Molecular Aspects of Chemotherapy

Molecular Aspects
of Chemotherapy

*Proceedings of the Third International Symposium
on Molecular Aspects of Chemotherapy
Gdańsk, Poland
June 19-21, 1991*

Edited by

David Shugar
Polish Academy of Sciences, University of Warsaw, Warsaw, Poland

Wojciech Rode
Polish Academy of Sciences, Warsaw, Poland

Edward Borowski
Technical University of Gdańsk, Gdańsk, Poland

Springer-Verlag Berlin Heidelberg GmbH

David Shugar
Institute of Biochemistry & Biophysics, Polish Academy of Sciences, 36 Rakowiecka St.,
02-532 Warszawa, and Department of Biophysics, Institute of Experimental Physics, University
of Warsaw, 93 Żwirki i Wigury St., 02-089 Warszawa, Poland

Wojciech Rode
Nencki Institute of Experimental Biology, Polish Academy of Sciences,
3 Pasteura St., 02-093 Warszawa, Poland

Edward Borowski
Department of Pharmaceutical Technology & Biochemistry, Technical University of Gdańsk,
80-952 Gdańsk, Poland

Distribution rights throughout the world with the exception of:
Albania, Bulgaria, Cuba, Czecho-Slovakia, Democratic People's Republic of Korea, Mongolia,
People's Republic of China, Poland, Romania, former USSR, Vietnam and Yugoslavia
granted to Springer-Verlag

ISBN 978-3-662-02742-4   ISBN 978-3-662-02740-0 (eBook)
DOI 10.1007/978-3-662-02740-0

25/3140-543210

Preface

While serendipity and random screening continue to fulfil a significant role in the search for new drugs, current remarkable advances in molecular biology and genetics are dictating to a profound extent the approaches employed in their development. Increasing attention is being devoted to investigations of the mechanisms of action of existing drugs, and the sources of undesired side effects, at the molecular level. The information so derived is now extensively applied, with the aid of broad inter-disciplinary approaches, both theoretical and experimental, to improvements in existing drugs, and the rational design of new ones.

The foregoing comprised the subject matter of the 3rd International Symposium on "Molecular Aspects of Chemotherapy", under the auspices of the International Society of Chemotherapy, and organized by the Committee on Drug Research, Polish Academy of Sciences, and the Department of Biotechnology and Biochemistry, Technical University of Gdańsk.

This volume includes the texts of the review lectures presented by invited participants on up-to-date achievements, and future perspectives, in molecular mechanisms of inhibition of cellular functions and metabolism, with emphasis on the design and mechanisms of action of chemotherapeutic agents for treatment of bacterial, neoplastic, viral and parasitic diseases. From the Contents, it will be seen that the range of disciplines represented was a broad one, including theoretical and experimental chemists, physicists, molecular biologists, biochemists, enzymologists, virologists, tumour biologists. Plenary sessions were supplemented by several poster sessions.

The next, 4th, Symposium is planned for June 23–25, 1993, under the auspices, and as a pre-symposium, of the 18th International Congress of Chemotherapy, Stockholm, June 27–July 2, 1993. While a list of invited speakers is in preparation, the organizers warmly welcome proposals from prospective participants for presentation of plenary lectures (of a review nature), which will be given due consideration.

David Shugar,
  Institute of Biochemistry & Biophysics,
  Polish Academy of Sciences (PAN), Warsaw
  and Department of Biophysics, Institute of
  Experimental Physics, University of Warsaw, Warsaw

Wojciech Rode,
  Nencki Institute of Experimental Biology,
  Polish Academy of Sciences (PAN), Warsaw

Edward Borowski,
  Department of Pharmaceutical Technology & Biochemistry,
  Technical University of Gdańsk, Gdańsk

Acknowledgements

We are indebted to the following for financial support: Sterling Drug Inc., Rensselaer, NY, USA; Societa Prodotti Antibiotici, Milan, Italy; F. Hoffmann-La Roche & Co., Basel, Switzerland; Faculty of Chemistry, Technical University of Gdańsk, Gdańsk, Poland; and the Ministry of National Education, Warszawa, Poland.

We are also grateful to Janina Jedlińska of PWN for her untiring patience and assistance in the technical editing of the texts.

Contents

Insight into Lactam Antibiotics and their Receptors from Computational Chemistry

D. B. Boyd and J. D. Snoddy

Lilly Research Laboratories, Eli Lilly and Company, Indianapolis, Indiana 46285, USA

Abstract. Molecular dynamics simulations have been performed on two penicillin-recognizing proteins. One protein is an enzyme from *Streptomyces* R61 and has transpeptidase and carboxypeptidase activity on peptide substrates terminating in D-alanyl-D-alanine. The DD-peptidase is a model for penicillin-binding proteins (PBPs) controlling bacterial peptidoglycan biosynthesis and is inhibited by various classes of natural and synthetic β-lactam antibiotics. The second protein is a β-lactamase from *Bacillus licheniformis* 749/C and deactivates penicillins and other β-lactam antibiotics by hydrolyzing them to inactive products. Crystallographic data for the two proteins have been refined by molecular mechanics energy minimizations, and the conformation space available to the proteins immersed in an 8-Å layer of water molecules has been explored by molecular dynamics simulations. The solid state model of the DD-peptidase is found to be much further from an energy-optimized structure than in the case of the β-lactamase. The DD-peptidase is discovered to have a flexible loop that can transiently close access to the active site serine. An overview of recent modeling studies of β- and γ-lactam antibiotics illustrate the type of information that can be obtained with modern methodologies of computational chemistry.

1. Introduction

An impressive array of β-lactam antibiotics has been produced by scientists in the last 50 years. Drugs of this type have many advantages in terms of potency, spectrum of activity, safety, resistance development, and administration compared to other classes of antibiotics. Indeed, the therapeutic value of these drugs has contributed to an improvement in the length and quality of life of patients.

However, in the process of natural selection among microorganisms for survival under the pressure of antibiotics, resistant strains have appeared. For instance, incidence of *Staphylococcus aureus* strains resistant to penicillin G is 80–90% worldwide. Incidence of *Staphylococcus aureus* infections caused by strains resistant to methicillin is as high as 85% in some urban and hospital settings. Also worrisome is the appearance in the last eight years of plasmid-mediated β-lactamases that counteract the effectiveness of powerful third generation cephalosporins. Evolving clinical challenges fuel a need for new, effective modes of antiinfective therapy.

Penicillins and cephalosporins have been the subject of many reviews, e.g.,

[1–3]. Medicinal chemists have synthesized tens of thousands of derivatives, probing the limits of structure-activity relationships. Meanwhile, microbiologists and molecular biologists have performed important experiments elucidating the penicillin-binding proteins (PBPs), which are the enzymes targeted by β-lactam antibiotics. Both the function and the primary amino acid sequence of many of these enzymes are known. Moreover, enzymes that mediate antibiotic resistance, namely β-lactamases and altered PBPs, have been sequenced and characterized [3–5]. The evolutionary relationship of the PBPs and β-lactamases has been discussed, e.g., [2].

There is a need to understand mankind's bacterial adversaries at the molecular and atomic level. Crystallographers have responded by providing information on the secondary and tertiary structures of β-lactamases and related enzymes [6–19]. Rational design of new specific inhibitors requires this kind of information. The present paper is a progress report of work aimed at obtaining energy-refined three-dimensional structures of two penicillin-recognizing proteins. One protein is the D-alanyl-D-alanine carboxypeptidase-transpeptidase from *Streptomyces* R61, the so-called R61 DD-peptidase (EC 3.4.16.-) [6, 7]. The other protein is a class A β-lactamase produced by *Bacillus licheniformis* 749/C (EC 3.5.2.6) [8].

Before describing our calculations to refine these structures, the philosophical underpinning of the computational approach is introduced. Then a brief survey of some examples of modeling studies of antibacterial agents will be given. From this foundation, we will then be in a position to describe our most recent computational experiments.

Molecular Structure Information. Ordinary X-ray diffraction experiments give a static, time-averaged picture of a molecular structure, which, in reality, is usually highly dynamic, undergoing continual conformational changes between energetically accessible states. Computers make it possible to model this dynamic behavior.

Computational chemistry encompasses several well known theoretical methodologies that quantitatively model chemical and biochemical phenomena [20, 21]. Molecular modeling can be used to rationalize experimental data and determine molecular phenomena even in cases where experiment is difficult or impossible.

The viability of computational chemistry rests on four cornerstones: (1) a theory of matter that has proven to model Nature realistically, (2) computer hardware capable of performing the computational experiments in a reasonable length of time, (3) graphics terminals for visualization of the input and output of the computations, and (4) software for efficient, multifunctional studies of molecular systems.

One of the important techniques of computational chemistry is molecular mechanics (MM) [22–24]. In this approach, a molecule can be thought of as a collection of masses held together by springs. The potential energy surface, which describes the shape of a molecule, is modeled by a combination of harmonic (bond stretching and bending), periodic (torsional), dispersive (van der Waals), and Coulombic (electrostatic) forces that tend to hold the atoms at their energetically favorable positions. The force field (Table 1) is used to compute a total energy E,

Table 1. Form of an empirical force field, Newton's equations of motion, and a common approximate scheme for solving the equations of motion. The total energy E is the sum of bond stretching, bond angle bending, torsional angle twisting, electrostatic, and Lennard-Jones 6–12 terms. The "improper" dihedral angle term is an expedient for keeping sp^2 atoms planar. The equations of motion relate the force on an atom to its mass, acceleration, and potential. The position of an atom is described by the vector r

Force Field

$$E = E_b + E_\theta + E_\phi + E_\tau + E_{el} + E_{vdw}$$

$$E_b = \sum_{bonds} k_b(r - r_0)^2 \qquad\qquad E_\theta = \sum_{angles} k_\theta(\theta - \theta_0)^2$$

$$E_\phi = \sum_{dihedrals} k_\phi[1 + \cos(n\phi - \delta)] \qquad E_\tau = \sum_{improper} k_\tau(\tau - \tau_0)^2$$

$$E_{el} = \sum_{i;j>i} \frac{q_i q_j}{4\pi\varepsilon r_{ij}} \qquad\qquad E_{vdw} = \sum_{i;j>i} \frac{A_{ij}}{r_{ij}^{12}} - \frac{B_{ij}}{r_{ij}^6}$$

Newton's Equations of Motion

$$F_i = m_i \frac{d^2 r_i(t)}{dt^2} \qquad\qquad F_i = -\frac{\partial V(r_i, ..., r_N)}{\partial r_i}$$

Verlet Integration Scheme

$$r_i(t + \Delta t) = 2r_i(t) - r_i(t - \Delta t) + \frac{F_i \Delta t^2}{m_i}$$

which a computer can evaluate rapidly. The energy reflects how much a geometrical arrangement of atoms deviates from an idealized bonding situation. Thus an energy minimization will bring the molecular structure to a "better" geometry.

, Molecular dynamics (MD) goes beyond molecular mechanics and aims to model the variation in a molecule's structure and energy over time [25, 26]. MD is a good method [26, 27] for sampling "conformational space" (the collection of accessible conformations) of proteins. The sampling is in the vicinity of the starting molecular geometry, but a longer simulation can sample farther from the initial conformation.

In MD simulations atoms move according to Newton's equations of motion (Table 1), so that atoms under the most strain move fastest and farthest. The Verlet integration approximation is used to extrapolate atomic positions for the next time step. Calculating the trajectories of atoms and forces as a molecule changes shape is computationally expensive. To adequately solve Newton's equations for each atom in a molecule requires extremely short time steps, typically only 1 femtosecond (10^{-15} s). The short time steps arise from the need to reevaluate the forces on each atom frequently, so that the trajectories can be updated before atoms collide with each other (which in the force field model would result in an infinite energy). Tracking the motion of thousands of atoms, such as constitute a protein, requires much computing and massive storage of data. A typical molecular dynamics run might simulate only 10–20 picoseconds of motion, even with the resources of the best supercomputers.

Synopsis of Structure-Activity Relationships and Computations on Antibacterial Agents. To convey an appreciation of the capabilities of computational chemistry, it may help to give examples of modeling studies of lactam antibiotics and to briefly review what is known about structure-activity relationships. We cannot take the space to review completely the applications of computational chemistry to antibiotics, so what follows in necessarily selective and abbreviated.

Opinions about the requirements for antibacterial activity have evolved with time. In 1972 when penicillins (Fig. 1(1)) and cephalosporins (Fig. 1(2)) were the

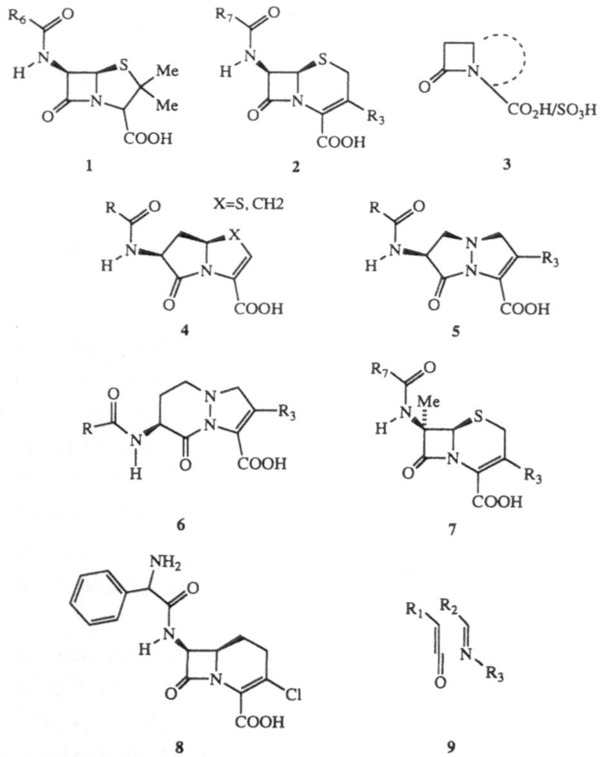

Fig. 1. Structures

two well known classes of lactam antibiotics, the features considered essential for activity were a diastereomer of a bicyclic β-lactam structure with an acidic group on the five- or six-membered ring and an acylamino group on the β-lactam ring [1]. Ten years later [2] with the discovery of the additional classes of β-lactams (Fig. 2), the pharmacophore, e.g., the structural elements responsible for activity, had simplified to a β-lactam ring and an acidic group (Fig. 1 (3)) [28].

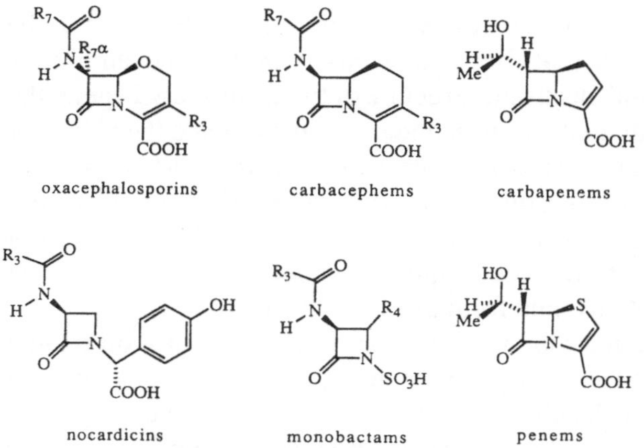

Fig. 2. Some classes of β-lactam antibiotics

Also by 1982, certain factors were known to be important for activity: (1) the interatomic distances defining the pharmacophoric group must be such that the molecule will fit in the receptor site and preferentially interact with the residues it encounters there, (2) the faces of the β-lactam ring must be sterically accessible because the antibiotic molecule must be able to approach close enough to the active site serine for an acylation reaction to occur, (3) the lactam amide bond must be sufficiently reactive to acylate the serine, but not so reactive as to hydrolyze readily, and (4) the scissle amide bond group must be in a ring (i.e., lactam) so that after the amide bond opens the fragments cannot easily dissociate away from the active site.

For a series of structurally related cephalosporins with modifications at the 3 position, it has been shown that *in vitro* activity against Gram-negative organisms can be correlated with chemical reactivity of the β-lactam ring [29]. Chemical reactivity can be computed quantum mechanically in terms of the ease of approach of a nucleophile to the β-lactam ring [28, 30] or measured experimentally from base-catalyzed hydrolysis rates or other properties. The various measures of reactivity were shown to be a consequence of the inductive effect of the substituent at the 3 position [31]. Hammett sigma values correlate with minimum inhibitory concentrations of the series of cephalosporins mentioned above.

As has been pointed out before, the correlation between reactivity and Gram-negative activity was observable because all the other factors that affect activity are constant in a series of closely related compounds [32].

It is important to note that, while correlations exist between the physicochemical properties and biological activities of lactam antibiotics, what is critical for inhibition of the PBPs is *the chemical reactivity of the compound in the milieu of the receptor site*. It is worth recalling that the mode of action of lactam antibiotics is a mechanism-based inhibition of an enzyme [33]. The function of an enzyme is, of course, to lower the transition state energy of a reaction from what it would be free in solution. Although the inherent chemical reactivity of the β-lactam ring is determined by the molecular and electronic structure of the antibacterial agent, the environment of the receptor site in the transpeptidases can modulate the ability of the agent to irreversibly acylate the active site serine. *Structurally dissimilar lactam antibiotics will fit in the receptor somewhat differently, and the forces between atoms of the inhibitor and receptor will be different.* Hence it is not surprising that the chemical reactivity range required for biological activity for one category of compounds will be different from that of another.

Another observation from microbiology is that within a series of related cephalosporins, those with good leaving groups at the 3 position [28] generally are more active than compounds with substituents that can only act inductively [31]. Again, trying to understand this fact from physical organic experiments can give an incomplete picture. It is necessary instead to understand what is happening in the active site of the target proteins. To eventually gain this detailed understanding of how a PBP perceives the ligands that enter its active site is one of the motivations for the present calculations.

A recent major revelation to the dogma of lactam antibiotics was that the β-lactam ring is not required for activity. Although studies dating from the 1940s had shown that γ-lactam analogues of penicillins and cephalosporins lacked any significant antibacterial activity, γ-lactam analogues of the penems and carbapenems (Fig. 1(4)) did show glimmers of activity [34–36]. These bicyclic nuclei are more chemically reactive than penams or cephems, and hence activity was observable.

A boost to the chemical reactivity and a quantum leap in activity was achieved by putting a second nitrogen in the γ-lactam ring. The bicyclic pyrazolidinones (Fig. 1(5)) have significant antibacterial properties comparable even to third generation cephalosporins [37–40]. Reactivity of the new compounds is close to or slightly greater than that of cephalosporins. Reactivity and activity of the bicyclic pyrazolidinones followed a pattern that was predictable from the properties of the substituents at the 3 position (R_3). Those substituents that were highly electron withdrawing and of intermediate hydrophilicity as measured on the Hammett σ and Hansch-Fujita π scales [41], respectively, were found to impart the best Gram-negative activity [42].

Experimentation revealed that further enlargement of the lactam ring to a δ-lactam ring essentially destroyed activity [43]. In an attempt to understand this, the base-catalyzed hydrolysis rates of the bicyclic tetrahydropyridazinones (Fig. 1(6)) were examined, but they were comparable to those of the bicyclic pyrazolidinones. Cell wall permeability and destruction by β-lactamases were not factors in the inactivity. The pharmacophoric structure of Fig. 1(6) was found to meet the geometrical requirements usually associated with potent antibacterial agents [44]. In other words, the distance between the acidic group and the lactam was neither too great or too short. Computational chemistry was used to compare the steric volume of the conformers of Fig. 1 (6) to that of the biologically inactive 7α-methylcephalosporins (Fig. 1(7)). It was discovered that the δ-lactam ring unavoidably occupies some of the same space as the 7α-methyl group on the β-lactam ring [43]. Hence the inactivity of the bicyclic tetrahydropyridazinones is due to them being unable to set properly in the receptor site for a productive interaction. This computational prediction was subsequently confirmed in PBP binding experiments when it was shown that concentrations of the compounds as high as 512 mcg/mL do not produce binding to any of the PBPs of *Escherichia coli.*

A class of β-lactam compounds that has received much recent interest is the carbacephems. In particular, loracarbef (Fig. 1(8)) has been found active at the 0.5–4 mcg/mL level against Gram-positive and Gram-negative bacteria, such as *Staphylococcus aureus, Streptococcus pyogenes, Streptococcus pneumoniae, Haemophilus influenziae, Branhamella catarrhalis, Escherichia coli, Klebsiella pneumoniae,* and *Proteus mirabilis* [45–50]. Loracarbef is orally bioavailable and is more stable than cephalosporins [51]. Computational chemistry has been used to compare the three-dimensional structure of this compound to cefaclor and to determine conformational preferences of derivatized pyridino substituents at the 3 position of other carbacephems [52].

One last example will be cited to show the potential of computational chemistry, pertaining to the chemical mechanisms used in a synthetic pathway to the carbacephems. The β-lactam ring can be formed by the well known Staudinger reaction [53]. Ordinarily organic chemistry envisions the 2+2 cycloaddition occurring from coplanar reactants as shown in Fig. 1(9). However, it was discovered from quantum mechanical energy optimization that the preferred relative orientation of the reactants forming the intermediate is nearly orthogonal as depicted in Fig. 3. After formation of the carbon-nitrogen bond, bonds in both the ketene and imine moieties rotate in a conrotatory manner to complete formation of the β-lactam ring with the substituents either both on the α face or both on the β face. The orthogonal geometry of the transition intermediate explains the asymmetric induction observed when a chiral auxiliary R_1 is on the ketene [54].

Limitations. Although computational methodologies can be used to answer many questions, they have not advanced to the point of being able to predict accurately the folding of a protein from an amino acid sequence. It has been pointed out that:

"no reliable approaches are as yet available for the prediction of de novo macromolecular structures, and there is no real alternative to experimental structure determination" [55].

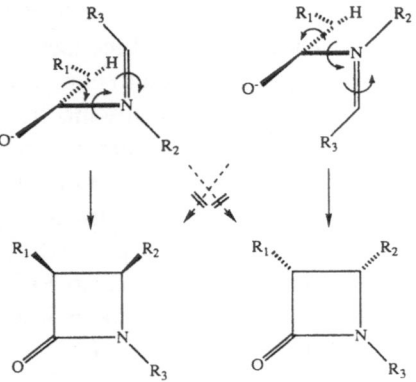

Fig. 3. Scheme for rotation about three bonds in the intermediate resulting in formation of the β-lactam ring with substituents at the 3 and 4 positions being cis and on the β or α face

In cases where no homologous structures are known, crystallography (or multi-dimensional NMR) is an indispensable tool for modeling the tertiary structure of a protein. By the same token, crystallography (or NMR) does not determine the one true conformation of a protein. A protein is in dynamic equilibrium between many conformations. Not uncommonly, an electron density map from X-ray diffraction data must be interpreted with some degree of subjectivity to "solve" a structure of a protein. If the protein is highly flexible, the X-ray data may be so blurred as to prevent reliable assignment. Again quoting from the recent literature:

"the structure reported by a protein crystallographer is not a universal constant: it is not perfectly refined, assumption-free, model-free, nor error-free, nor is it invariant to the environment" [56].

The *combination* of experimental determination and molecular modeling is the best approach to obtaining a reliable protein structure. For example, one can start with a proposed experimental model and then optimize bond lengths and angles according to a force field. A variant of MD called simulated annealing has rapidly become a tool crystallographers routinely employ to refine macromolecular structures [57].

Toward the goal of eventually being able to understand the biochemical mechanisms that occur in the active sites of the PBPs and β-lactamases, we next describe computations on the two β-lactam-recognizing proteins modeled here.

2. Methods

In the first molecular modeling study of an entire β-lactam-recognizing protein [58], the starting DD-peptidase structure had four loops that were ill-defined in the X-ray data, and these segments had to be built computationally. The present computations are based on a starting structure resulting from further crystallographic refinement. The new X-ray data provided a complete set of atomic coordinates of R61. Also Tyr-159, which is in a flexible loop at the lip of the active site, is in the position the crystallographers earlier presumed to be occupied by Phe-164. In other words, the interpretation of the electron density map has changed such that the residues in this loop were shifted five positions compared to previously published models [6,7]. The new interpretation of the experimental data is another step toward an improved model of a protein that has yielded somewhat grudgingly to elucidation.

The second protein modeled here, the *Bacillus licheniformis* β-lactamase has yielded to high resolution and refinement by crystallographic studies [8]. The atomic coordinate set used for *B.l.* is reported elsewhere [8]; the sequence for modeling starts with Asp-31 and runs through Lys-295. Both the R61 DD-peptidase and the *B.l.* β-lactamase have the signature Ser-Xaa-Xaa-Lys sequence in their active sites and fold into a similar overall tertiary structure [12, 13]. Experimental data on the R61 and *B.l.* proteins are summarized in Table 2.

Table 2. X-ray data on the two proteins

	DD-peptidase	*B.l.* β-lactamase
MW (kDa)	37.5	29.5
Residues	349	264
Heavy atoms	2629	2033
Resolution	1.7	2.0
R factor (%)	35	22

The molecular dynamics methodology employed here is similar to that recently reported [58]. The QUANTA/CHARMm molecular simulation software suite [59–62] was used for minimizations and MD. The SYBYL molecular modeling program [62–65] was used for computer graphics.

For the energy minimizations, all atoms, including hydrogens, were treated explicitly. Version 21 (1990) of the CHARMm force field was used. Conjugate gradient minimization was run until the change in total energy between cycles was less than 0.0001 kcal/mole (about 3000 iterations). The resulting structure was solvated with 8-Å layer of TIP3P (transferable intermolecular potential-functions, three-point) waters [66, 67]. This layer is as thick as possible with the software which has a limit of 10000 atoms. There were 1529 waters in the R61 model increasing the total number of atoms in the computational model from 5182 to 9769. There were 954 waters in the *B.l.* model bringing the total number of atoms from 4083 to 6945. After solvation, the models were minimized for another 5000 iterations until the gradient criterion mentioned above was satisfied.

The simulations on the minimized, solvated models were run with CHARMm. Time steps of 1 fs were used. Generally the program was run with defaults supplied by QUANTA. Thus the SHAKE algorithm was applied to maintain bond distances to hydrogens [25, 27, 55, 68]. The criterion for including atom pairs in the nonbonded interaction table (updated every 50 iterations) was a distance of 8 Å or less; the cutoff for computing nonbonded interaction energies was 7.5 Å. A shifting function was applied to the nonbonded energy contribution to eliminate discontinuities at the cutoff distance in the electrostatic and van der Waals terms. A dielectric constant of 1.0 was used. A harmonic constraining potential was applied to the oxygen of each water molecule $E_{con} = \sum_{waters} 0.001(r-r_0)^2$. The effect of these potentials is to discourage the waters from drifting away from the protein (evaporating) during the simulation. Recall that the water positions have been energy minimized. Because no constraints are on the hydrogens of the waters, the solvent molecules are free to rotate to seek favorable hydrogen bonding arrangements.

The protocol for the MD calculations was as follows: (a) heating from 0 to 300 K through a 3 ps period, (b) equilibration at 300 K for 10 ps, and (c) simulation at 300 K for 100 ps. Velocities were rescaled in a Gaussian distribution every 1000 steps whenever the averaged temperature was not within 10° of 300 K. Atomic coordinate trajectories were recorded every 50 fs during the simulation period and used to analyze the results.

To maximize sampling of conformational space, relatively long simulations were run on each protein. Our 100 ps simulations are quite long for such large systems. Experiments like this are impossible without high performance computing. For R61, the computations required 150 h of central processing unit time on a Silicon Graphics 4D240 Power Series workstation, plus 80 h on Eli Lilly and Company's Cray-2S/128 supercomputer. For *B.l.*, the entire simulation was run on the Cray-2 and required 108 h.

Titration curves were estimated according to approximate theory [69, 70] by iterating until the calculated mean net charge on the surface of a sphere representing a protein at a given pH was consistent with the ionization states of the amino acids in the protein.

3. Results

Often there will be counterions associated with a protein in solution, but these may not be located unless the X-ray data are of very high quality and refinement has proceeded well. Most computer simulations neglect counterions because of the additional complexity involved and the often arbitrariness of placing a finite number of ions. Some proteins as extracted from the Brookhaven Data Bank bear very high positive or negative charges. Before modeling, it is therefore helpful to consider the titration curve of a protein to check its net charge. As seen in Fig. 4,

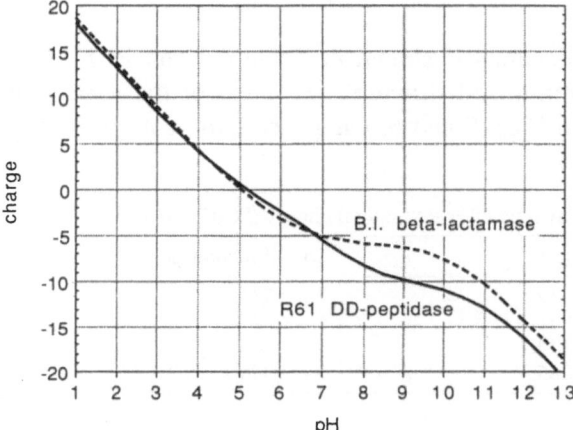

Fig. 4. Titration curves for R61 and *B.l.* as estimated from the amino acid composition of the protein models

Table 3. Amino acid composition of the modeled proteins

Amino acid	DD-peptidase	*B.l.* β-lactamase
Ala	34	26
Arg	*14*	*15*
Asn	16	13
Asp	**23**	**24**
Cys	2	0
Gln	19	7
Glu	**8**	**19**
Gly	30	15
His	*9*	*1*
Ile	9	14
Leu	33	27
Lys	*7*	*22*
Met	8	4
Phe	12	7
Pro	12	11
Ser	27	11
Thr	40	20
Trp	2	3
Tyr	14	6
Val	30	15
Computed MW	37347	28885

the models of the R61 and *B.l.* proteins (with the amino acid compositions as given in Table 3) carry a charge between −5 and −6 at neutral pH. These charges are not so large as to invalidate a simulation without counterions. It would be of interest, however, to include 5 or 6 sodium ions in future simulations. (By way of comparison, a *Staphylococcus aureus* β-lactamase [18] bears a net charge of about +15 at neutral pH). The isoelectric points of R61 and *B.l.* are computed to be at pH 5.2 and pH 5.0, respectively.

Progress of the energy minimizations is shown in Table 4. Data in Table 5 compare the X-ray and energy minimized structures. Progress of the molecular dynamics simulations is shown in Figs. 5 and 6, where the temperature of each

Table 4. Progress of minimizations as shown by CHARMm total energy (kcal/mol)

	DD-peptidase	*B.l.* β-lactamase
Start	+26,600,012.	−3,533.
Final	−8,821.	−8,275.
Solvated by 8-Å layer	−5,379.	−8,416.
Final with 8-Å layer	−30,065.	−22,048.

Table 5. Root mean square and maximum atomic displacements (Å) comparing starting X-ray structure and structure after minimizations with water layer

	DD-peptidase		*B.l.* β-lactamase	
	rms	maximum	rms	maximum
α carbons	1.86	4.98	1.36	5.44
All heavy atoms	2.20	8.63	1.70	7.84

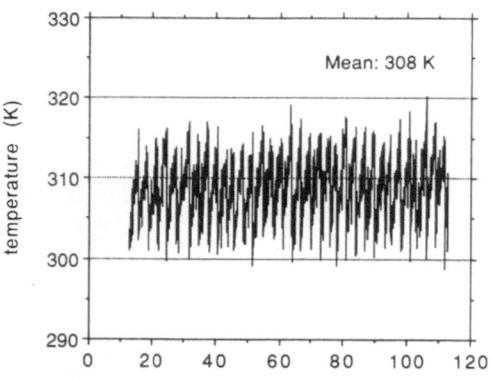

Fig. 5. Behavior of the theoretical temperature during the production phase of the molecular dynamics simulation of the solvated R61 DD-peptidase

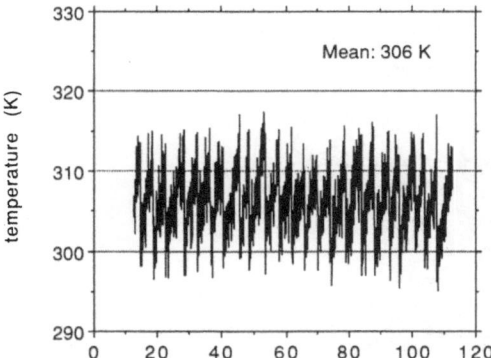

Fig. 6. Behavior of the theoretical temperature during simulation of the solvated *B.l.* β-lactamase

system is plotted. Velocity rescaling is a common technique in molecular dynamics to prevent the temperature from rising excessively due to mathematical approximations and to conformational relaxation and the consequent conversion of potential to kinetic energy. The fact that the velocities had to be rescaled frequently and the total energy of the system had wide fluctuations suggests it would be desirable to use longer nonbonded cutoffs in the future as has been proposed with the CHARMm force field [71]. For sampling conformational space and thereby obtaining mean interatomic distances in the active site, the present computations should nevertheless be useful.

Comparisons of the final geometries of the (solvated) minimized proteins and the final geometries after the MD simulations are given in Table 6. Molecular graphics of the proteins are shown in Figs. 7 and 8. Mean interatomic distances around the active site of R61 and *B.l.* are shown in Figs. 9 and 10.

Table 6. Root mean square and maximum atomic displacements (Å) comparing the structure minimized with the water layer and the final structure from the simulation

	DD-peptidase		*B.l.* β-lactamase	
	rms	maximum	rms	maximum
α-carbons	2.64	8.75	1.66	5.39
All heavy atoms	3.14	10.37	1.98	6.72

4. Discussion

The energy of the R61 structure (Table 4) begins dramatically high. The main reason for this is that bond lengths and angles in the X-ray model were far from optimum. Many interatomic distances were too short or too long. Residue Asp-84,

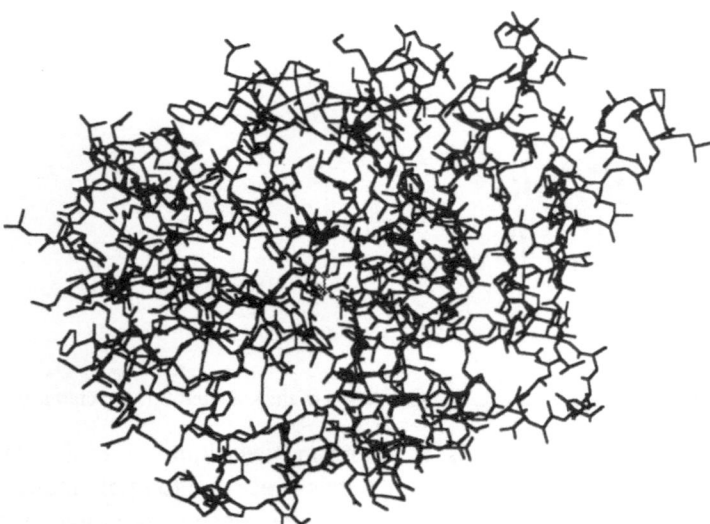

Fig. 7. Structure of R61 DD-peptidase at 93 ps into the production phase of the simulation. At this point, no hydrogen bond between Tyr-159 and Thr-301 exists and the entrance to the receptor site (center) is open. Hydrogens and waters of solvation are not shown for sake of clarity, but were included in the modeling. The active site serine is in gray. The *N*- and *C*-termini are at the upper right

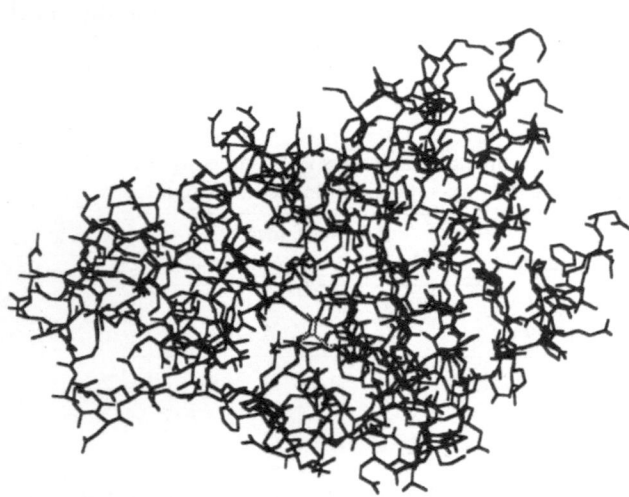

Fig. 8. Structure of *B.l.* β-lactamase at 73 ps into the simulation. The β sheet region domain is to the right, and the active site serine is in gray

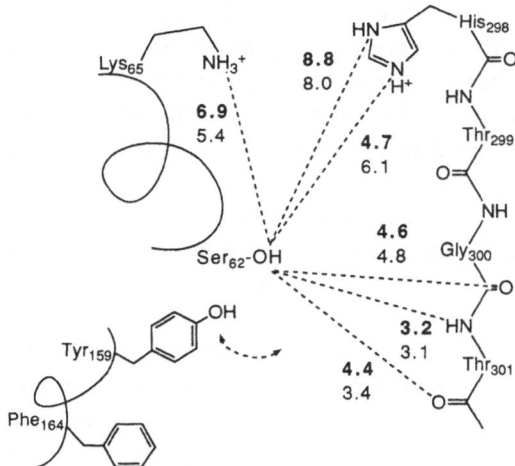

Fig. 9. Mean interatomic distances (Å) around the active site serine of R61 DD-peptidase. The values from the conclusion of the molecular dynamics simulation are given in bold face; the other values are from the X-ray data used as the starting point for the computations. The motion of Tyr-159 that can close the entrance to the receptor site is denoted schematically. Standard deviation in the distances from the MD simulations is ± 0.4 Å

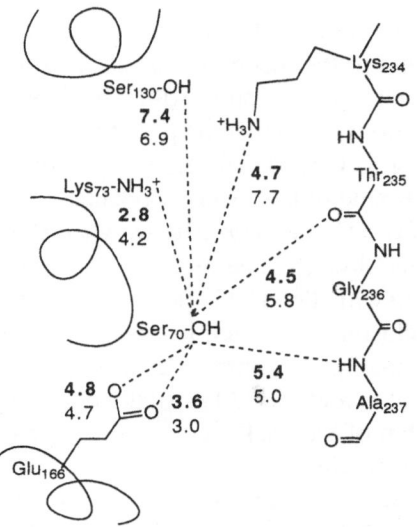

Fig. 10. Mean interatomic distances (Å) around the active site serine of *B.l.* β-lactamase. The values from the conclusion of the molecular dynamics simulation are given in bold face; the other values are from the X-ray data used as the starting point for the computations. Standard deviation in the distances from the MD simulations is ± 0.6 Å

in particular, was far from residues Leu-83 and Ala-85. In contrast, the *B.l.* starting structure has a much lower initial energy. This is due to its higher level of refinement and to the fact that it has already been energy-refined by simulated annealing calculations [8]. Further minimizations after solvation with the 8-Å layer of waters lowers the energies of both proteins still more. (The total force field energies of R61 and *B.l.* (Table 3) are not directly comparable because the two proteins are of different size and composition.)

The poorer starting geometry of R61 also is seen (Table 5) in the fact that the atomic displacements are generally larger than in *B.l.* Our results are consistent with the crystallographic *R* factors (Table 2); i.e., the starting atomic coordinates of the *B.l.* structure were more refined. The root mean square displacements can be viewed from two perspectives. From a computational chemist's point of view, the data in Tables 5 and 6 show the effects of conformational fredom compared to the solid state. On the other hand, from a crystallographer's point of view, the data show how well (or poorly) the force field parameters are able to match experimental data. Readers are free to adopt their own perspective.

Minimizations and simulations show that the R61 DD-peptidase has more flexible regions than does the *B.l.* β-lactamase. The loops of R61 can undergo continual conformational change. The floppiness of these loops is no doubt responsible for the difficulties in refinement of the X-ray data of R61. R61 has less ordered secondary structure: about 10% of the residues are in β strands, 30% in helices, and 60% in loops. In contrast, *B.l.* has 25% of its residues in β strands, 45% in helices, and only 30% in loops.

A discovery from the simulations is that a loop in the helical domain near the active site of R61 DD-peptidase can reduce access to the receptor. The active site Ser-62 lies at the base of a cone; the serine's hydroxyl oxygen is 10–11 Å below the surface of the globular protein. The computations show that the para-hydroxyl group of Tyr-159 can move into a position to hydrogen bond to the hydroxyl group of Thr-301, which is in the β strand on the other side of the receptor pocket. The hydroxyl oxygen of Ser-62 lies 5–6 Å under Tyr-159 and Thr-301, so when the hydrogen bond is formed, access to the serine is blocked. The receptor does not stay closed during the whole simulation; the loop with Tyr-159 is flexible enough that it only transiently closes the entrance. The closing of the cleft is seen dramatically in a motion picture visualization of the simulation. In contrast, the active site of the *B.l.* β-lactamase stays open. Its cleft is wider, and Ser-70 is totally open to the surface. The loop in the β-lactamase at the left of the pocket corresponding to the one with Tyr-159 in R61 is shorter and not as flexible.

We suggest that the flexibility of any loops near the portals of the active site of D-alanyl-D-alanine transpeptidases and carboxypeptidases may be important to biochemical function. The substrate of these enzymes is the pentapeptide tail on a glycan polymer of *N*-acetylmuramic acid and *N*-acetylglycosamine [2]. The size of this substrate would require a pocket large enough to accommodate the peptide tail. In the case of the transpeptidases, a second peptide tail must also be accommodated because crosslinking involves formation of a peptide bond between two peptidoglycan strands. Flexibility of loops near the active site of transpeptidases allows them to move out of the way of an incoming substrate.

Our results from the 100 ps simulations (Figs. 9 and 10) show an interesting phenomenon. Both enzymes have the Ser-Xaa-Xaa-Lys sequence in the active site. The ammonium group of the invariant lysine is probably essential for attracting a ligand into the receptor and for orienting the groups in the active site. To the "northeast" of the serine is a basic residue, Lys-234 in *B.l.* and His-298 in R61. In both proteins, a side chain nitrogen of these residues came closer to the serine's hydroxyl group than observed in the solid state. Proximity of these basic groups on the innermost β strand would help increase the nucleophilicity of the serine's hydroxyl oxygen.

The R61 enzyme, although it has transpeptidase activity, is not a PBP. Its function in *Streptomyces* remains unclear. Functionally and structurally it is most closely related to class C β-lactamases [5]. In the family tree of penicillin-recognizing proteins [5], R61 is next most closely related to the class A β-lactamases, whereas all these categories of enzymes are only distantly related to the high molecular weight PBPs. This fact is consistent with earlier findings that the kinetics of reactions between the R61 enzyme and β-lactam antibiotics cannot be directly related to antibacterial potency of these compounds [72]. The reasons for this are that (1) the target transpeptidases in the periplasmic space of bacteria possess a transglycosylase domain and so there may be conformational differences, (2) there are multiple transpeptidases, which are known to compensate for one another to some extent when one is blocked, and (3) *in vitro* potency depends on additional factors such as β-lactamase susceptibility and outer membrane permeability [73]. Nevertheless, the R61 DD-peptidase can serve as a model system for studying interactions between an inhibitor and an enzyme in isolation.

Based on the computations done to date, we have two energy-refined models of penicillin-recognizing proteins. These can be used in ongoing research to examine mechanistic questions and to explore effects of site specific mutations. As mentioned, crystallography gives a time-averaged picture of a structure, and, for highly disordered regions, no definition of structure may be possible. Computational chemistry can extend the information from crystallography to show the dynamic nature of proteins. X-ray data provides a basis for hypothesizing the mechanism of action of an enzyme, while molecular modeling can actually explore possible reaction paths and transitory species.

We have discussed several applications of computational chemistry in the arena of antibacterial research. As with any tool of research, it is not applicable to every question. However, many questions can be addressed. It is not exaggerating to state that computational chemistry is now indispensable to modern drug discovery research. To date, one antibiotic and several other biologically active compounds that have reached commercial use owe their development in part to computer-assisted molecular design [74]. Applications in antiinfective research will continue to grow.

Acknowledgements. Professor James R. Knox (University of Connecticut) furnished crystallographic data, Dr. James E. Shields (Lilly Research Laboratories) made available his program for estimating titration curves of proteins, Dr. Thalia I. Nicas provided many helpful suggestions on the manuscript, Dr. Richard J. Loncharich, Dr. B. L. Barnett, and Dr. W. D. Laidig helped clarify questions on the molecular dynamics calculations, and Dr. Riaz F. Abdulla supplied his unflagging enthusiasm.

References

1. Flynn EH, ed, *Cephalosporins and Penicillins; Chemistry and Biology*. Academic Press, New York, 1972.
2. Morin RB and Gorman M, eds, β-*Lactam Antibiotics; Chemistry and Biology*. Academic Press, New York, 1982, Vol. 1–3.
3. Umezawa H, ed, *Frontiers of Antibiotic Research*. Academic Press, Tokyo, 1987.
4. Sing MD, Wachi M, Doi M, Ishino F and Matsuhashi M, Evolution of an inducible penicillin-target protein in methicillin-resistant *Staphylococcus aureus* by gene fusion. *FEBS Lett.* **221**: 167–171, 1987.
5. Joris B, Ghuysen JM, Dive G, Renard A, Dideberg O, Charlier P, Frère JM, Kelly JA, Boyington JC, Moews PC and Knox JR, The active-site-serine penicillin-recognizing enzymes as members of the *Streptomyces* R61 DD-peptidase family. *Biochem J.* **250**: 313–324, 1988.
6. Kelly JA, Knox JR, Zhao H, Frère JM and Ghuysen JM, Crystallographic mapping of β-lactams bound to a D-alanyl-D-alanine peptidase target enzyme. *J. Mol. Biol.* **209**: 281–295, 1989.
7. Kelly JA, Knox JR and Zhao H, Studying enzyme-β-lactam interactions using X-ray diffraction. *J.Mol. Graphics* **7**: 87–92, 1989.
8. Moews PC, Knox JR, Dideberg O, Charlier P and Frère JM, β-lactamase of *Bacillus licheniformis* 749/C at 2 Å resolution. *Proteins* **7**: 156–171, 1990.
9. Knox JR, Moews PC, Kelly JA and DeLucia ML, R-TEM β-lactamase secondary structure prediction and X-ray analysis at 4 Å resolution. In: β-*Lactamases* (eds Hamilton-Miller JMT and Smith JT), pp. 127–140. Academic Press, New York, 1979.
10. Charlier P, Dideberg O, Frère JM, Moews PC and Knox JR, Crystallographic data for the β-lactamase from *Enterobacter cloacae* P99. *J. Mol. Biol.* **171**: 237–238, 1983.
11. Kelly JA, Knox JR, Moews PC, Hite GJ, Bartolone JB, Zhao H, Joris B, Frère JM and Ghuysen JM, 2.8-Å structure of penicillin-sensitive D-alanyl carboxypeptidase-transpeptidase from *Streptomyces* R61 and complexes with β-lactams. *J. Biol. Chem.* **260**: 6449–6458, 1985.
12. Kelly JA, Dideberg O, Charlier P, Wéry JP, Libert M, Moews PC, Knox JR, Duez C, Fraipont C, Joris B, Dusart J, Frère JM and Ghuysen JM, On the origin of bacterial resistance to penicillin: Comparison of a β-lactamase and a penicillin target. *Science* **231**: 1429–1431, 1986.
13. Samraoui B, Sutton BJ, Todd RJ, Artymiuk PJ, Waley SG, and Phillips DC, Tertiary structural similarity between a class A β-lactamase and a penicillin-sensitive D-alanyl carboxypeptidase-transpeptidase. *Nature* **320**: 378–380, 1986.

14. Sutton BJ, Artymiuk PJ, Cordero-Borboa AE, Little C, Phillips DC, and Waley SG, An X-ray-crystallographic study of β-lactamase II from *Bacillus cereus* at 0.35 nm resolution. *Biochem. J.* **248**: 181–188, 1987.

15. Dideberg O, Charlier P, Wéry JP, Dehottay, Dusart J, Erpicum T, Frère JM, and Ghuysen JM, The crystal structure of the β-lactamase of *Streptomyces albus* G at 0.3 nm resolution. *Biochem. J.* **245**: 911–913, 1987.

16. Herzberg O and Moult J, Bacterial resistance to β-lactam antibiotics: Crystal structure of β-lactamase from *Staphylococcus aureus* PC1 at 2.5 Å resolution. *Science* **236**: 694–701, 1987.

17. Oefner O, D' Arcy A, Daly JJ, Gubernator K, Charnas RL, Heinze I, Hubschwerlen C, and Winkler FK, Refined crystal structure of β-lactamase from *Citrobacter freundii* indicates a mechanism for β-lactam hydrolysis. *Nature* **343**: 284–288, 1990.

18. Herzberg O, Refined crystal structure of β-lactamase from *Staphylococcus aureus* PC1 at 2.0 Å resolution. *J. Mol. Biol.* **217**: 701–719, 1991.

19. Knox JR and Moews PC, β-lactamase of *Bacillus licheniformis* 749/C: Refinement at 2 Å resolution and analysis of hydration. *J. Mol. Biol.* **220**: 435-453, 1991.

20. Lipkowitz KB and Boyd DB, eds, *Reviews in Computational Chemistry,* Vol. 1, VCH Publishers, New York, 1990.

21. Boyd DB, The computational chemistry literature. In: *Reviews in Computational Chemistry* (eds Lipkowitz KB and Boyd DB), Vol. 2, pp. 461–479. VCH Publishers, New York, 1991.

22. Boyd DB and Lipkowitz KB, Molecular mechanics: The method and its underlying philosophy. *J. Chem. Educ.* **59**: 269–274, 1982.

23. Bowen JP and Allinger NL, Molecular mechanics: The art and science of parameterization. In: *Reviews in Computational Chemistry* (eds Lipkowitz KB and Boyd DB), Vol. 2, pp. 81–97. VCH Publishers, New York, 1991.

24. Dinur U and Hagler AT, New approaches to empirical force fields. In: *Reviews in Computational Chemistry* (eds Lipkowitz KB and Boyd DB), Vol. 2, pp. 98–164, VCH Publishers, New York, 1991.

25. Lybrand TP, Computer simulation of biomolecular systems using molecular dynamics and free energy perturbation methods. In: *Reviews in Computational Chemistry* (eds Lipkowitz KB and Boyd DB), Vol. 1, pp. 295–320, VCH Publishers, New York, 1990.

26. Troyer JM and Cohen FE, Simplified models for understanding and predicting protein structure. In: *Reviews in Computational Chemistry* (eds Lipkowitz KB and Boyd DB), Vol. 2, pp. 57–80. VCH Publishers, New York, 1991.

27. Leach AR, Survey of methods for searching the conformation space of small and medium-sized molecules. In: *Reviews in Computational Chemistry* (eds Lipkowitz KB and Boyd DB), Vol. 2, pp. 1–55. VCH Publishers, New York, 1991.

28. Boyd DB, Theoretical and physicochemical studies of β-lactam antibiotics. In: *β-Lactam Antibiotics: Chemistry and Biology* (eds Morin RB and Gorman M), Vol. 1, pp. 437–545. Academic Press, New York, 1982.

29. Boyd DB, Substituent effects in cephalosporins as assessed by molecular orbital calculations, nuclear magnetic resonance, and kinetics. *J. Med. Chem.* **26**: 1010–1013, 1983.

30. Boyd DB, Quantum mechanics in drug design: Methods and applications. *Drug. Inf. J.* **17**: 121–131, 1983.

31. Boyd DB, Electronic structure of cephalosporins and penicillins. 15. Inductive effect of the 3-position side chain in cephalosporins. *J. Med. Chem.* **27**: 63–66 (1984).

32. Boyd DB, Elucidating the leaving group effect in the β-lactam ring opening mechanism of cephalosporins. *J. Org. Chem.* **50**: 886–888, 1985.

33. Neuhaus FC and Georgopapadakou N, Strategies in β-lactam design. In: *Emerging Targets in Antibacterial and Antifungal Chemotherapy* (eds Sutcliffe J and Georgopapadakou N). Chapman and Hall, New York, 1991.

34. Boyd DB, Elzey TK, Hatfield LD, Kinnick MD, and Morin JM Jr., γ-lactam analogues of the penems. *Tetrahedron Lett.* **27**: 3453–3456, 1986.

35. Boyd DB, Foster BJ, Hatfield LD, Hornback WJ, Jones ND, Munroe JE, and Swartzendruber JK, γ-lactam analogues of carbapenems. *Tetrahedron Lett.* **27**: 3457–3460, 1986.

36. Allen NE, Boyd DB, Campbell JB, Deeter JB, Elzey TK, Foster BJ, Hatfield LD, Hobbs JN Jr., Hornback WJ, Hunden DC, Jones ND, Kinnick MD, Morin JM Jr., Munroe JE, Swartzendruber JK, and Vogt DG, Molecular modeling of γ-lactam analogues of β-lactam antibacterial agents: Synthesis and biological evaluation of selected penem and carbapenem analogues. *Tetrahedron* **45**: 1905–1928, 1989.

37. Jungheim LN, Sigmund SK, Jones ND, and Swartzendruber JK, Bicyclic pyrazolidinones, steric and electronic effects on antibacterial activity. *Tetrahedron Lett.* **28**: 289–292, 1987.

38. Indelicato JM and Pasini CE, The acylating potential of γ-lactam antibacterials: Base hydrolysis of bicyclic pyrazolidinones. *J. Med. Chem.* **31**: 1227–1230, 1988.

39. Allen NE, Hobbs JN Jr., Preston DA, Turner JR, and Wu CYE, Antibacterial properties of the bicyclic pyrazolidinones. *J. Antibiot.* **43**: 92–99, 1990.

40. Ternansky RJ and Draheim SE, [3.3.0]Pyrazolidinones: An efficient synthesis of a new class of synthetic antibacterial agents. *Tetrahedron Lett.* **31**: 2805–2808, 1990.

41. Hansch C and Leo A, *Substituent Constants for Correlation Analysis in Chemistry and Biology.* Wiley-Interscience, New York, 1979.

42. Boyd DB and Seward CM, The substituent parameter database: A powerful tool for QSAR analysis. In: *QSAR; Rational Approaches to the Design of Bioactive Compounds* (eds Silipo C and Vittoria A), pp. 167–170. Elsevier, Amsterdam, 1991.

43. Jungheim LN, Boyd DB, Indelicato JM, Pasini CE, Preston DA, and Alborn WE Jr., Synthesis, hydrolysis rates, supercomputer modeling, and antibacterial activity of bicyclic tetrahydropyridazinones. *J. Med. Chem.* **34**: 1732–1739, 1991.

44. Boyd DB, Eigenbrot C, Indelicato JM, Miller MJ, Pasini CE, and Woulfe SR, Heteroatom activated β-lactam antibiotics: Considerations of differences in the biological activity of [[3(*S*)-(acylamino)-2-oxo-1-azetidinyl]oxy]acetic acids (oxamazins) and the corresponding sulfur analogues (thiamazins). *J. Med. Chem.* **30**: 528–536, 1987.

45. Hirata T, Matsukuma I, Mochida K, and Sato K, KT3777 (LY163892), a new oral carbacephem antibiotic; synthesis and chemistry. *Abstracts of 27th Interscience Conference on Antimicrobial Agents and Chemotherapy*, No. 1187. New York, NY, Oct. 4–7, 1987.

46. Counter FT, Ensminger PW, Alborn WE, Preston DA, and Turner JR, Microbiological evaluation of LY163892 (KT3777), a new orally absorbed β-lactam antibiotic. *Abstracts of 27th Interscience Conference on Antimicrobial Agents and Chemotherapy.* No. 1188. New York, NY, Oct. 4–7, 1987.

47. Gray G, Ramotar K, Krulicki W, and Louie TJ, Stability and *in vitro* activity of LY163892 and cefaclor. *Abstracts of 27th Interscience Conference on Antimicrobial Agents and Chemotherapy*, No. 1200. New York, NY, Oct. 4–7, 1987.

48. Sato K, Okachi R, Mochida K, and Hirata T, KT3777 (LY163892), a new orally active carbacephem antibiotic: antibacterial activity and pharmacokinetics in animals. *Abstracts of 27th Interscience Conference on Antimicrobial Agents and Chemotherapy,* No. 1203. New York, NY, Oct. 4–7, 1987.

49. Turner JC, Sullivan HR, Quay JF, Finch LS, and Stucky JF, Absorption, distribution, metabolism, and elimination of LY163892 (KT3777), an orally absorbed β-lactam, in laboratory animals. *Abstracts of 27th Interscience Conference on Antimicrobial Agents and Chemotherapy,* No. 1204. New York, NY, Oct. 4–7, 1987.

50. Quay JF, Coleman DL, Finch LS, Indelicato JM, Pasini CE, Shoufler JR, Sullivan HR, and Turner JC, Pharmacokinetics of orally absorbed β-lactam antibiotics LY163892 (KT3777) and

LY213735 (KT3799) in laboratory animals. *Abstracts of 27th Interscience Conference on Antimicrobial Agents and Chemotherapy*, No. 1205. New York, NY, Oct. 4–7, 1987.

51. Blaszczak LC, Brown RF, Cook GK, Hornback WJ, Hoying RC, Indelicato JM, Jordan CL, Katner AS, Kinnick MD, McDonald JH III, Morin JM Jr., Munroe JE, and Pasini CE, Comparative reactivity of 1-carba-1-dethiacephalosporins. *J. Med. Chem.* **33**: 1656–1662, 1990.

52. Cook GK, McDonald JH III, Alborn W Jr., Boyd DB, Eudaly JA, Indelicato JM, Johnson R, Kasher JS, Pasini CE, Preston DA, and Wu CYE, 3-Quaternary-1-carba-1-dethiacephalosporins. *J. Med. Chem.* **32**; 2442–2450, 1989.

53. Cooper RDG, Daugherty BW, and Boyd DB, Chiral control of the Staudinger reaction. *Pure Appl. Chem.* **59**: 485–492, 1987.

54. Boyd DB, Computer-assisted molecular design studies of β-lactam antibiotics. In: *Frontiers of Antibiotic Research* (ed Umezawa H), pp. 339–356. Academic Press, Tokyo, 1987.

55. van Gunsteren WF and Berendsen HJC, Computer simulation of molecular dynamics: Methodology, applications, and perspectives in chemistry. *Angew. Chem. Int. Ed. Engl.* **29**: 992–1023, 1990.

56. Robson B and Platt E, Comparison of the X-ray structure of baboon α-lactalbumin and the tertiary predicted computer models of human α-lactalbumin. *J. Comput.-Aided Mol. Design.* **4**: 369–379 (1990).

57. Brünger AT and Karplus M, Molecular dynamics simulations with experimental restraints. *Acc. Chem. Res.* **24**: 54–61, 1991.

58. Boyd DB, Snoddy JD, and Lin HS, Molecular simulations of DD-peptidase, a model β-lactam-binding protein: Synergy between X-ray crystallography and computational chemistry. *J. Comput. Chem.* **12**: 635–644, 1991.

59. Brooks BR, Bruccoleri RE, Olafson BD, States DJ, Swaminathan S, and Karplus M, CHARMM: A program for macromolecular energy, minimization, and dynamics calculations. *J. Comput. Chem.* **4**: 187–217, 1983.

60. *Manuals to QUANTA/CHARMm*, Polygen Corporation, Waltham, Massachusetts.

61. Momany FA, Klimkowski VJ, and Schäfer L, On the use of conformationally dependent geometry trends from *ab initio* dipeptide studies to refine potentials for the empirical force field CHARMM. *J. Comput. Chem.* **11**: 654–662, 1990.

62. Boyd DB, Compendium of software for molecular modeling. In: *Reviews in Computational Chemistry* (eds Lipkowitz KB and Boyd DB), Vol. 1, pp. 383–392. VCH Publishers, New York, 1990.

63. Van Opdenbosch N, Cramer R III, and Giarrusso FF, SYBYL, the integrated molecular modelling system. *J. Mol. Graphics.* **3**: 110–111, 1985.

64. *Manual to SYBYL Molecular Modeling Software*, Tripos Associates, St. Louis, Missouri.

65. Clark M, Cramer RD III, and Van Opdenbosch N, Validation of the general purpose Tripos 5.2 force field. *J. Comput. Chem.* **10**: 982–1012, 1989.

66. Jorgensen WL, Revised TIPS for simulations of liquid water and aqueous solutions. *J. Chem. Phys.* **77**: 4156–4163, 1982.

67. Jorgensen WL, Chandrasekhar J, Madura JD, Impey RW, and Klein ML, Comparison of simple potential functions for simulating liquid water. *J. Chem. Phys.* **79**: 926–935, 1983.

68. Ryckaert JP, Ciccotti G, and Berendsen HJC, Numerical integration of the Cartesian equations of motion of a system with constraints: molecular dynamics of *n*-alkanes. *J. Comput. Phys.* **23**: 327–341, 1977.

69. Tanford C, *Physical Chemistry of Macromolecules*. Wiley, New York, 1961, pp. 548–564.

70. Martin RB, *Introduction to Biophysical Chemistry*. McGraw-Hill, New York, 1964, pp. 79–97.

71. Loncharich RJ and Brooks BR, The effects of truncating long-range forces on protein dynamics. *Proteins: Structure, Function, and Genetics* **6**: 32–45, 1989.

72. Boyd DB and Ott JL, Lack of relevance of kinetic parameters for exocellular DD-peptidases to cephalosporin MICs. *Antimicrob. Agents Chemother*. **29**: 774–780, 1986.
73. Boyd DB and Ott JL, Examination of model enzyme and penetration systems in relation to antibacterial activity. *J. Antibiotics,* **39**: 281–185, 1986.
74. Boyd DB, Successes of computer-assisted molecular design. In: *Reviews in Computational Chemistry* (eds Lipkowitz KB and Boyd DB), Vol. 1, pp. 355–371. VCH Publishers, New York, 1990.

New Methods in Molecular Shape Analysis to Identify and Characterize Active Conformations

B. J. Burke*, K. Rowberg†, M. G. Cardozo*, M. G. Koehler* and A. J. Hopfinger*†

Departments of Medicinal Chemistry and Pharmacognosy* and Chemistry†, University of Illinois at Chicago, Chicago, IL 60680, USA

Abstract. This presentation focuses upon developing methods to postulate active conformations/shapes in a series of flexible analogs when the receptor geometry is unknown. In one study of a set of 3-X substituted triazines which inhibit dihydrofolate reductase (DHFR), the size and flexibility of the substituents were both too large, and the number of analogs and substituent structural changes each too small, to permit a reliable inference of the active substituent conformation. However, use of template conformations from ligands bound to DHFR, along with minimum energy substituent conformations, permitted us to postulate that the bonding topology of the substituent, and not its conformation, controls the SAR. In another study of a series of β-hydroxylase inhibitors it was possible to construct a molecular shape descriptor that employs loss in intramolecular conformational energy as part of the shape comparison. In both cases significant 3D-QSARs were generated and the methods developed were added to our molecular shape analysis (MSA) formalism.

1. Introduction

A major goal of any computer-assisted drug design investigation is to determine the *active conformation* of a bioactive compound, or series of analogs. The active conformation gives rise to one, or more, spatial descriptors that can be used to hypothesize a basis for biological activity. For a series of analogs, the measures of these spatial descriptor(s) can be correlated with the corresponding measured biological activities as part of a quantitative structure-activity relationship (QSAR).

The active conformation is usually thought of as that conformation which corresponds to the preferred molecular shape of a molecule when it binds to the receptor. While this does not have to be the case, the *active conformation-active molecular shape* interrelationship has proven a useful concept in drug design. The practical implementation of a strategy to identify the active conformation in a series of analogs has evolved. Succinctly, the goal is to discover which stable *intramolecular* conformation is common to active analogs, but is a high-energy, unstable intramolecular conformer state for inactive analogs. Analogs of "medium" activity are usually not considered in the search for the active conformation. Their reduced activity is often found from a QSAR to be due to nonconformational properties, and/or modest losses in stability of the active conformer. Also, an analog series may not contain key members which provide the structure-activity information to uniquely identify the active conformation. Rather, two, or more, conformations can be valid candidates for the active conformer state. In these situations the relative statistical significance of each of the QSARs can sometimes be used to postulate the most likely active conformation. However, if such an approach does not work, the investigation must look to further synthesis and testing of new analogs to complete the identification of the active conformation.

Recently, we have come upon two distinct analog series where the best QSAR equations are based upon active conformations which are energetically available, on an intramolecular basis, to both active and inactive analogs [1, 2]. That is, even though there may be conformer states of high intramolecular energy for inactive analogs that are available to active analogs, none of these conformer states lead to QSARs that are as statistically significant as a QSAR based upon a conformation available to all analogs in the data set. In retrospect, it appears that the chemical bonding topology (configuration) imparted to an analog by the substituent modification over-rides conformation in dictating the active molecular shape and generating a QSAR. The purpose of this paper is to discuss these two studies and the combined treatment of chemical bonding topology and conformation in determining active molecular shapes.

2. A Molecular Shape Analysis and Quantitative Structure-Activity Relationship Investigation of Some Triazine-Antifolate Inhibitors of *Leishmania* Dihydrofolate Reductase

2.1. *Data set*

Booth et al. [3], as part of a linear free energy QSAR analysis, have reported the structure and corresponding Michaelis inhibition constant, K_i, for the inhibition of *Leishmania major* dihydrofolate reductase (DHFR) by a series of 4,6-diamino-1,2-dihydro-2,2-dimethyl-1-(3-substituted-phenyl)-s-triazines (Fig. 1).

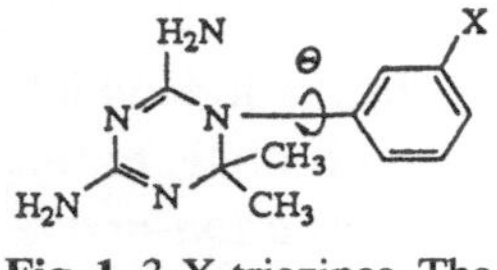

Fig. 1. 3-X-triazines. The structures and K_i are reported as part of Table I

We have investigated the inhibition of DHFR from different species (4–9) using a method called *Molecular Shape Analysis* (MSA) [10] which uses explicit three-dimensional information on the shape of a molecule, as determined from confor-

Table I. The set of 3-X triazine analogs, corresponding observed DHFR binding constants, QSAR descriptor values, predicted DHFR inhibition potencies, and corresponding residuals

No.	X	$\log(1/K)_{obs}$	$\Delta\pi_{3'}$	Shape reference compound 45 and Eq. (3)			Shape reference compound 45 and Eq. (4)		
				$V_0(\text{Å}^3)$	$\log(1/K)_{pred}$	$\Delta\log(1/K)$	$V_0(\text{Å}^3)$	$\log(1/K)_{pred}$	$\Delta\log(1/K)$
1	SO_2NH_2	3.44±0.08	−3.48	173.3	3.78	−0.34	174.5	3.80	−0.36
2	$CONH_2$	3.93±0.06	−3.15	170.5	3.97	−0.04	170.5	3.98	−0.05
3	$COOCH_2CH_3$[a]	3.95±0.03	−1.15	151.6	4.81	−0.86	157.3	4.93	−0.98
4	OH	4.27±0.05	−2.33	153.5	4.22	0.05	153.5	4.24	0.03
5	DL-$CH(OH)C_6H_5$[b]	4.37±0.03	−1.12	165.8	5.07	−0.70	148.8	4.80	−0.43
6	$COCH_3$	4.40±0.02	−2.21	161.5	4.43	−0.03	161.5	4.45	−0.05
7	OC_2H_5	4.42±0.12	−1.28	170.9	4.90	−0.48	172.3	5.13	−0.71
8	OCH_3	4.64±0.05	−1.68	158.5	4.67	−0.21	158.5	4.69	−0.05
9	H	4.90±0.06	−1.66	136.6	4.30	0.60	136.6	4.33	0.57
10	NO_2	5.00±0.04	−1.94	167.5	4.69	0.31	167.5	4.71	0.29
11	CN	5.22±0.02	−2.23	154.5	4.30	0.92	154.5	4.32	0.90
12	$O(CH_2)_{13}CH_3$	5.23±0.08	5.33	213.7	5.79	−0.56	219.3	5.82	−0.59
13	$C(CH_3)_3$	5.24±0.02	0.32	183.5	5.86	−0.62	180.7	5.82	−0.58
14	$O(CH_2)_5CH_3$	5.40±0.03	1.01	198.6	5.82	−0.42	205.3	6.37	−0.97
15	OCH_2-L-adamantyl	5.51±0.04	1.41	188.3	6.11	−0.60	176.4	5.91	−0.40
16	CF_3	5.66±0.04	−0.78	168.2	5.25	0.41	168.2	5.27	0.39
17	C_2H_5	5.71±0.08	−0.63	175.6	5.43	0.28	175.6	5.45	−0.26
18	$OCH_2C_6H_5$	5.72±0.03	0.00	210.8	6.25	−0.53	215.3	6.33	−0.61
19	Cl	5.92±0.03	−0.95	169.5	5.20	0.72	169.5	5.22	0.70
20	$O(CH_2)_{10}CH_3$	6.01±0.04	3.71	222.8	6.52	−0.51	219.3	6.43	−0.42
21	$CH_2NHC_6H_4$-4'-SO_2NH_2	6.12±0.04	−0.66	224.9	6.28	−0.16	228.5	6.43	−0.22
22	$CH_2NHC_6H_4$-4'-Cl	6.18±0.04	−0.66	223.1	6.25	−0.07	226.7	6.31	−0.13
23	$CH_2OC_6H_4$-3'-CH_2OH	6.19±0.03	0.00	242.5	6.80	−0.61	231.4	6.61	−0.42
24	$CH_2NHC_6H_3$-3'-5'$(CONH_2)_2$	6.29±0.03	−0.66	247.8	6.68	−0.39	233.5	6.43	−0.14
25	$CH_2OC_6H_4$-3'-CH_3	6.37±0.04	0.00	233.7	6.65	−0.28	230.6	6.59	−0.22
26	$CH_2SC_6H_5$	6.39±0.09	0.64	195.7	6.14	0.25	191.4	6.07	0.32
27	$CH_2OC_6H_4$-4'$(CH_2)_4CH_3$	6.47±0.03	0.00	213.5	6.30	0.17	220.6	6.42	0.05
28	$O(CH_2)_8CH_3$	6.52±0.01	2.63	208.3	6.45	0.07	219.3	6.63	0.11
29	$CH_2Sec_6H_5$[c]	6.56±0.02	0.71	209.3	6.39	0.17	198.8	6.21	0.35
30	$O(CH_2)_2OC_6H_4$-3'-CF_3	6.60±0.06	0.02	193.6	5.96	0.65	190.5	5.91	0.69
31	$CH_2OC_6H_4$-3'-Cl	6.65±0.14	0.00	248.8	6.91	−0.26	241.7	6.78	−0.13
32	S-CH_2-C_6H_5	6.65±0.05	0.64	230.0	6.73	−0.08	219.6	6.55	0.10
33	$CH_2OC_6H_4$-3'-CN	6.75±0.07	0.00	234.9	6.67	0.08	239.1	6.74	0.01
34	$CH_2OC_6H_4$-3'$CH(CH_3)_2$	6.76±0.08	0.00	244.8	6.84	−0.08	250.3	6.93	−0.17
35	$CH_2OC_6H_4$-3'-OCH_3	6.82±0.04	0.00	235.3	6.68	0.14	231.6	6.61	0.21
36	$CH_2OC_6H_4$-3'-C_2H_5	6.90±0.05	0.00	239.4	6.75	0.15	233.0	6.63	0.27
37	$SCH_2C_6H_4$-4'-Cl	6.96±0.02	0.64	218.9	6.54	0.42	221.5	6.58	0.38
38	$CH_2OC_6H_4$-3'-$NHCONH_2$	7.04±0.09	0.00	246.5	6.87	0.17	242.4	6.79	0.24
39	$CH_2OC_6H_4$-3'-$NHCOCH_3$	7.12±0.05	0.00	246.5	6.87	0.25	242.4	6.80	0.32
40	$O(CH_2)_4OC_6H_4$-3'-CF_3	7.12±0.05	1.05	218.2	6.60	0.52	229.3	6.78	0.34
41	$(CH_2)_{11}CH_3$	7.14±0.08	4.75	235.4	6.42	0.73	235.1	6.35	0.79
42	$CH_2OC_6H_2$-2',4',5'-Cl_3	7.16±0.04	0.00	251.3	6.96	0.20	259.9	7.09	0.07
43	CH_2O-1-naphthyl	7.40±0.08	0.00	277.9	7.42	−0.02	286.6	7.55	−0.15
44	$(CH_2)_8CH_3$	7.68±0.05	3.13	244.3	7.02	0.66	245.7	7.01	0.67
45	$CH_2OC_6H_4$-3'-C_6H_5	7.77±0.10	0.00	302.9	7.86	−0.09	302.9	7.83	−0.06

mational analysis [11], in construction of a QSAR. Shape comparisons for a series of molecules are performed by computing the common overlap steric volume between each molecule in the data set and a common reference molecule. The measure of the common overlap steric volume is used as a correlation descriptor, along with other molecular properties, like $\log P$, to establish a QSAR. Each molecule in the set is used, in turn, as the reference molecule. The objective is to search for the reference molecule which leads to the statistically most significant QSAR.

All of the MSA-based QSARs we have constructed are of the general form

$$\text{Biological activity} = f(\text{molecular shape, substituent lipophilicity, DBSV}) \tag{1}$$

where DBSV stands for a data base specific variable which is sometimes needed to aid in explaining the activity of a small number of compounds in a particular data set. The DBSV is usually found to be an explicit three-dimensional feature related to conformation. The general QSAR form embodied in Eq. (1) holds not only for the inhibition of DHFR from different species but also for different chemical classes of DHFR inhibitors. In fact, we have been able to use the molecular shape of a triazine, as a reference, to generate a QSAR for a set of quinazoline DHFR inhibitors [7]. This consistency among MSA-QSARs across both DHFR species and inhibitor chemical classes is in contrast to DHFR QSARs developed by other investigators. These latter QSARs differ from one another in the set of independent variables, both as a function of DHFR species and the chemical class of the inhibitors. Even QSARs for 3-X-triazine DHFR inhibitors vary markedly from one another in the set of independent variables for different structure-activity data bases. Thus, we thought it would be of interest to carry out a MSA-QSAR analysis of the compounds in Table I to see if the general form of Eq. (1) holds up for this data set, and to further develop MSA as a method to "marry" three-dimensional molecular modeling and classic QSAR analysis.

The conformational behavior of the 3-X-triazines with respect to torsion angle, θ, defined in Fig. 1, is well characterized [4]. In fact, a number of the compounds in Table I having smaller and/or relatively rigid side chains have been included in

Fig. 2. A space-filling representation of the active conformation, with respect to θ, of X = 3,4-Cl$_2$

prior conformational analyses [4, 5]. This information along with the postulated "active" conformation with respect to θ, and shown in Fig. 2 for X=3, 4-Cl$_2$, were used in the present analysis. That is, the conformation assigned to all of the triazines in Table I, with respect to θ, is as shown in Fig. 2. The basis for this assignment of conformation is described in Refs. [4] and [5]. It is possible that constraints of species and specificity might alter the postulated active conformation. However, there is a high correlation between the inhibition potencies reported in Ref. [5], using the bovine liver I$_{50}$ values, and the Michaelis inhibition constants of the same compounds reported in Table I:

$$\log(1/K_i) = 0.876 \log(1/I_{50}) - 0.727 \tag{2}$$
$$N = 11, \ R = 0.956, \ SD = 0.16, \ F = 96.1$$

This correlation suggests similar binding modes in the two DHFR enzymes which we assume includes indentical inhibitor conformations with respect to θ. Consequently, we focused our efforts on the conformational behavior of the X substituent, especially highly flexible groups of the forms:

$$X = -(CH_2)_n - Z - R, \quad -Z - (CH_2)_n - R, \quad \text{and} \quad -(CH_2)_n - CH_3.$$

Unfortunately, there are not sufficient types of structural variations among the substituents in Table I to identify an active substituent conformation using MSA. Hence, an alternate strategy was derived to quantitatively characterize the role of substituent conformation upon the QSAR. The conformation of methotrexate, as bound to *Escherichia coli* DHFR in the crystal structure [12], was used as a structural comparison template for the 3-X-triazines in Table I. A hydrogen-suppressed illustration of the methotrexate conformation bound to *E.coli* DHFR is shown in Fig. 3. The criteria for comparison were to superimpose the fragments common to methotrexate and each of the triazines, as well as to freeze θ at the postulated active triazine conformation.

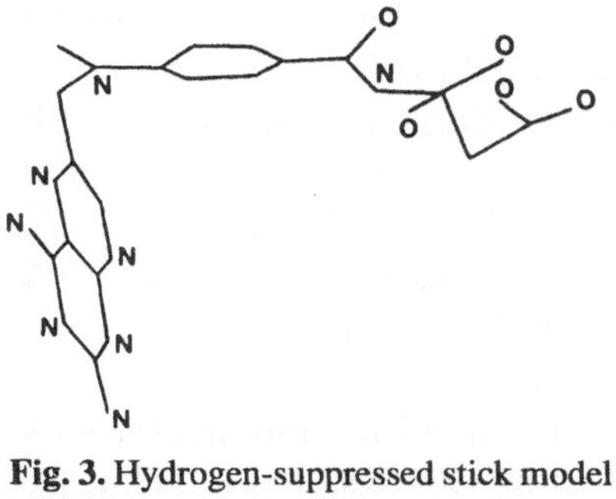

Fig. 3. Hydrogen-suppressed stick model representation of the *E.coli* DHFR-bound conformation of methotrexate

The substituent torsion angles were then varied to optimize the pairwise common overlap steric volume between each of the compounds in Table I and the inhibitor-bound conformation of methotrexate. The resulting conformation of each flexible substituent was then used as a starting point in a conformational energy minimization of each compound with respect to the torsion angles. The resulting minimum energy conformers were used in the MSA analysis to generate a QSAR.

This approach to assigning the active substituent conformation assumes that the 3-X-triazines bind to *L.major* DHFR in an intramolecular minimum energy conformation which is close to the bound conformation of methotrexate to *E. coli* DHFR. While this is a reasonable approach, it is of interest to know how dependent the resulting MSA-QSAR is on the substituent conformations. Thus, an alternate set of minimum energy substituent conformations, generated from a different common conformational template, were also considered as potential active substituent conformations. In this case all substituents were initially placed in planar conformations such that each substituent was coplanar to the 3-substituted phenyl ring.

All substituent torsion angles were selected to generate an extended conformation. For substituents containing an aromatic ring, the plane of the extended conformation was taken to be the same as the plane of the substituent ring. Three structural classes of substituents require further definition within the context of an extended conformation. These three structural classes are characterized by compound 23 of Table I for class I, compound 27 for class II, and compound 12 for class III. Schematic illustrations of the extended conformations for each of these three classes are given in Fig. 4. The solid circles indicate the pattern of atoms composing the bonding topology used to define the extended conformation for each structural class. The extended conformation for compound 45 of Table I, the most active inhibitor, is shown in Fig. 5.

The 3-X-triazines were again frozen into the postulated active conformation with respect to θ. The conformational energy of each compound was minimized as a function of the substituent torsion angles using the extended conformation as a starting point. The resulting minimum energy conformations were then considered in a second set of statistical analyses to generate a MSA-QSAR.

The conformational energy minimizations were performed using the MMFF option of CHEMLAB-II [13] which is based upon the Allinger MM2 force field [14]. Atomic charge densities were computed using the CNDO/2 method [15]. The pairwise common overlap steric volumes were computed using the MSA option in CHEMLAB-II which is based upon a three-dimensional numerical volume integration algorithm [13].

Other physicochemical descriptors were also considered in conjunction with pairwise common overlap steric volume in formulating QSARs. In particular, we focused upon various combinations of substituent lipophilicity because of our

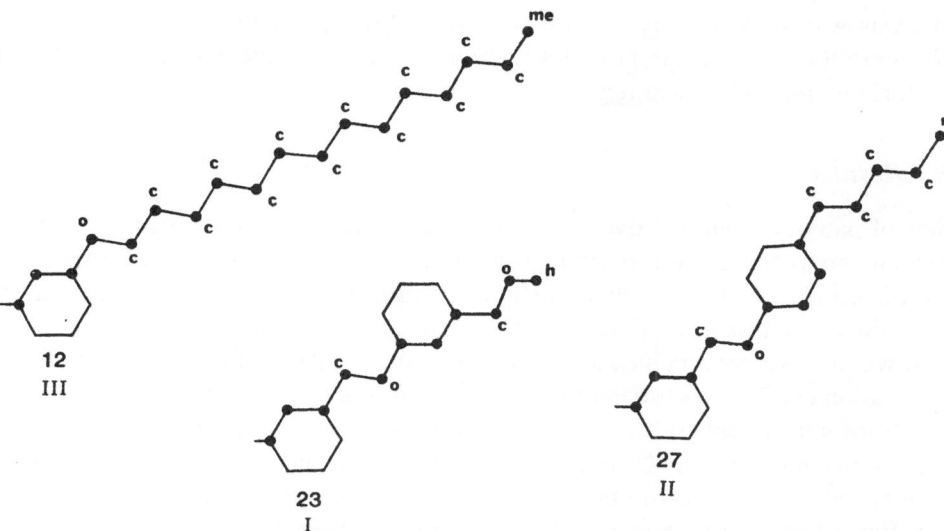

Fig. 4. Schematic representation of the extended planar conformations of the three classes of large substituents contained in Table I. Class I, compound 23 shown. Class II, compound 27 shown. Class III, compound 12 shown

previous general findings, and because of the findings of Booth et al. in their analysis of this data set (Table I) [1]. Some of the other descriptors considered are differences in the conformational energies between relative intramolecular minima, the minimum energy conformer having the largest overlap volume with methotrexate, net dipole moment of the molecule, and total conformational entropy of the X substituent.

The multivariational matrix containing inhibitor potency and molecular descriptors as columns and compounds as rows was preanalyzed by using two block partial least-squares principal component analysis, CPLS2 [16]. A program to perform this analysis is part of SIMCA, a statistical software package for pattern recognition studies [17]. Combinatorial descriptor set, multiple linear regression

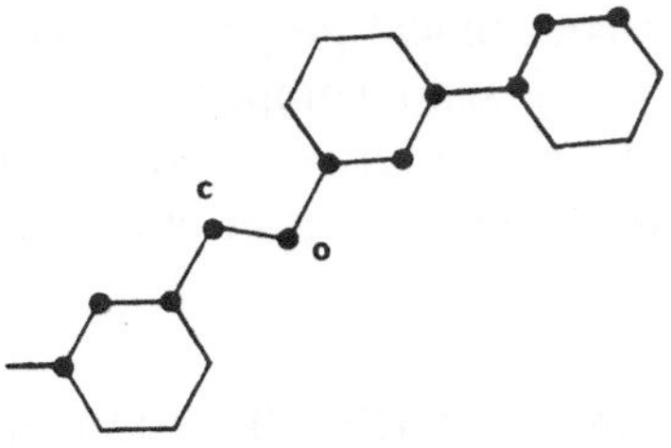

45

Fig. 5. Schematic representation of the extended planar conformation of compound 45, the shape reference standard used to construct the QSARs

analysis was subsequently performed, with SIMCA, to compute MSA-QSARs. No more than three descriptors were used at a time as independent variables in performing multiple linear regression analyses.

2.2. Results

Sets of pairwise common overlap steric volumes were generated using the minimum energy conformers derived from both the methotrexate conformational template and from the extended substituent conformational state. The results based upon the methotrexate conformational template are presented first.

It was necessary to include a lipophilicity term in the QSARs in order to achieve correlation coefficients greater than 0.80. In particular, setting the lipophilicity of $Y = 0$ for substituents of the form $X = - CH_2ZC_6H_4 -Y$, as done by Booth et al. [3], yielded the optimum QSARs. The lipophilicity values, $\Delta\pi_{3'}$, reported in Table I, and used in QSAR construction, are based upon the values given by Booth et al. [3]. For internal consistency in data base construction, the lipophilicity of each shape reference standard is defined as zero. This corresponds to subtracting a constant from each absolute lipophilicity value of each compound for each choice in the shape reference standard. Scaling an independent variable by a constant does not affect the significance of fit in multidimensional linear regression analysis.

No descriptors, other than overlap volume and relative lipophilicity, were found to be significant. The optimum MSA-QSAR has the most active inhibitor, compound 45 of Table I, as the shape reference standard:

$$\log(1/K_i) = 0.017 \ (\pm 0.005)V_0 + 0.273 \ (\pm 0.114)\Delta\pi_{3'} - 0.069 \ (\pm0.030)\Delta\pi_{3'}^2 +$$
$$+ 2.58 \ (\pm 1.04)$$
$$N = 45, \ R = 0.910, \ SD = 0.47, \ F = 66.1, \ \Delta\pi_{3'} \ (\text{opt}) = 1.98$$

(3)

The alternate MSA-QSAR analyses were carried out using the substituent minimum energy conformations determined from the extended planar substituent conformations. Once again, various additional physicochemical properties were used in conjunction with V_0 to establish a QSAR. The same set of compounds used to derive Eq. (3) were individually considered as the shape reference standard compound in constructing sets of V_0 descriptors for the linear regression analyses. The optimum MSA-QSAR found is

$$\log(1/K_i) = 0.017 \ (\pm 0.005)V_0 + 0.276 \ (\pm 0.111)\Delta\pi_{3'} - 0.072 \ (\pm0.030)\Delta\pi_{3'}^2 +$$
$$+ 2.65 \ (\pm 0.99)$$
$$N = 45, \ R = 0.913, \ SD = 0.47, \ F = 68.4, \ \Delta\pi_{3'} \ (\text{opt}) = 1.92$$

(4)

The shape reference standard for Eq. (4) is, again, compound 45 of Table I, the most active analog. $\Delta\pi_{3'}$ is the same as in Eq. (3), and accounts for a very substantial amount of the variance in the observed binding constants of the 45 compounds in Table I. The predicted binding constants, and their respective differences from observed values, using Eq. (4), are also given in Table I.

2.3. Discussion

A drawback to MSA-QSAR analyses has been the difficulty in characterizing the shape of large flexible substituents. There is generally not enough information available in the structure-activity table to permit identification, or, at least, construction of a reasonable hypothesis for the active conformation. Moreover, the large number of conformational degrees of freedom can make the conformational search impractical.

Our strategy for the MSA-QSAR analysis of large, flexible substituents is to first select a common conformational class, such as the extended conformation. This conformation is then assigned as the starting point in an intramolecular energy minimization for each analog being investigated. The resulting set of minimum energy conformations are then used in the MSA-QSAR analysis. Another general conformational class is then selected, and the process is repeated.

A comparison of the resulting MSA-QSARs provides information regarding how critical explicit substituent conformation is to activity. This is in contrast to the role of size and/or other physicochemical properties of the substituent upon biological activity.

In this particular study we had the relatively rare advantage of knowing the binding conformation of methotrexate to DHFR. This structure was used to define a conformational class of substituent conformations for MSA-QSAR analysis. The inherent assumption in using this conformational class is that the 3-X-triazines bind to *L.major* DHFR with substituent conformations mimicking the binding conformation of methotrexate. A significant MSA-QSAR, Eq. (3), was constructed from this conformational hypothesis. However, Eq. (4), generated from a different active substituent conformation model, is as significant a MSA-QSAR as Eq. (3). This suggests that the binding conformations of the substituents may not necessarily mimic the methotrexate crystal structure binding conformation. Rather, the intrinsic shape of the substituent, owing to its chemical bonding topology, may control binding potency. If this were not the case, it is reasonable to expect a major difference in statistical significance between Eqs. (3) and (4). These two MSA-QSARs are predicated upon very different substituent binding conformations.

The finding that different common substitutent conformational templates can be used to construct significant MSA-QSARs may also indicate relatively non-specific conformational binding modes for the larger 3-X-triazine substituents.

3. Molecular Shape and QSAR Analysis of Some 1-X-Benzylimidazole-2-Thione Inhibitors of Dopamine β-Hydroxylase

3.1. *Data set*

Kruse et al. have reported inhibition potencies of a set of 52 1-X-benzylimidazole-2-thiones (Fig. 6) as multisubstrate inhibitors of dopamine β-hydroxylase (DβH) [18]. These inhibitors effectively reduce blood pressure in adult male Okamoto-Aoki spontaneously hypertensive rats using oral or intraperitoneal dosing. DβH is

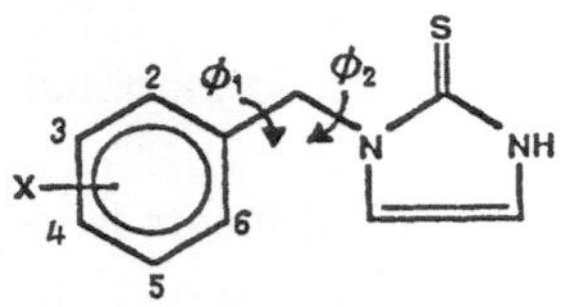

Fig. 6. 1-X-benzylimidazole-2-thiones

a copper-containing mixed-function oxidase that catalyzes the conversion of dopamine to norepinephrine. As such, this DβH-inhibitor system represents an intervention endpoint for treatment of cardiovascular disorders related to hypertension.

Two different measures of DβH inhibition were reported by Kruse et al. [18]. The less active analogs (compounds 1–15 of Table II) have inhibition reported in terms of percent inhibition at a fixed inhibitor concentration of 1.0×10^{-4} M. The inhibition potencies of the rest of the analogs are given as actual IC_{50} values. We attempted to put these two inhibition measurements on a common scale so that all 52 compounds could be considered in MSA-QSAR analyses.

The following relationship was assumed in order to combine the two activity scales:

$$\text{Activity} = -\log(IC_{50}) = \log\left(\frac{X \% \text{ inhibition} \times 10^4 \text{ M}}{50\% \text{ inhibition}}\right) \qquad (5)$$

where X is the percent inhibition reported at 1.0×10^{-4} M. The set of inhibition potencies, based upon the IC_{50} measure, for all 52 compounds is given in Table II.

3.2. *Methods*

The compounds selected for the SAR database were built using standard bond lengths and angles with the CHEMLAB-II molecular modeling package [13]. The geometries were optimized by free valence molecular mechanics using the MMFF

Table II. General SAR table for some X-substituted 1-aralkylimidazole-2-thiones

No.	X	V_0	$Q_{3,4,5}$	Obs $-\log(IC_{50})$	Pred[&] $-\log(IC_{50})$	Diff(Obs-Pred) $-\Delta\log(IC_{50})$
1	4-CO$_2$H					
2	*2,6-Me$_2$	0.8158	−0.03	3.00	3.12	−0.12
3	4-CH$_2$OH					
4	*2,6-Cl$_2$	0.8419	0.06	3.15	3.31	−0.16
5	3-SO$_2$NH$_2$, 4-OMe					
6	2,6-(OMe)$_2$	0.7475	−0.03	3.30	3.62	−0.32
7	*2-Cl	0.9084	0.04	3.45	3.65	−0.20
8	2-Me	0.8961	0.05	3.47	3.56	−0.09
9	3,4-(OMe)$_2$	0.8239	0.28	3.47	3.83	−0.36
10	4-CF$_3$	0.8937	0.07	3.70	3.58	0.12
11	3-CF$_3$, 4-OMe	0.8550	0.10	3.76	3.43	0.33
12	2,6-Cl$_2$, 4-OMe	0.7633	0.17	3.81	3.91	−0.10
13	4-CH$_3$	0.9435	0.07	3.83	4.18	−0.35
14	4-Br	0.9171	0.16	3.94	4.03	−0.09
15	3-Br, 4-OMe	0.8764	0.30	4.08	3.99	0.09
16	*3-F, 4-OMe	0.8974	0.31	4.13	4.18	−0.05
17	2-OMe	0.8826	0.01	4.13	3.36	0.77
18	3-Me, 4-OMe	0.8853	0.16	4.16	3.73	0.43
19	2-OH	0.9470	0.01	3.24	4.10	−0.86
20	*3-NO$_2$, 4-OMe	0.8827	0.19	3.45	3.78	−0.33
21	*4-OMe	0.8980	0.10	3.69	3.69	0.00
22	3-OMe	0.9002	0.20	3.80	3.95	−0.15
23	3-OH	0.9861	0.20	3.83	—	—
24	3-CF$_3$, 4-OH	0.9195	0.11	3.92	3.94	−0.02
25	2,4,6-Cl$_3$	0.7980	0.17	3.99	3.65	0.34
26	*2,5-Cl$_2$	0.9075	0.12	4.01	3.83	0.18
27	4-Cl	0.9480	0.16	4.02	4.46	−0.44
28	2,6-Cl$_2$, 4-OH	0.8271	0.17	4.12	3.57	0.55
29	2,3,5,6-F$_4$, 4-OH	0.9508	0.42	4.21	5.11	−0.90
30	4-NO$_2$	0.9419	0.13	4.28	4.30	−0.02
31	*2,3-Cl$_2$	0.9041	0.12	4.28	3.80	0.48
32	3-Me, 4-OH	0.9635	0.16	4.31	4.73	−0.42
33	4-F	0.9890	0.17	4.33	5.26	−0.93
34	3,5-Cl$_2$, 4-OMe	0.9016	0.30	4.33	4.19	0.14
35	3,5-F$_2$, 4-OMe	0.8995	0.50	4.44	4.64	−0.20
36	*H-	0.9480	0.02	4.48	4.14	0.34
37	3-NO$_2$, 4-OH	0.9517	0.22	4.51	4.66	−0.15
38	*3,4-Cl$_2$	0.9505	0.19	4.55	4.57	−0.02
39	*2,4-Cl$_2$	0.8744	0.14	4.77	—	—
40	3-Br, 4-OH	0.9589	0.02	4.92	4.32	0.60
41	3-Cl	0.9862	0.15	4.92	5.16	−0.24
42	3-F	0.9905	0.25	5.25	5.48	−0.28
43	#3,5-F$_2$					
44	4-OH	0.9892	0.13	5.59	5.17	0.42
45	*3,5-Cl$_2$	0.9931	0.19	5.62	5.40	0.22
46	*3,4-(OH)$_2$	0.9831	0.28	5.66	5.39	0.27
47	#3,5-Cl$_2$					
48	*3-Cl, 4-OH	0.9986	0.19	5.70	5.53	0.17
49	*3-F, 4-OH	0.9999	0.30	5.82	5.81	0.01
50	3,5-F$_2$	0.9905	0.43	5.92	5.90	0.02
51	*3,5-Cl$_2$, 4-OH	1.0000	0.27	6.17	5.74	0.43
52	*3,5-F$_2$, 4-OH	1.0000	0.50	7.13	6.28	0.85

& = Values of $-\log(IC_{50})$ predicted by Eq. (8).
* = Compounds used to develop trial QSARs.
# = Rings connected by three methylene units, not congeneric in series.

option in CHEMLAB-II; this is a version of Allinger's MM2 program [14] with extended parameterization and force field function generalization.

The CHEMLAB-II modeling package option SCAN was used to perform a fixed valence conformation energy scan at ten degree increments of ϕ_1 and ϕ_2. The reference conformation for $\phi_1 = \phi_2 = 0°$ corresponds to the coplanar ring defined by Fig. 6. Flexible side chains were also scanned at ten degree increments about the principal bonds. A molecular mechanics force field composed of dispersion/steric, electrostatic, and, where applicable, hydrogen bonding contributions were used to estimate the conformational energy. The nonbonded steric MMFF parameters from CHEMLAB-II were used to compute the dispersion/steric interactions.

The electrostatic interactions were calculated using a Coulombic functional representation with the dielectric constant equal to 3.5 and atomic charges calculated by the CNDO/2 method [15]. When hydrogen bonding atoms were available for bonding, the hydrogen bonding potential developed by Hopfinger [11] was used. The global conformational energy minimum was used to define the relative stability of each conformational state sampled.

The QSAR conformation of an analog refers to that conformation whose shape measure, relative to the shape reference compound, yields the most significant QSAR. Conformations within ΔE_u^* kcal/mol of the global energy minimum for each analog u were considered as candidates for the QSAR conformation. Three separate ΔE_u^* values (1, 3 and 6 kcal/mol) were considered in this study.

The shape reference compound is the compound to which all others in the analog series are compared. Each analog in the data set is evaluated as possibly being the shape reference compound. Selection of the shape reference compound is ultimately dictated by maximizing the statistical significance of the resulting QSAR. However, the choice for the shape reference compound also depends upon the conformations assigned to a candidate reference compound. In this investigation conformations within $\Delta E_\upsilon^* = 1.0$ kcal/mol of the global energy minimum for analog υ, were selected as possible conformers for the shape reference compound.

The geometric criterium for pairwise analog molecular superposition was to place the N–C–N of the thione rings of each pair of molecules identically upon one another. Each analog, υ, was considered as the shape reference compound and assigned a conformation from the set of conformations satisfying $\Delta E_\upsilon^* = 1.0$ kcal/mol. Each analog, u, in the data set was then assigned the conformations consistent with the ΔE_u^* constraint and compared to υ. The criterium for selecting a unique conformation for u in a trial QSAR was to maximize the shape similarity measure (see below) between u and υ in terms of the conformation of u.

The common overlap steric volume, V_0, between each analog u in the data set and the reference compound, υ, was determined as

$$V_0 = V_u \cap V_\upsilon \tag{6}$$

where V_x represents the spatial occupancy of compound x [4, 9, 10]. V_0 was computed using a numerical integration scheme in CHEMLAB-II [13], and was used as a measure of how similar in shape the analogs are to the reference compound. V_0 does not contain information regarding the intramolecular stability of each conformation of the pair of compounds from which they are derived. Loss in conformational stability in order to realize particular measures in V_0 was taken into account by defining the shape commonality index, I_c:

$$I_c = S\,(u, \upsilon, w) - \left[\frac{\Delta E_u}{[\Delta E_u^*(\Delta E_u^* + \Delta E_\upsilon^*) + \varepsilon]^{1/2}} + \frac{\Delta E_\upsilon}{[\Delta E_\upsilon^*\,(\Delta E_u^* + \Delta E_\upsilon^*) + \varepsilon]^{1/2}} \right] \tag{7}$$

where $S\,(u, \upsilon, w)$ is wV_0, and ΔE_x^* is the difference in conformational energy between the global minimum and the conformation used to compute V_0 for analog $x = u$ and/or υ [19]. The parameter w is a weighting factor between shape similarity and loss in intramolecular conformational stability. The MSA-QSAR is optimized as a function of w.

In addition to molecular shape, other molecular properties were considered in the construction of the MSA-QSAR. These properties included various measures of lipophilicity, partial charge densities, dipole moment parameters, and conformational entropy.

3.3. Results

The conformational profiles of the analogs can be characterized by three conformational (ϕ_1, ϕ_2) energy maps shown in Fig. 7. The three maps represent analogs having di-ortho substituents (2,6-diCl), Fig. 7a, mono-ortho substitution (2-Cl), Fig. 7b, and non-ortho substituents (4-Cl), Fig. 7c. To facilitate the presentation of the results, we divide the maps into four quadrants as defined in Fig. 7a. All three maps have common conformational energy minima in quadrants II and III at energies within 1 kcal/mol of the global minimum. Quadrant IV contains a relatively small space where minima are found in all three maps. However, the minima of the mono- and di-ortho compounds are relatively high energy conformers (about 3 kcal/mol above the global minimum). Location of conformational energy minima in quadrant I vary from map to map.

The optimum QSAR was found using compound 51, of Table II, the X = (3,5-Cl$_2$, 4-OH) analog. The shape reference compound from past investigations has usually been either the most active, or largest analog in the data set. Compound 51 is the second most active analog and one of the largest. The conformation of compound 51, as the shape reference compound, corresponds to (ϕ_1, ϕ_2) = ($-120°$, $-120°$) which is shown in Fig. 8. This conformation is the global energy minimum

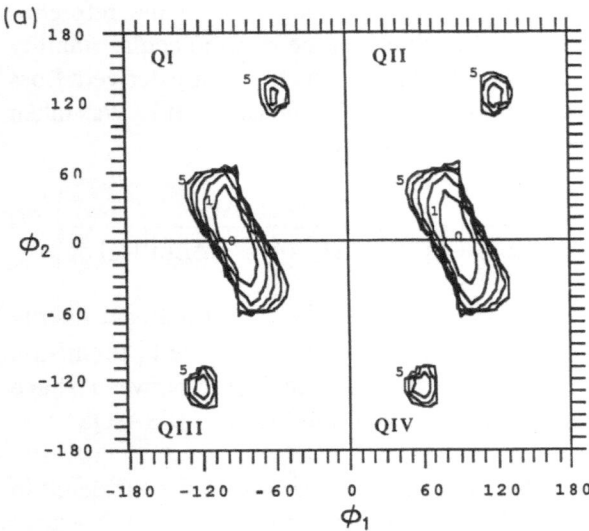

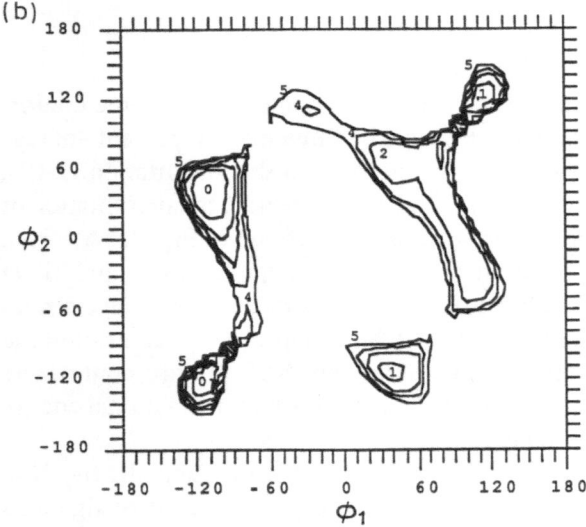

Fig. 7a–c. Conformational energy maps of three DβH inhibitors with MMFF potentials: (**a**) *di-ortho* subsituted analog, X = 2, 6-Cl$_2$, –log(IC$_{50}$) = 3.15; (**b**) *mono-ortho* substituted analog, X = 2, 5-Cl$_2$, log(IC$_{50}$) = –log 4.01; (**c**) *non-ortho* substituted analog, X = 3, 5-Cl$_2$, –log (IC$_{50}$) = 5.62. Quadrants are listed in (**a**)

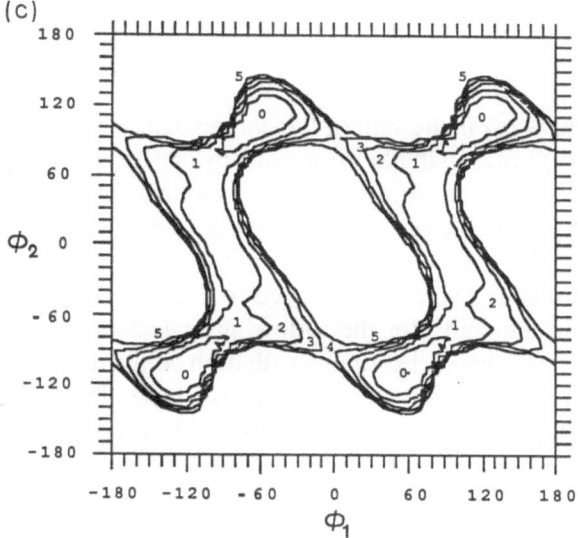

Fig. 7c

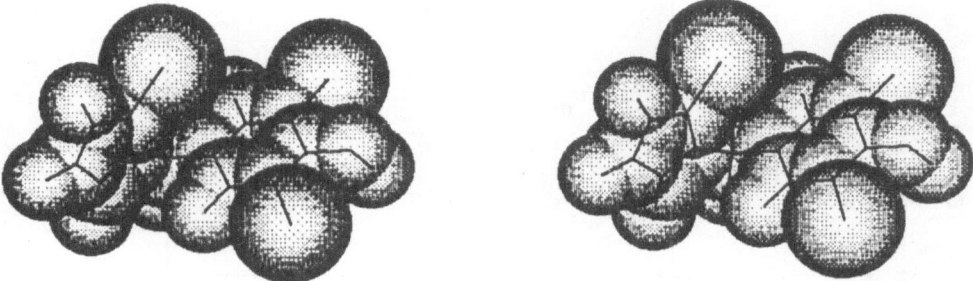

Fig. 8. A space-filling steric representation of compound 51 (X = 3,5-Cl$_2$, 4-OH) for ϕ_1, ϕ_2 = -120°, -120°

for compound 51. It must be stressed that (ϕ_1, ϕ_2) = (-120°, -120°) is not necessarily the "active" conformation since inactive analogs can adopt this conformation as a stable minimum energy state. Rather, the shape reference compound conformational state corresponds to the conformer that optimizes the QSAR from the set of conformations energetically available to the shape reference compound.

The optimum QSAR in terms of the F-statistic measure is (for shape reference compound 51 of Table II):

$$- \log(IC_{50}) = -117.4\ (\pm 21.9)V_0 + 70.4\ (\pm 12.3)V_0^2 +$$
$$+ 2.33\ (\pm 0.51)Q_{3,4,5} + 52.12 \tag{8}$$
$$N = 45,\ R = 0.90,\ F = 57.4,\ S = 0.41$$

where V_0 is defined by Eq. (6) and $Q_{3,4,5}$ is the sum of the charge densities on C_3, C_4, and C_5 (see Fig. 6). Equation (8) states that inhibition potency can be increased by analogs which have the steric shape of compound 51, as shown in Fig. 8, resulting from substituents which make $-(-C_3 - C_4 - C_5 -)-$ a positive charge domain.

Forty-seven of the 52 compounds in Table II were used to construct Eq. (8). Five analogs were not included in the study for the following reasons: two had longer chains than all of the others, two had substituents with high conformational flexibility and one was potentially a charged species. Each of these classes of substituents represent possible complications to the application of MSA.

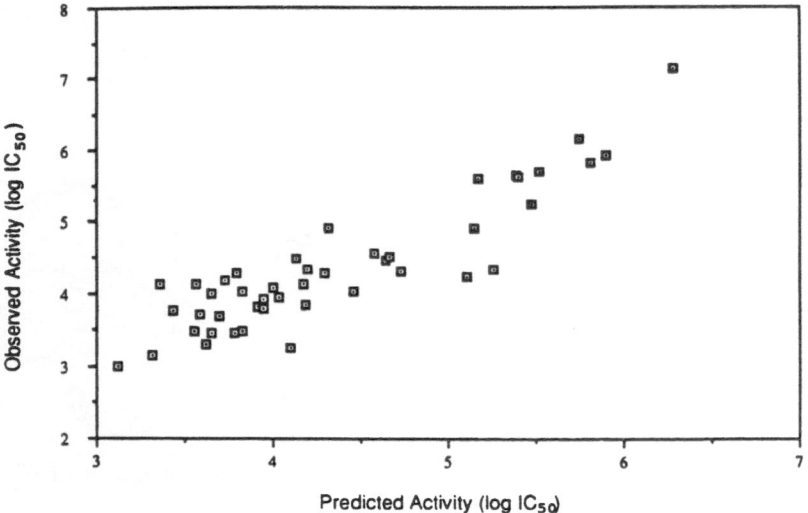

Fig. 9. A plot of observed activity ($-\log(IC_{50})$) vs. predicted activity ($-\log(IC_{50})$) as predicted by Eq. (8)

There are two outliers for Eq. (8). Compound 39 is an outlier, unless lipophilicity is included in the QSAR. Compound 23 is the second outlier. Predicted and residual, $-\Delta\log(IC_{50})$, inhibition potencies are given in Table II along with V_0 and $Q_{3,4,5}$ values. A plot of observed versus predicted $-\log(IC_{50})$ values, using Eq. (8), are shown in Fig. 9.

We considered the effect of loss in intramolecular stability in order to achieve a particular measure of V_0 using Eq. (7). A surprise finding is that any attempt to penalize the shape fit, V_0, for loss in intramolecular stability *decreases* the significance of the corresponding QSAR relative to Eq. (8).

3.4. *Discussion*

Perhaps the most interesting aspect of this study, as it relates to the MSA formalism, is postulating the "active" conformation. In most MSA studies the hypothesized active conformation corresponded to that conformer state for the analogs whose intramolecular stability decreases as biological activity decreases. However, for the 1-X-benzylimidazole-2-thione DβH inhibitors, we conclude the arrangement of the substituents on the phenyl ring, that is the shape of the molecule due to its *configuration* (chemical bonding topology), over-rides conformational effects due to torsion angles ϕ_1 and ϕ_2 in specifying activity. Of course, conformation is important in that only a limited set of low energy conformation states are available to a particular configuration. Still, both active and inactive thione analogs can energetically adopt the postulated "active" conformations.

The possible active conformations were arrived at by determining which set of V_0, as a function of ϕ_1, ϕ_2 and ΔE_u^*, yields the optimum MSA-QSAR. Low energy conformers in quadrants II and III, see Fig. 7a, are the best active conformation candidates. Quadrants II and III comprise two sets of conformers which are enantiomeric and are not distinguishable in MSA. We arbitrarily decided to use conformers in quadrant III in this investigation.

The shape commonality index, I_c, can provide a means of determining the balance between ligand molecular shape and ligand conformational stability on inhibition potency. In so far as ligand molecular shape can be construed as a representation of ligand-receptor binding, I_c can be viewed as a composite measure of ligand stability and ligand binding potential. The inclusion of I_c into MSA overcomes many of the restrictions and limitations of comparing the shape similarity among *flexible* molecules. A major thrust in future research will be to further explore and define the shape commonality index.

Acknowledgements. MC is a Fogarty International NIH Postdoctoral Fellow and BJB is a Fellow of the American Foundation of Pharmaceutical Education (AFPE). Funds from the Laboratory of Computer-Aided Molecular Modeling and Design at UIC were used in support of this work.

References

1. Koehler MG, Rowberg-Schaefer K and Hopfinger AJ, A molecular shape analysis and quantitative structure-activity relationship investigation of some triazine-antifolate inhibitors of *Leishmania* dihydrofolate reductase. *Arch. Biochem. Biophys.* **266**: 152–161, 1988.

2. Burke BJ and Hopfinger AJ, 1 (substituted-benzyl) imidazole-2 (3H)-thione inhibitors of dopamine β-hydroxylase. *J. Med. Chem.* **33**: 274–281, 1990.

3. Booth RG, Selassie CD, Hansch C and Santi DV, Quantitative structure-activity relationship of triazine-antifolate inhibition of *Leishmania* dihydrofolate reductase and cell growth. *J. Med. Chem.* **30**: 1318–1334, 1987.

4. Hopfinger AJ, A QSAR investigation of dihydrofolate reductase inhibition by Baker triazines based upon molecular shape analysis. *J. Amer. Chem. Soc.* **102**: 7196–7206, 1980.

5. Hopfinger AJ, A general QSAR for dihydrofolate reductase inhibition by 2,4-diaminotriazines based upon molecular shape analysis. *Arch. Bioch. & Bioph.* **208**: 153–163, 1981.

6. Hopfinger AJ, Inhibition of dihydrofolate reductase: Structure-activity correlations of 2,4-diamino-5-benzylpyrimidines based upon molecular shape analysis. *J. Med. Chem.* **24**: 818–822, 1981

7. Battershell C, Malhotra D and Hopfinger AJ, Inhibition of dihydrofolate reductase: Structure-activity correlations of quinazolines based upon molecular shape analysis. *J. Med. Chem.* **24**: 812–818, 1981.

8. Hopfinger AJ, Theory and application molecular potential energy fields in molecular shape analysis: A quantitative structure-activity relationship study of 2,4-diamino-5-benzylpyrimidines as dihydrofolate reductase inhibitors. *J. Med. Chem.* **26**: 990–996, 1983.

9. Mabilia M, Pearlstein RA and Hopfinger AJ, Molecular shape analysis and energetics-based intermolecular modelling of benzylpyrimidine dihydrofolate reductase inhibitors. *Eur. J. Med. Chem.* **20**: 163–174, 1985

10. Mabilia M, Pearlstein RA and Hopfinger AJ, Computer graphics in molecular shape analysis. In: *Molecular Graphics and Drug Design* (Burgen ASV, Roberts GCK and Tute MS, eds), Elsevier, Amsterdam, pp.157–182, 1986.

11. Hopfinger AJ, *Conformational Properties of Macromolecules*, Academic Press, New York, 1973.

12. Matthews DA, Bolin TJ, Burridge JM, Filman DJ, Volz KW, Kaufman BT, Beddell CR, Champness JN, Stemmers DK and Kraut J, X-ray crystallography of *L. Casei* dihydrofolate reductase. *J. Biol. Chem.* **281**: 260–267, 1985.

13. *CHEMLAB-II Users Guide*, Chemlab, Inc., Lake Forest, IL 60045, 1990.

14. Burkert U and Allinger NL, *Molecular Mechanics*, American Chemical Society, Washington DC, 1982.

15. Pople JA and Beveridge DC, *Approximate Molecular Orbital Theory*, McGraw-Hill, New York, 1970.

16. Wold S, Hellber S and Dunn WJ III, Statistical approaches to analyze toxicological databases. *Acta Pharmacol.Toxicol.* **52**: 158–169, 1983.

17. Dunn WJ III and Wold S, Regression analysis in structure-activity studies. *Bioorg.Chem.* **9**: 505–512, 1980.

18. Kruse LI, Kaiser C, DeWolf WE Jr, Frazee JS, Ross ST, Wawro J, Ezekiel M, Ohlstein EH and Berkowitz BA, Multisubstrate inhibitors of dopamine β-hydroxylase. 2. Structure-activity relationships at the phenethylamine binding site. *J. Med. Chem.* **30**: 486–494, 1987.

19. Hopfinger A J and Burke B J, Molecular shape analysis of structure-activity tables. In: *Quantitative Structure-Activity Relationships in Drug Design* (Fauchere JL, ed), Alan R. Liss, New York, pp.151–159, 1989.

Molecular Mechanisms for the Sequence Recognition of DNA by (+)-CC-1065

C. H. Lin* and L. H. Hurley*†

Drug Dynamics Institute, College of Pharmacy† and Department of Chemistry*,
University of Texas at Austin, Austin, Texas 78712, USA

Abstract. (+)-CC-1065 is an extremely potent antitumor antibiotic produced by *Streptomyces zelensis*. We have previously postulated that a sequence dependent catalytic activation and/or conformational flexibility are/is responsible for the DNA sequence selectivity of (+)-CC-1065. In this review article we demonstrated that both of these factors are likely to be involved. Using a 12-mer DNA duplex containing a unique (+)-CC-1065 bonding site within a highly reactive 5'AGTTA* (where* denotes the covalent modification site) sequence in combination with high field proton NMR, we have examined the structure of both the duplex and its covalent adduct. First, we demonstrate the involvement of a bridging water molecule between a phenolic proton on the alkylating subunit of (+)-CC-1065 and an anionic oxygen in the phosphate on the noncovalently modified strand of DNA. This structure has important implications for catalytic activation of the covalent reaction between (+)-CC-1065 and DNA and, consequently, the molecular basis for sequence selective recognition of DNA by the alkylating subunit of (+)-CC-1065. Second, we illustrate the importance of bending and associated conformational flexibility of DNA in the sequence selectivity of the covalent reaction with (+)-CC-1065.

1. Introduction

(+)-CC-1065 is an antitumor antibiotic with a unique structure [1–2] and mechanism of action (Fig. 1) [3–4]. Previous studies have demonstrated that this antibiotic is extraordinary for both its base and DNA sequence specificity [1, 5–7]. An analog of (+)-CC-1065 designed and synthesized by Upjohn scientists was recently introduced into phase I clinical trials [8]. Structurally, (+)-CC-1065 consists of three repeated pyrroloindole subunits (A, B, and C in Fig. 1) attached via amide linkages that are approximately 15° out of plane, providing the drug molecule with a right-hand twisted banana shape [2, 9]. Subunit "A" contains the DNA reactive cyclopropane ring that alkylates N3 of adenine when it binds within certain reactive sequences (Fig. 1) [10–11]. Since only adenines in certain sequence contexts react with (+)-CC-1065, this drug has sequence selectivity [5]. Surprisingly, the "A" subunit alone contains sufficient structural information to encode the primary molecular basis for sequence selectivity [5], and this subunit is also essential for antitumor activity [4]. However, as we have previously demonstrated, the noncovalent binding interactions of the B and C subunits with

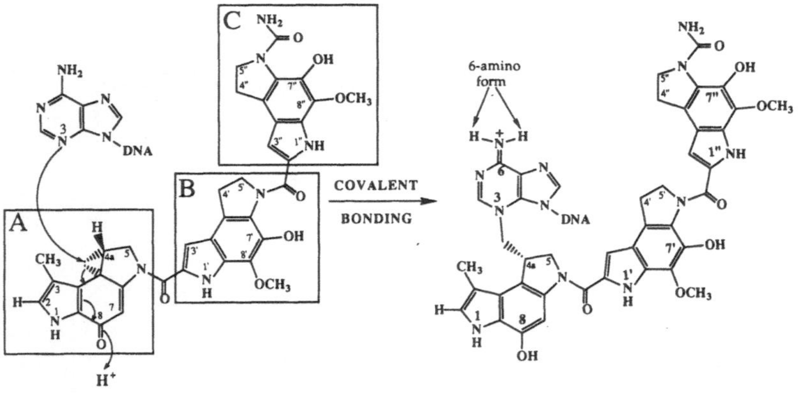

Fig. 1. Reaction of (+)-CC-1065 with double-stranded DNA to form the (+)-CC-1065-(N3 adenine)-DNA adduct. The covalently modified adenine is in the quaternized 6-amino form [10]

DNA can modulate or fine tune this sequence selectivity [5]. We have previously suggested that the primary basis for sequence selectivity is through *a sequence dependent catalytic activation and/or a sequence dependent conformational flexibility* [12]. In this article, we review the evidence for both catalytic activation and sequence dependent conformational flexibility. These results have been published previously [13–14].

2. Material and Methods

2.1. *Chemicals*

(+)-CC-1065 was obtained from The Upjohn Company and used without further purification. ^{17}O-water (60 and 45 atom% ^{17}O) was purchased from Cambridge Isotope Laboratories. Reagents used to prepare for NMR buffer, sodium phosphate (99.99%), and sodium chloride (99.99%) were purchased from Aldrich Co. HPLC, water and methanol were purchased from Baxter Scientific and Fisher Co., respectively. Hydroxylapatite used to purify the 12-mer duplex and the 12-mer adduct was purchased from CalBiochem Co., Sephadex G-25 (superfine) was purchased from Pharmacia Co.

2.2. *Preparation and Purification of the 12-mer Duplex and the (+)-CC-1065-12-mer Duplex Adduct*

The non-self-complementary 12-mer duplex (Fig. 2) for NMR studies was synthesized in house on a 10 μmol scale by using the solid-phase cyanoethyl phosphoramidite approach [15] on an Applied Biosystem automated DNA synthesizer, Model 381A. The general procedures for synthesis, deprotection, drug bonding,

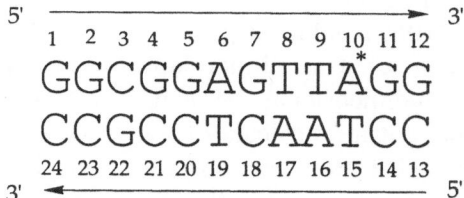

Fig. 2. Sequence and numbering scheme of the 12-mer duplex which contains a highly reactive bonding sequence 5′AGTTA* for (+)-CC-1065, where* denotes the covalently bonding site of (+)-CC-1065

HPLC,[1] and chromatography purification of the 12-mer duplex and the (+)-CC-1065-12-mer duplex adduct have been reported previously [10, 13–14].

2.3. High-field NMR Spectroscopy

One- and two-dimensional 500 MHz ^{1}H NMR data sets in 90% H_2O/10% D_2O or 99.96% D_2O buffered solution containing 10 mM sodium phosphate and 100 mM sodium chloride at pH 6.85 were recorded on a General Electric GN-500 FT NMR spectrometer. Proton chemical shifts were recorded in parts per million (ppm) and referenced relative to external TSP[1] (1 mg/ml) in D_2O (HOD signal was set to 4.751 ppm). Two-dimensional NMR data sets were recorded according to the procedures described in the recent publications [16–17]. Suppression of the water signal was achieved with 1–3̄–3̄–1̄ pulse sequence [18] with a delay of 120 ms. One-dimensional NOE[1] difference experiments were performed at 23°C.

2.4. T_1 Inversion-Recovery Experiments

T_1 measurements in 90% H_2O/10%D_2O were made on a General Electric GN-500 NMR instrument by the conventional inversion-recovery method executed with a 1–3̄–3̄–1̄ selective excitation pulse sequence. The 1–3̄–3̄–1̄ routine was optimized to provide exact 90- and 180-degree flip angles on the resonances of interest. The pulse repetition delay is 5 s.

3. Results and Discussion

The non-self-complementary 12-mer duplex sequence (Fig. 2) is part of the early promoter region of SV40 DNA and is contained within the 21 bp repeat region.

---

[1] Abbreviations: CPI, cyclopropylpyrroloindole; BPDE, benzapyrenediolepoxide; HPLC, high performance liquid chromatography; NMR, nuclear magnetic resonance; NOE, nuclear overhauser effect; NOESY, two-dimensional NOE correlated spectroscopy; ppm, parts per million; TSP, 3-(trimethylsilyl) propionate.

This 21 bp repeat region contains the Sp1 protein binding sites 5'GGGCGG as well as two identical highly reactive (+)-CC-1065 bonding sequences 5'AGTTA* (where* denotes the covalently modified adenine). This 12-mer duplex has been thoroughly characterized by one- and two-dimensional ^{1}H and ^{31}P NMR, hydroxyl-radical footprinting, and molecular dynamics calculations and exists as an overall right-handed B-form DNA duplex [16]. However, this 12-mer duplex contains a numbers of unique features which are unusual for a B-form DNA.

3.1. ^{1}H NMR of the ^{17}O-labeled Water and Phosphate of the (+)-CC-1065-12-mer Duplex Adduct

(+)-CC-1065 was reacted with the non-self-complementary 12-mer duplex that contains one of the preferred bonding sequences 5'AGTTA*. The duplex and its (+)-CC-1065-modified sequences were characterized by one- and two-dimensional ^{1}H NMR [10, 14]. A partial ^{1}H-NMR spectrum (downfield region) of the non-isotopically labeled (+)-CC-1065 12-mer duplex adduct is plotted in Fig. 3A. The assignments of the exchangeable protons in the 12-mer DNA duplex adduct have been made previously [10, 14] and are noted in Figs. 3A and 3C. Of particular significance is the upfield shifted ^{10}A-^{15}T imino proton (2.07 ppm) and the downfield shifted 6-amino protons ($\approx$2 ppm) of ^{10}A (the covalently modified adenine) that occur at 9.19 and 9.08 ppm in the duplex adduct relative to the duplex alone. These latter signals were assigned to the ^{10}AH6 hydrogen-bonded amino proton (H6$_b$) and ^{10}AH6 external amino proton (H6$_e$) respectively.[2] Confirmation of the assignments for the ^{10}AH6 amino protons was made by synthesizing (6-^{15}N)^{10}A-deoxyadenosine-labeled (+)-CC-1065 12-mer duplex adduct [10] (Fig. 3B). As expected, the ^{10}AH6$_b$ and ^{10}AH6$_e$ resonance signals are split into doublets due to coupling with the ^{15}N nucleus located at N6. Upon heating the sample to 45°C, we found, contrary to our expectation, that the resonance signal assigned to the ^{10}AH6$_b$ proton *exchanged more rapidly* than that assigned to the ^{10}AH6$_e$ proton (unpublished results). Since the 6-amino group of adenine is in the doubly protonated form [10] and the ^{10}A-^{15}T imino proton is shifted upfield, which is indicative of reduced hydrogen bonding strength, we considered the possibility that the observed rapid exchange of the ^{10}AH6$_b$ proton might be due to a facile exchange with an ordered and judiciously positioned hydrogen-bonded water molecule. To evaluate this possibility, the (6-^{15}N)^{10}A-deoxyadenosine-labeled (+)-CC-1065 12-mer duplex adduct was dissolved in ^{17}O-labeled water to attain a 40.5% overall enrichment. To our surprise, not only was the doublet for the ^{10}AH6$_b$ proton broadened, but the (+)-CC-1065 8-phenolic proton of the A subunit was also broadened relative to the equivalent resonance signals in ^{16}O-water

---

[2] The assignments of the H6$_b$ and H6$_e$ protons were made on the basis of comparison of coupling constants to the equivalent protons in the (6-^{15}N) ^{10}A-deoxyadenosine labeled duplex [9].

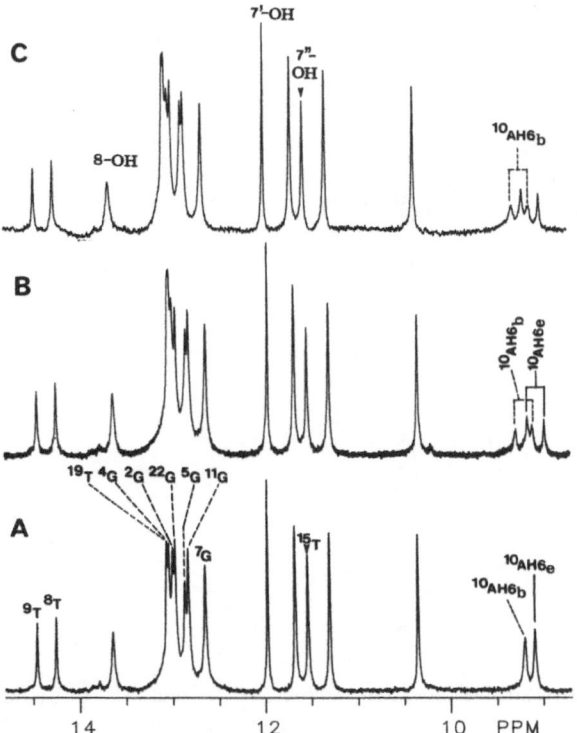

Fig. 3 A, B, C. 500-MHz proton NMR spectrum (8.5–15 ppm, downfield region) of the non-isotopically labeled (+)-CC-1065 12-mer duplex adduct (1 equivalent of (+)-CC-1065 per 12-mer duplex) in 0.5 ml 90% H_2O/10% D_2O buffer containing 10 mM NaH_2PO_4, 100 mM NaCl, pH 6.85 at 25°C. The assignments of the 12-mer imino and the covalently modified adenine 6-amino protons are based on one-dimensional NOE difference and two-dimensional NOESY experiments. 500-MHz ^{1}H-NMR spectrum (8.5–15 ppm, downfield region) of the (6[15]N)deoxyadenosine-labeled (+)-CC-1065 12-mer duplex adduct in (**B**) ^{16}O-water and in (**C**) 40.5% [17]O-enriched water. The [15]T imino and 7"-OH proton resonances overlap

(compare Figs. 3B and 3C).[3] An expansion of the regions containing the broadened proton NMR signals in comparison with the ^{16}O-water sample is shown in Fig. 4. Most likely, the broadening of these protons is due to their interaction with the [17]O nucleus. Oxygen-17, with a spin $I=5/2$, possesses an unsymmetrical charge distribution and therefore an electronic quadrupole moment. The proximity of the 8-phenolic and [10]AH6$_b$ protons to these fluctuating magnetic dipoles enhan-

---

[3] The sample in [17]O–H_2O was prepared by lyophilizing the (6-^{15}N)deoxyadenosine labeled (+)-CC-1065 12-mer duplex adduct in ^{16}O–H_2O solution to complete dryness and then 0.45 mL 45% [17]O–H_2O/0.05 mL D_2O mixture was added to the above sample. So, the only difference between two samples for Figs. 3B and 3C is the [17]O–H_2O content. Other factors such as DNA concentration, amount of metal, and volume of the sample are exactly the same.

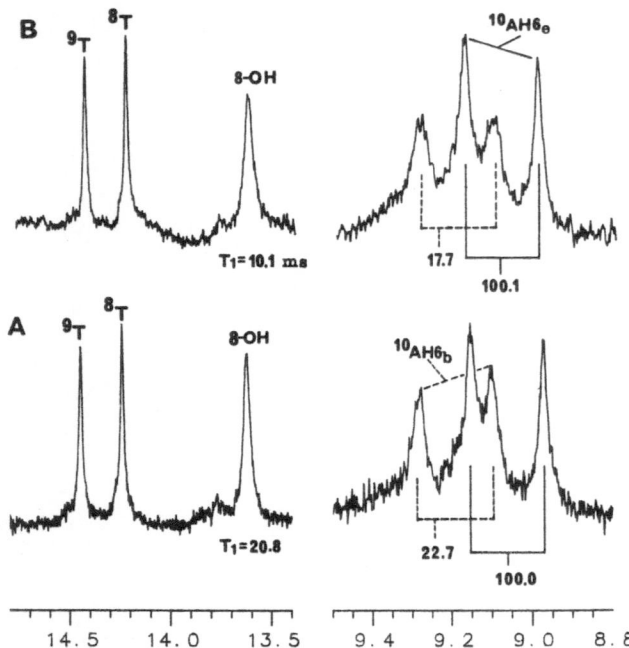

Fig. 4 A, B. 500-MHz ^{1}H-NMR spectrum (expanded downfield regions, 8.7–9.5 and 13.4–14.8 ppm) of the [6-^{15}N]deoxyadenosine-labeled (+)-CC-1065 12-mer duplex adduct in (**A**) regular water (**B**) 40.5% ^{17}O-enriched water NMR buffer containing 10 mM NaH$_2$PO$_4$, 100 mM NaCl, pH 6.85 at 25°C. T_1 measurements were made on a General Electric GN-500 NMR instrument by the conventional inversion-recovery method executed with a 1–3–3–$\bar{1}$ selective excitation pulse sequence. The 1–3–3–$\bar{1}$ routine was optimized to provide exact 90- and 180-degree flip angles on the resonances of interest. Both 8-phenolic proton of (+)-CC-1065 and ^{10}AH6$_b$ of the covalently modified adenine have smaller T_1 values in ^{17}O-enriched water solution. The T_1 relaxation time of ^{10}AH6$_e$ remained the same, regardless of the solvent

ces their spin-lattice (and presumably also spin-spin) relaxation rates, correspondingly reducing their T_1 values. These comparative measurements in ^{16}O-water and ^{17}O-water solutions are shown in Fig. 4. For the 8-phenolic proton in ^{17}O-water, a 50% reduction in T_1 relaxation time is noted in comparison with the non-isotopically-labeled water, and while the T_1 of the ^{10}AH6$_b$ proton in ^{17}O-water is also reduced (22%), it is not so significantly reduced as the 8-phenolic proton. For comparison we also measured the T_1 values for the two other phenolic protons and imino protons in the B and C subunits of (+)-CC-1065 and other exchangeable protons, including the H1 in the A subunits of (+)-CC-1065 in both ^{16}O- and ^{17}O-labeled water. Significant differences between T_1 values in ^{16}O and ^{17}O-labeled water were not observed (unpublished results).

A two-dimensional ^{1}H-NMR study of the (+)-CC-1065 12-mer duplex adduct shows that there is a discontinuity between the ^{9}T-^{16}A and ^{10}A-^{15}T base pairs, which is characterized by a highly propeller twisted ^{10}A-^{15}T [17]. The ^{10}A deoxyadenosine has an unusual sugar conformation and the ^{10}A-^{15}T imino proton resonance is shifted upfield by 2.07 ppm, indicative of a considerably weakened hydrogen-bonding interaction. This discontinuity may well be responsible for the bending of DNA induced by (+)-CC-1065, which appears to be at least superficially similar in magnitude, direction, and in structural origin to that associated with intrinsic A-tract bending [13]. In view of the discontinuity between ^{9}T-^{16}A and ^{10}A-^{15}T, it is perhaps not too surprising that a water molecule may be required to bridge the ^{10}AH6$_b$ proton, and we suspect the O4 of thymine on the opposite strand in the same base pair to stabilize the otherwise badly distorted duplex.

In previous publications, molecular modeling studies from both the French group [19] and the Upjohn-Texas group [6] have suggested that the anionic oxygen of the ^{17}A-^{18}C phosphate on the non-covalently modified strand that is two base pairs to the 5' side of the covalently modified adenine may be involved in a hydrogen bonding interaction with the 8-phenolic proton of the A subunit of (+)-CC-1065-12 mer adduct. It was also suggested that this interaction might be involved in general acid catalysis of the covalent bonding reaction and be important in the observed sequence selectivity [6]. The ≈ 2 ppm downfield shift of the 8-phenolic proton of the A subunit relative to the corresponding B and C subunit protons is in accord with this idea (Fig. 3C). To further evaluate this proposal, we prepared three individual samples of the 12-mer duplex used in this study with ^{17}O labels in the anionic oxygens of the ^{11}G-^{12}G, ^{16}A-^{17}A, and ^{17}A-^{18}C phosphates.[4] The position of the ^{17}O labels in each sample was confirmed by broadening of the corresponding phosphate resonances in the ^{31}P-NMR (unpublished results). Unexpectedly, we found it was the ^{17}O labeled *^{16}A-^{17}A phosphate sample* rather than the ^{17}O-labeled *^{17}A-^{18}C phosphate sample* that produced a very significant sharpening of the resonance signal of the 8-phenolic proton of the A subunit (Fig. 5). A comparison of the T_1 relaxation times of the 8-phenolic proton for the three anionic oxygen ^{17}O-labeled phosphate samples is shown in Fig. 5. While the ^{11}G-^{12}G and ^{17}A-^{18}C ^{17}O-phosphate labeled samples did not reveal any significant change in the T_1 relaxation time of the 8-phenolic proton, the ^{16}A-^{17}A ^{17}O-phosphate labeled sample (Fig. 5C) showed a significant increase (20.8 to 37 ms) in T_1 relaxation time in comparison to the non-isotopically enriched sample. The sharpening of the 8-phenolic proton resonance signal and associated 78% *increase in* T_1 relaxation time is presumably due to the competing dipolar coupling of the ^{17}O nucleus in the phosphate for the relaxation potential of the bridging water

---

[4] In addition to the ^{17}O label at ^{17}A-^{18}C, we decided to evaluate ^{11}G-^{12}G and ^{16}A-^{17}A as possible alternative phosphates involved in stabilization of the adduct because modeled results showed the distances between the anionic-oxygen in each phosphate to the A-subunit quinine-oxygen were 7.35, 4.69, and 5.46 Å in the covalently bound complex and 6.81, 7.88, and 4.26 Å in the nonconvalently bound complex for ^{11}G-^{12}G, ^{16}A-^{17}A, and ^{17}A-^{18}C, respectively (unpublished results).

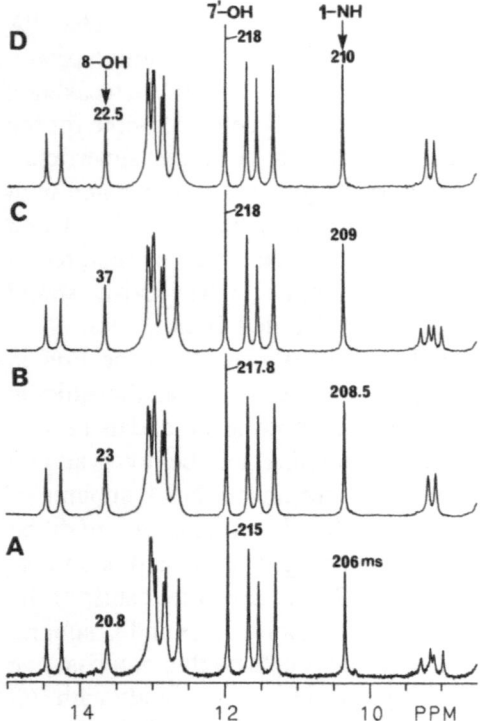

Fig. 5A. 500-MHz ^{1}H-NMR spectrum (8.5–15 ppm, downfield region) of the non-isotopically-labeled (+)-CC-1065 12-mer duplex adduct as described in the legend for Fig. 3.   **B, C** and **D** are the 500 MHz ^{1}H-NMR spectra of the ^{17}O-phosphate-labeled ^{11}G–^{12}G, ^{16}A–^{17}A, and ^{17}A–^{18}C samples

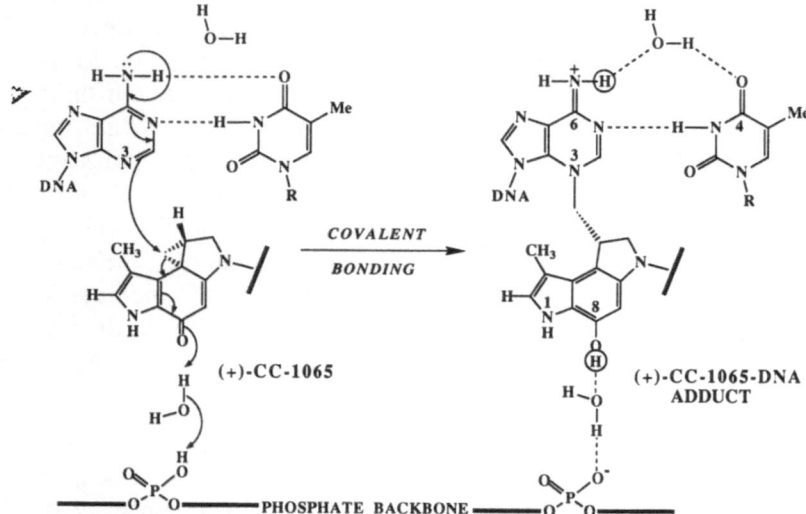

Fig. 6. Proposed mechanism for the catalytic activation of the reaction of the (+)-CC-1065 with DNA involving two strategically placed water molecules

molecule located between the anionic oxygen of the phosphate at ^{16}A-^{17}A and the 8-phenolic proton of the A-subunit of the (+)-CC-1065-DNA adduct.

Based on the results of this study, we are now able to further refine the proposed reaction mechanism that results in covalent modification of N3 of adenine in DNA by (+)-CC-1065 (Fig. 6). The important additions are the two water molecules, one of which is demonstrated to bridge the 8-phenolic proton of (+)-CC-1065 and the phosphate between ^{16}A-^{17}A, and a second that is hydrogen bonded to ^{10}AH6$_b$ and perhaps bridges to O4 of ^{15}T, although we lack direct data that would support the O4 bridging position. The observation of significantly reduced T_1 relaxation times caused by quadrupole induced relaxation of protons by specifically ^{17}O-labeled water molecules suggests these water molecules are ordered and have considerable residence times as part of the (+)-CC-1065-DNA adduct structure. To the best of our knowledge, the critical importance of ordered water molecules in relaying the catalytic activation of covalent bond formation or stabilizing the resulting DNA adducts has not previously been demonstrated. However, their importance in noncovalent complexes of drugs with DNA, such as beneril [20], and in DNA structure [21–22] has been documented. Just as we propose here that an ordered water molecule may relay the general acid catalysis of CPI in its covalent reaction with N3 of adenine, the proposed general acid catalysis of BPDE hydrolysis by an acidic phosphate group [23] may also involve an ordered water molecule. How general this involvement of ordered water molecules may be in providing bridging hydrogen bonds in covalent or noncovalent complexes of drug/carcinogen-DNA adducts remains to be seen. Moreover, their importance in mechanisms of molecular recognition between drug/carcinogens and DNA remains largely unrecognized. Based upon the example described here, the use of ^{1}H-NMR with ^{17}O-labeled water or phosphates may be a powerful probe for detecting such water bridging systems in complexes, as well as in catalytic processes that occur on enzymes and DNA.

3.2. ^{1}H-NMR Studies on the (+)-CC-1065-12-mer Duplex Adduct to Evaluate Groove Width

Using the gel electrophoresis, we have shown that following the covalent reaction with (+)-CC-1065 the 12-mer duplex becomes bent with the bending being between the 8T and 9T nucleotides within the 5'AGTTA sequence [13]. The bend is about 17–22° toward the minor groove of DNA and structurally bears many similarities to a bent DNA structure that is intrinsically associated with A-tract [13, 24]. In this paper two different ^{1}H-NMR parameters, which have been previously associated with narrowing of the minor groove in A-tracts, are reported on this (+)-CC-1065-12 mer duplex adduct.

(1) *Interproton NOEs between adenine H2 and H1' protons on the opposite strands displaced one nucleotide to the 3' side of the adenine.* The relative

intensity of the NOEs between the H2 of adenine and the H1' proton of a
5'-neighboring residue on the complementary strand can provide qualitative and
quantitative information on minor groove width [25]. The 3' one base-pair offset
is necessary to compensate for the right-handed twist of B form DNA. There are
four such NOEs available for this type of assessment in the 12-mer duplex and its
(+)-CC-1065 duplex adduct (i.e., $6AH_2$ to $20CH1'$, $10AH_2$ to $16AH1'$, $16AH_2$ to
$10AH1'$, and $17AH_2$ to $9TH1'$). In the duplex alone, at 250 ms mixing time all four
NOEs can be detected and show qualitatively approximately the same intensities
[16]. In contrast, at both 120 and 60 ms mixing times the AH_2 to H1' protons in the
(+)-CC-1065 duplex adduct show quite dramatic variations in NOE intensities
between the four sites (Fig. 7). At 120 ms the NOE between $17AH_2$ and $9TH1'$ is

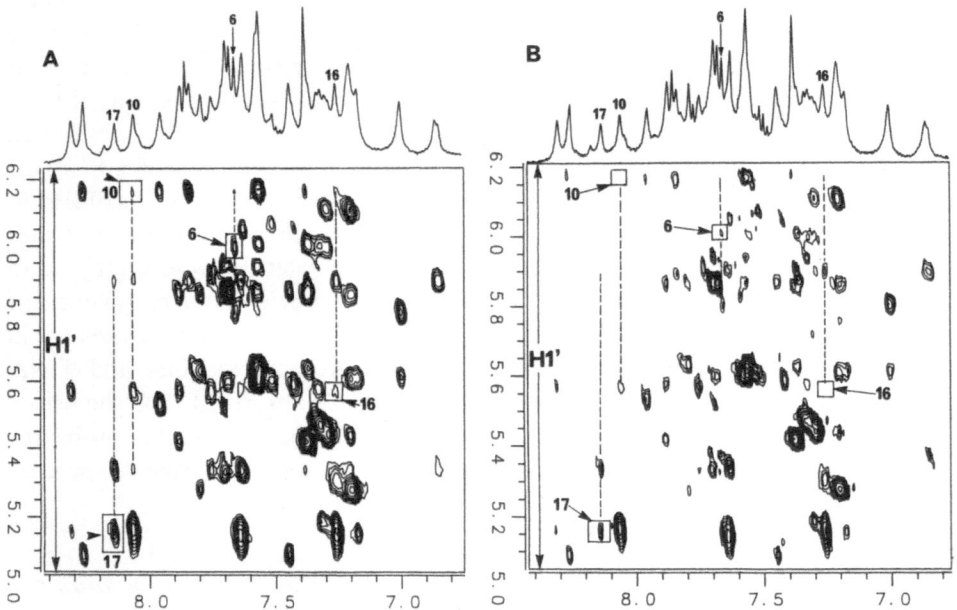

Fig. 7A. Phase sensitive NOESY (120 ms mixing time) expanded contour plot of the PuH8 and the
PyH6 to the H1' sugar protons region of the 12-mer duplex adduct. **B.**Phase sensitive NOESY (60
ms mixing time) expanded contour plot of the PuH8 and PyH6 to the H1' sugar protons of the 12-mer
duplex adduct in 0.5 ml 10 mM NaH_2PO_4, 100 mM NaCl, pH 6.85 at 23°C

significantly enhanced in intensity relative to the same NOE in the duplex alone,
and in comparison to the remaining three NOEs within the (+)-CC-1065 duplex
adduct (Fig. 7A). In contrast, the NOEs between $10AH_2$ and $16AH_2$ and their
corresponding H1' protons on the complementary strands are weak. The $6AH_2$ to
$20CH1'$ NOE has an intermediate intensity that is about the same as that of the

equivalent NOE in the duplex alone. At 60 ms mixing time (Fig. 7B), neither of the weak NOEs for 10AH$_2$ and 16AH$_2$ to their corresponding H1′ on the complementary strand, which were evident at 120 ms, are visible, while the remaining two NOEs are still visible but somewhat weaker than at 120 ms mixing time. At this shorter mixing time, any effect due to spin diffusion should be minimized. These results are consistent with compression of the minor groove around 8T and 9T with widening on either side and particularly abruptly at 10A. These are reminiscent of that found with A-tracts that show bending.

(2) *Up- and down-field shifts of the A-T imino protons in the vicinity of the covalent bonding site.* Up- and down-field shifts of A-T imino protons are associated with widening and narrowing of the minor groove of DNA respectively [26]. A comparison of the exchangeable proton regions that include the imino

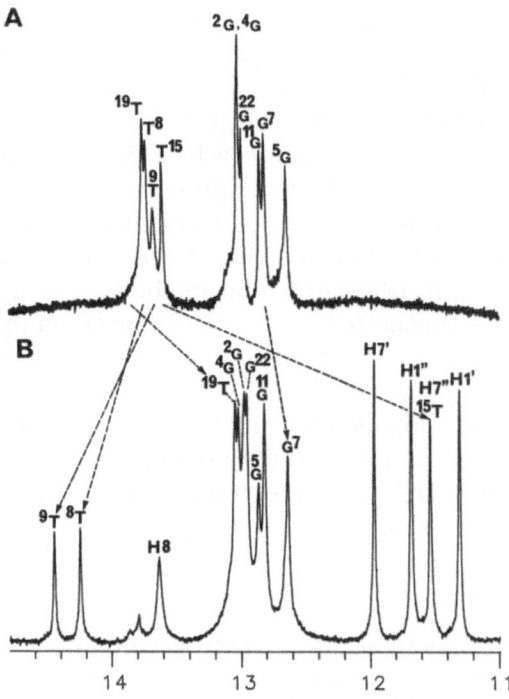

Fig. 8 A, B. One-dimensional ^{1}H-NMR in 90% H$_2$O at 23°C, showing the exchangeable protons region of: **A** the 12-mer duplex and **B** the (+)-CC-1065-12-mer duplex adduct

proton of the 12-mer-duplex and (+)-CC-1065 duplex adduct is show in Fig. 8. While the imino protons for 8T-17A and 9T-6A move downfield by 0.25 and 0.77 ppm respectively, the 10A-15T moves upfield by 2.07 ppm, the latter being indicative of base-pair opening at the covalent bonding site [10, 14]. These results are consistent with compression of the groove at the 8T to 9T step and widening on both sides, but especially abruptly at 10A, the covalent modification site.

3.3. Comparison of (+)-CC-1065 Induced Bending and Intrinsic Bending Associated with A-tracts

While the precise structural details of the bends produced by (+)-CC-1065 and A-tracts are likely to be different, there are a number of general similarities, such as direction of bending, its magnitude, and compression of the minor groove. In both cases the bending is associated with a discontinuity at the 3′ end of bent sequences. Furthermore, for this discontinuity to arise in both A-tracts and at a (+)-CC-1065 bonding site, the sequence to the 5′ side of this junction is crucial to permit the DNA flexibility and associated bent DNA structure. We suggest that this may be the sequence dependent conformational flexibility that is associated with the sequence selectivity of DNA alkylation by (+)-CC-1065. It is interesting to speculate what might be the biological significance of (+)-CC-1065-induced or entrapped structures in DNA. It is well known that bent DNA structures which can naturally occur as a consequence of sequence or be induced by DNA binding proteins are important in processes such as control of gene expression or replication [27], although the precise manner in which these effects occur is still unclear. The stabilization of a naturally occuring bend or the induction of a bend in DNA at a location not normally associated with a bent structure might have quite significant biological consequences. Experiments to test these ideas are currently being carried out.

Acknowledgements. This research was supported by grants from the Public Health Service (CA-49751), the Welch Foundation, and the Burroughs Wellcome Scholars Program. We are grateful to Steve D. Sorey for technical assistances.

References

1. Hanka LJ, Dietz A, Gerpheide SA, Kuentzil SL and Martin DG, CC-1065 (NSC-298223), a new antitumor antibiotic. Production *in vitro* biological activity, microbiological assays, and taxonomy of the producing microorganism. *J. Antibiot.* **31**: 1211–1217, 1978.
2. Chidester CG, Krueger WC, Mizak SA, Duchamp DJ and Martin DG, The structure of CC-1065, a potent antitumor agent and its binding to DNA, *J. Am. Chem. Soc.* **103**: 7629–7635, 1981.
3. Hurley LH, Reynolds BL, Swenson DH and Scahill T, Reaction of the antitumor anitbiotic CC-1065 with DNA: Structure of a DNA adduct with DNA sequence specificity. *Science* **226**: 843–844, 1984.
4. Warpehoski MA, Gebhard I, Kelly RC, Krueger WC, Li LH, McGovren JP, Prairie MD, Wicnienski N and Wierenga W, Stereoelectronic factors influencing the biological activity and

DNA interaction of synthetic antitumor agents modeled on CC-1065. *J. Med. Chem.* **31**: 590–603, 1988.

5. Hurley LH, Lee C-S, McGovren JP, Mitchell M, Warpehoski MA, Kelley RC and Aristoff PA, Molecular basis for the DNA sequence specificity of CC-1065. *Biochemistry* **27**: 3886–3892, 1988.

6. Hurley LH, Warpehoski MA, Lee C-S, McGovren JP, Scahill TA, Kelley RC, Mitchell M, Wicnienski NA, Gebhard I, Johnson PD and Bradford VS, Sequence specificity of DNA alkylation by the unnatural enantiomer of CC-1065 and its synthetic analogues. *J. Am. Chem. Soc.* **112**: 4633–4649, 1990.

7. Reynolds VL, Molineux IJ, Kaplan D, Swenson DH and Hurley LH, Reaction of the antitumor antibiotic CC-1065 with DNA, location of the site of therminally induced strand breakage, and analysis of DNA sequence specificity. *Biochemistry* **24**: 6228–6237, 1985.

8. McGovren JP, Upjohn Co., personal communication, 1990.

9. McGovren JP, Clarke GL, Pratt EA and Deckoning TF, Preliminary toxicity studies with the DNA-binding antibiotic, CC-1065. *J. Antibiot.* **37**: 63–70, 1984.

10. Lin CH and Hurley LH, Determination of the major tautomeric form of the covalently modified adenine in the (+)-CC-1065-DNA adduct by ^{1}H- and [15]-NMR studies. *Biochemistry* **29**: 9503–9507, 1990.

11. Scahill TA, Jensen RM, Swenson DH, Hatzenbuhler NT, Petzold G, Wierenga W and Brahme ND, An NMR study of the covalent and noncovalent interactions of CC-1065 and DNA. *Biochemistry* **29**: 2852–2860, 1990.

12. Warpehoski MA and Hurley LH, Sequence selectivity of DNA covalent modification. *Chem. Res. in Toxicology* **1**: 315–333, 1988.

13. Lin CH, Sun D and Hurley LH, (+)-CC-1065 produces bending of DNA that appears to resemble adenine/thymine tracts. *Chem. Res. Toxicol.* **4**: 21–26, 1991.

14. Lin CH, Beale JM and Hurley LH, Structure of the (+)-CC-1065-DNA adduct: Critical role of ordered water molecules and implications for involvement of phosphate catalysis in the covalent reaction. *Biochemistry* **30**: 3597–3602, 1991.

15. Gait MJ, ed. In: *Oligonucleotide Synthesis-A Pratical Approach*. IRL, Oxford, England, 1984.

16. Lin CH, Hill CG and Hurley LH, Characterization of a 12-mer duplex d(GGCGGAGTTAGG)·d(CCTAACTCCGCC) containing a highly reactive (+)-CC-1065 sequence by ^{1}H and ^{31}P NMR, hydroxyl-radical footprinting, and molecular dynamics calculations. *Chem. Res. in Toxicology*, submitted for publication, 1991.

17. Lin CH and Hurley LH, Determination of the solution structure of a (+)-CC-1065-12-mer DNA duplex adduct by one- and two-dimensional ^{1}H and ^{31}P NMR. *J. Am. Chem. Soc.*, in preparation, 1991.

18. Hore PJ, Solvent suppression in Fourier transform Nuclear Magnetic Resonance. *J. Magn. Reson.* **55**: 283–300, 1983.

19. Zakrzewska K, Randrianarivelo M and Pullman B, Theoretical study of the sequence specificity in the covalent binding of the antitumor drug CC-1065 to DNA. *Nucl. Acids Res.* **15**: 5775–5785, 1987.

20. Brown DG, Sanderson MR, Skelly JV, Jenkins TC, Brown T, Garman E, Stuard DI and Neidle S, Crystal structure of a berenil-dodecanucleotide complex: The role of water in sequence-specific ligand binding. *EMBO J.* **9(4)**: 1329–1334, 1990.

21. Drew HR and Dicherson RE, Structure of a B-DNA dodecamer: III. Geometry of hydration. *J. Mol. Biol.* **151**: 535–556, 1981.

22. Kopa ML, Fratini AV, Drew HR and Dickerson RE, Ordered water structure around a B-DNA dodecamer: A quantitative study. *J. Mol. Biol.* **163**: 129–146, 1982.

23. Gupta SC, Iskim NB, Nhalen DL, Yagi H and Jarina DM, Bifunctional catalysis in the nucleotide-catalyzed hydrolysis of (±)-7β, 8α-dihydroxy-9α, 10α-epoxy-7,8,9,10-tetra-hydrobenzo(a)-pyrene. *Org. Chem.* **52**: 3812–3815, 1987.
24. Lee C-S, Sun D, Kizu R and Hurley LH, Determination of the structure features of (+)-CC-1065 that are responsible for bending and winding of DNA. *Chem. Res. in Toxicology* **4**: 203–213, 1991.
25. Katahira M, Sugeta H and Kyogoku Y, A new model for the bending of DNAs containing the oligo(dA) tracts based on NMR observations. *Nucleic Acids Res.* **18**: 613–618, 1990.
26. Nadeau J and Crothers DM, Structural basis for DNA bending. *Proc. Natl. Acad. Sci. USA* **86**: 2622–2626, 1989.
27. Travers AA, Why bend DNA? *Cell* **60**: 177–180, 1990.

A Common Structural Pattern among Many Biologically Active Compounds of Both Natural and Synthetic Origin – A Novel Approach to the Design of Antineoplastic Agents

C. C. Cheng

Drug Development Laboratory and Department of Pharmacology, Toxicology & Therapeutics
University of Kansas Medical Center, Kansas City, Kansas 66160-7419, USA

Abstract. A tricyclic chemical structural pattern, consisting of a phenyl ring attached to the 2-position of a naphthalene nucleus, or composed of various heterocyclic units with similar molecular structural arrangements, is observed among a large number of biologically and pharmacologically active compounds. Possible relationships between this structural pattern and certain essential biomolecules are discussed. The pattern *per se* is not sufficient for attaining biological activity. It is believed that with proper substituents attached to specific positions of both ring units, compounds with desired biological activity could be rationally designed.

The fact that the structural framework of the antineoplastic agent coralyne [1] (structure **1**) resembles that of the carcinogen 7,12-dimethylbenz[*a*]anthracene [2] (structure **2**, DMBA), and that the structural framework of the antineoplastic alkaloid nitidine [3] (structure **3**) is similar to that of the carcinogen 5-methylchrysene [4] (structure **4**) prompted us to conduct a number of studies, including the evaluation of mutagenicity and carcinogenicity of these antineoplastic agents and related compounds [5] as well as a comparison of the structural characteristics of carcinogenic polynuclear hydrocarbons.

When the ring skeleton of DMBA (**5**) is placed directly over that of chrysene (**6**), two distinct features are noticed. One is that the resulting pentacyclic ring, benzo[*a*]pyrene (**7**), is a well-known potent carcinogen; the other is that a common structural pattern can be observed among these carcinogens in the superimposed, doubly shaded area, which depicts the structure of 2-phenylnaphthalene (**8**).

The same structural pattern can also be observed with the structures of coralyne (**1**) and nitidine (**3**) except that the skeleton of the "2-phenylnaphthalene unit" is not entirely comprised of carbon atoms, rather one of the carbon atoms is replaced by a nitrogen atom. Nevertheless, this isosteric replacement is by no means the cause of opposite pharmacological actions between the antineoplastic and the carcinogenic compounds, since DMBA itself was found to exhibit good antineoplastic activity in the U.S. National Cancer Institute experimental animal tumor screens [5] against adenocarcinoma CA755, sarcoma S180 and leukemias P388 and L1210.

In order to explore the generality of this structural pattern, an examination of the relationship between the chemical structure and the reported carcinogenicity of unsubstituted polycyclic aromatic hydrocarbons with four to six condensed ring systems was conducted. Existing data [6] indicated that triphenylene (**9**), perylene (**10**), pentacene (**11**), dibenzo[c,g]phenanthrene (**12**), dibenzo[e,l]pyrene (**13**) and coronene (**14**) possess very little or no carcinogenicity, whereas benzo-[b]triphenylene (**15**), benzo[a]triphenylene (**16**), dibenz[a,j]anthracene (**17**), dibenz[a,h]anthracene (**18**), benzo[a]pyrene (**7**), dibenzo[def,p]chrysene (**19**) and dibenzo[b,def]chrysene (**20**) are potent carcinogens. An inspection of these structures reveals the existence of at least one "2-phenylnaphthalene" structural pattern among these potent carcinogens (see the shaded areas). This pattern is not observed in the less or noncarcinogenic polycyclic compounds **9–14**.

A list of many polycyclic aromatic hydrocarbons with different degrees of

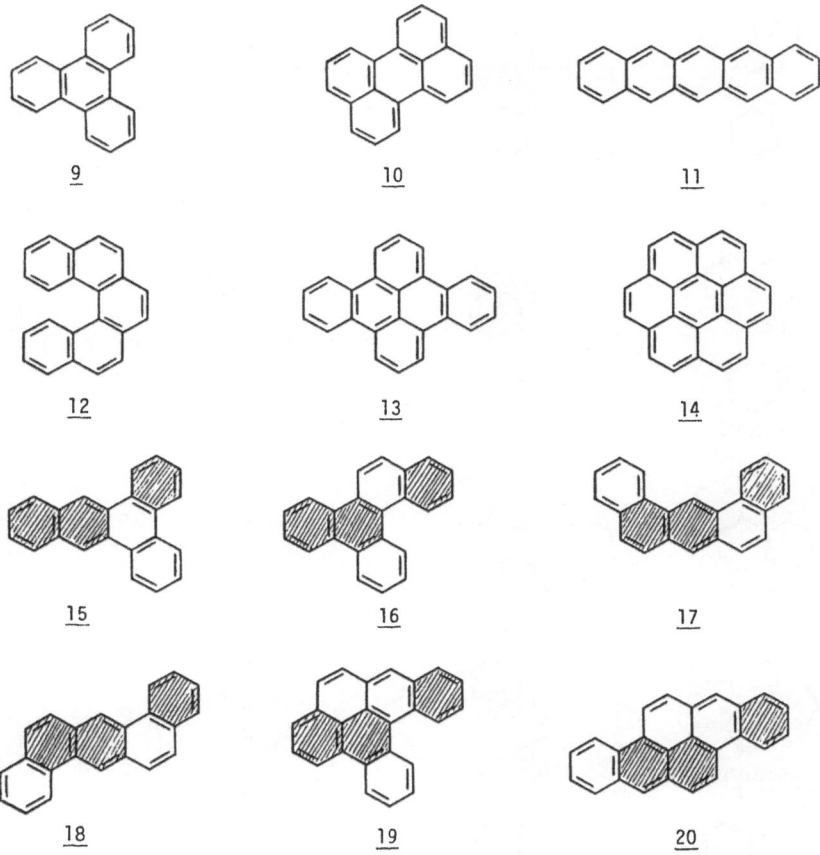

9   10   11

12   13   14

15   16   17

18   19   20

carcinogenicity and mutagenicity was published in 1986 [7]. The data also strongly support the proposed structural pattern, as shown in Figs. IA. B.

The intriguing information prompted an extensive search of biologically, pharmacologically,   and clinically active compounds of both natural and synthetic origin. Following are additional examples.

The antibacterial and antineoplastic antibiotic chartreusin (**21a**), isolated from *Streptomyces chartreusis* and other streptomyces cultures [7–9], possesses a condensed coumarin complex containing a D-digitalose and a D-fucose unit. The presence of a 2-phenyl-substituted naphthalene in its aglycone portion can be readily seen. Related antibiotics such as elsamicin A (**21b**) and elsamicin B (**21c**) only differ in side chain glycoside substitution [10, 11]. The antitumor antibiotics gilvocarcin V (toromycin, chrysomycin A, **22a**), gilvocarcin M (chrysomycin B, **22b**), and gilvocarcin E (**22c**) contain a similar structural pattern [12–14]. The fact that these antibiotics and many other naturally occurring analogs of gilvocarcin V

(**A**)

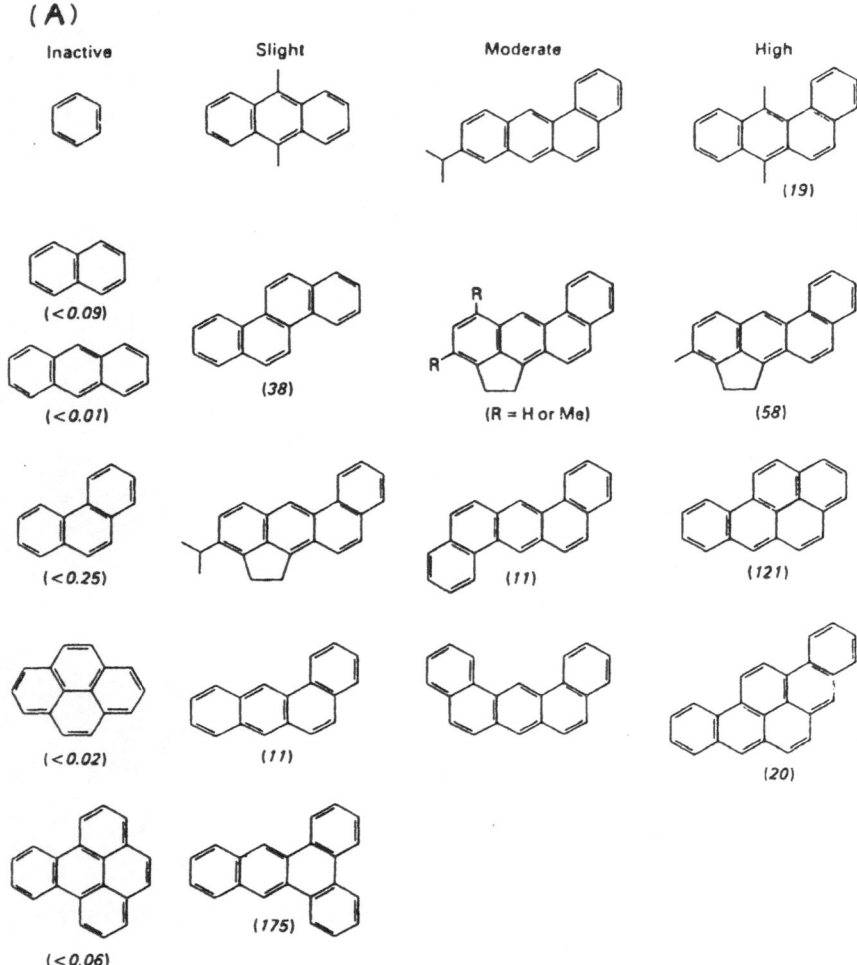

with modified glycoside moieties, such as defucogilvocarcin V (**22d**) or ravi-domycin (**22e**), still retain the original antitumor activity [15, 16] suggests that the C-glycoside moiety could be modified considerably without destruction of biological activity as long as the basic aglycone skeleton remains unchanged.

A large number of the angular isotetracene antibiotics (the angucyclines), such as SS-228Y [17] (structure **23a**) and its glycoside-substituting derivative capoamycin [18] (contains a β-4-(2,4-decadienoyl)olivoside C-glycoside side chain substituted at the C-9 position) as well as many glycoside-substituting derivatives of the related "hydrated" structure **24** including aquayamycin [19] vineomycin A [20], the sakyomicins [21], the kerriamycins [22], the sa-

(B)

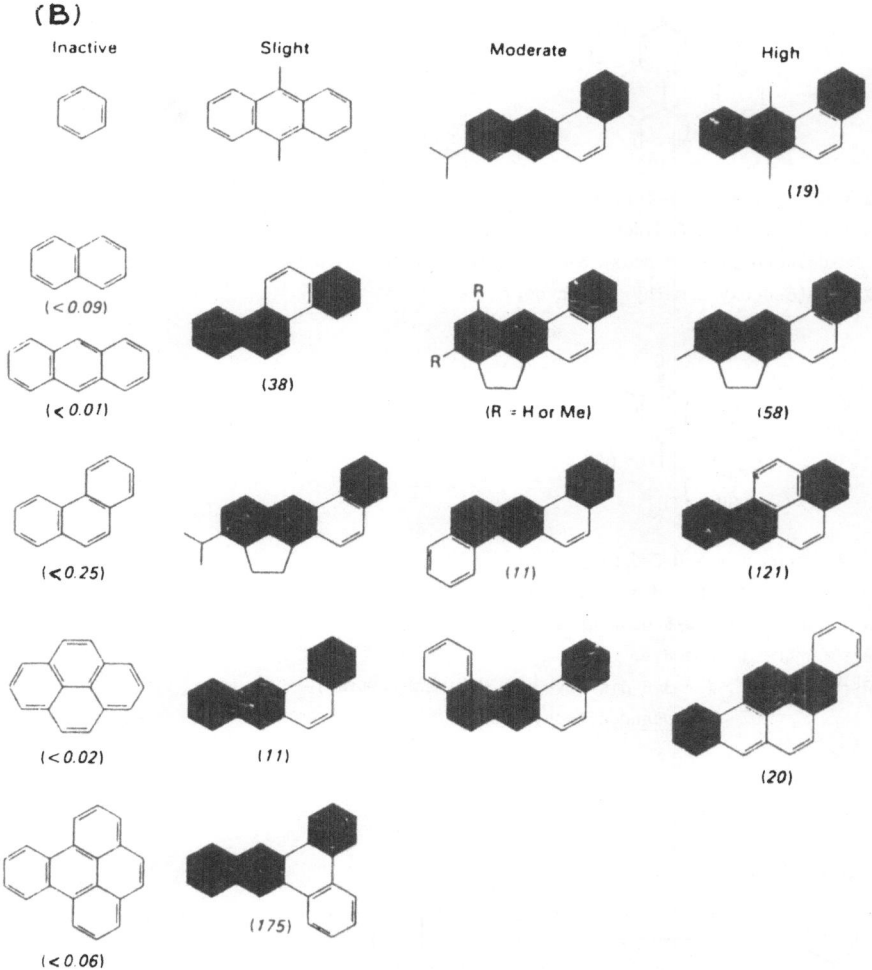

Fig. IA, B. Degree of carcinogeneity of some polycyclic aromatic hydrocarbons. **A.** From Blackburn GM and Kellard B, *Chem. Ind.* **687**, 1986. **B.** Shaded area added by CCC

quayamycins [23] and the urdamycins [24], showed inhibitory activity against bacteria and tumor cells. Many of these antibiotics readily undergo structural rearrangements in the presence of heat, light, dilute acid or dilute base to the biologically less active linear tetracenones [18] (for example SS-228Y to SS-228R, structure **23b**). These antibiotics and related angucyclines, such as tetran-gomycin [25] (structure **25a**), rabelomycin [26] (structure **25b**), tetrangulol [25] (structure **26a**), the aglycones for benzanthrins A and B [27] (structure **26b**), fujianmycins A and B [28] (structures **27a** and **27b**, respectively), and PD-116740

21a X = D-digitalosyl-D-fucosyl
  b X = D-(2-amino-2,6-dideoxy-3-O-methyl)-
        galactosyl-D-(6-deoxy-3-methyl)galactosyl
  c X = D-(6-deoxy-3-methyl)galactosyl

22a R = CH=CH$_2$; X = D-6-deoxygalactosyl
  b R = CH$_3$; X = D-6-deoxygalactosyl
  c R = C$_2$H$_5$; X = D-6-deoxygalactosyl
  d R = CH=CH$_2$; X = H
  e R = CH=CH$_2$; X = 3,6-dideoxy-3-(N,N-dimethylamino)-
              pseudocitropyrenosyl

23a   23b

[29] (structure **28**) can be considered as on various stages of the reduced or hydrated forms of benz[*a*]anthraquinones.

Two groups of structurally related oxygen-containing naphthoquinone antibiotics also exhibited interesting biological activities. WS-5995 A (structure **29**) can efficiently protect coccidia infection [30] and the pluramycin (iyomycin) antibiotics, represented by pluramycin, neopluramycin, hedamycin, kidamycin, griseorubicin, largomycin, etc. possess a common chromophore **30**. Many have been reported to exhibit cytotoxic and antineoplastic action.

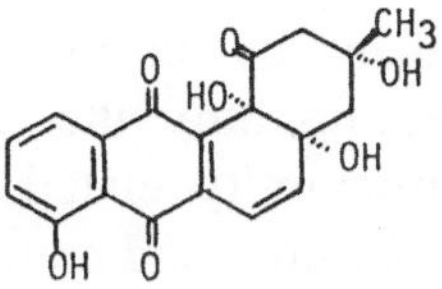

24

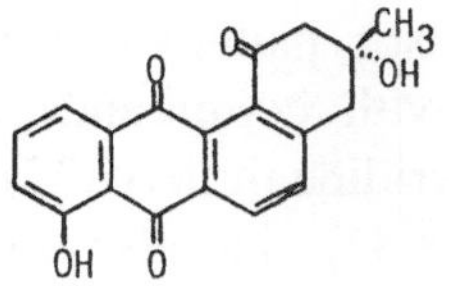

25a X = H
  b X = OH

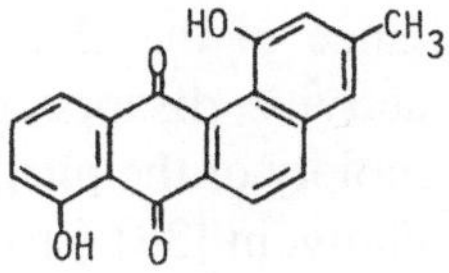

26a X = H
  b X = OH

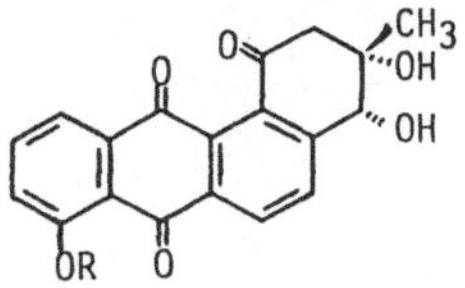

27a X = H
  b X = CH$_3$

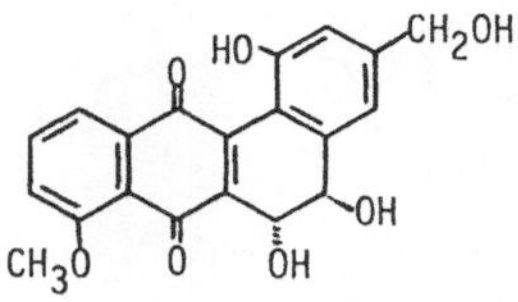

28

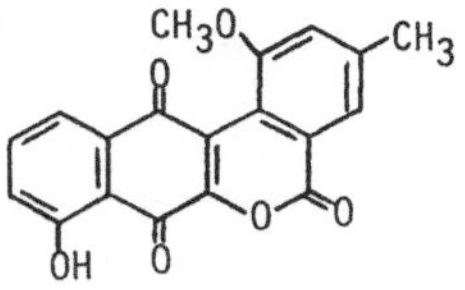

29

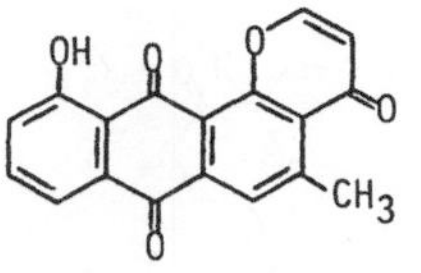

30

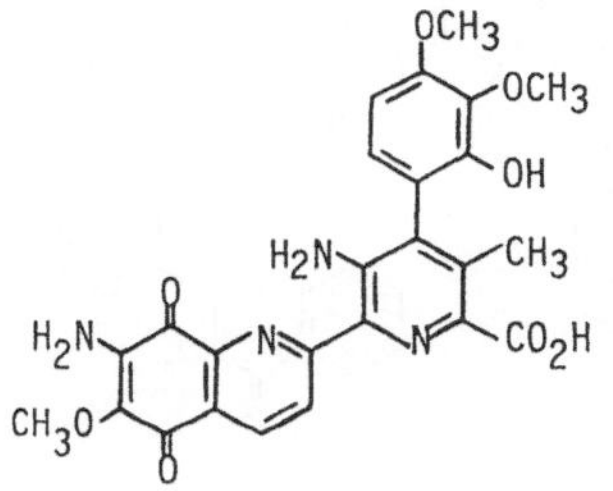

31

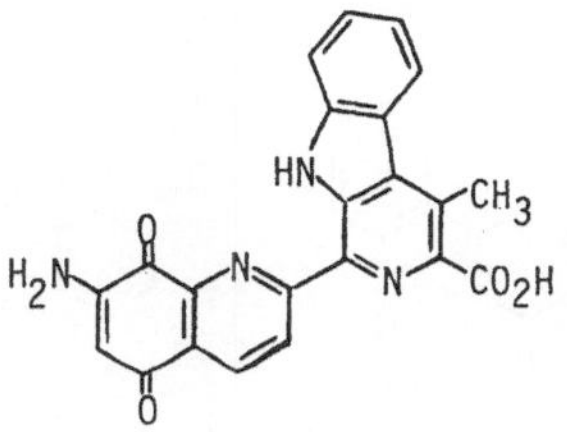

32

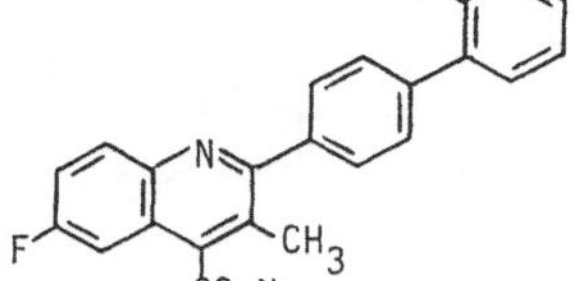

33

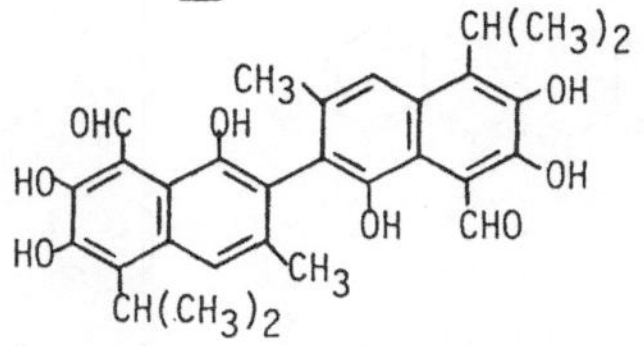

34

Streptonigrin [30] (structure **31**) is another antitumor antibiotic, but possesses a different and rather unique structure. The pyridine ring and the quinoline unit constitute a familiar "2-phenylnaphthalene" pattern except that it is composed of nitrogen heterocycles. X-ray diffraction study of streptonigrin revealed that the pyridyl unit is very nearly coplanar with the quinoline portion [31]. The importance of the characteristic pattern can be realized by the fact that the original activity disappeared with compounds possessing only the quinolinequinone moiety or the phenylpyridine moiety [32]. A structurally related antibiotic, lavendamycin [33] (structure **32**), has also shown antitumor and antibacterial activity.

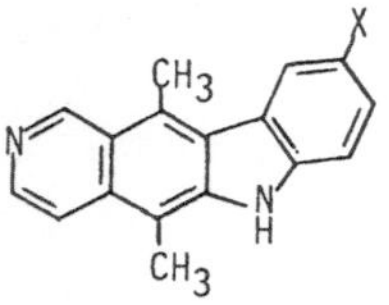

35a X = H
  b X = OCH₃
  c X = OH

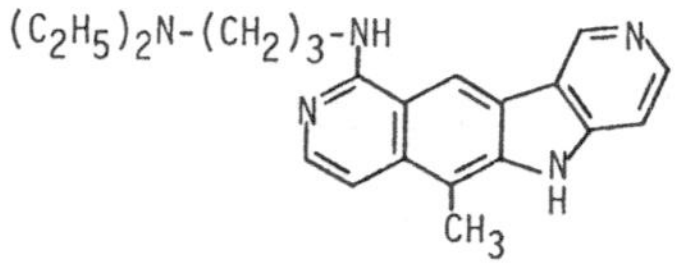

36

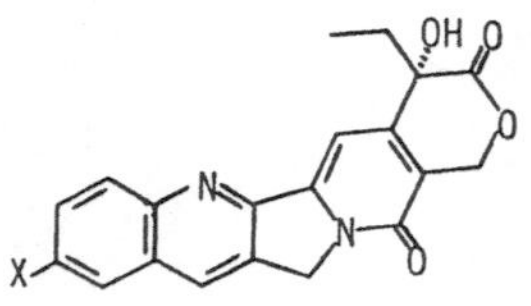

37a X = H
  b X = OH

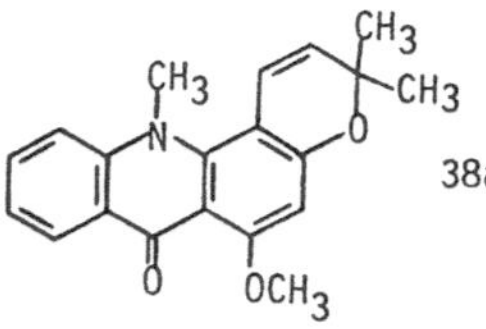

38a

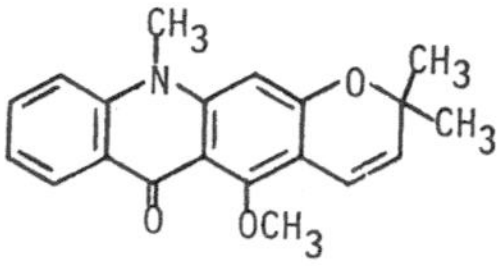

38b

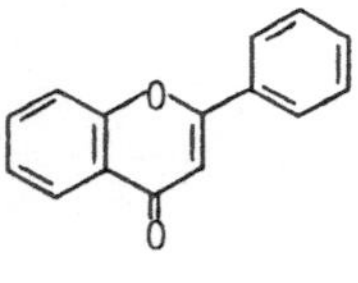

39

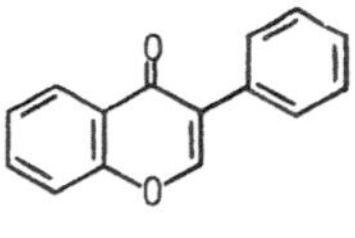

40

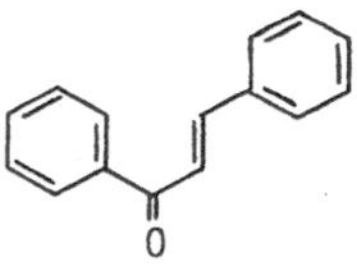

41

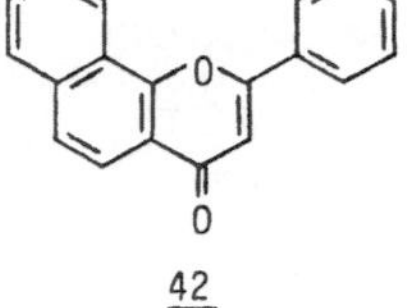

42

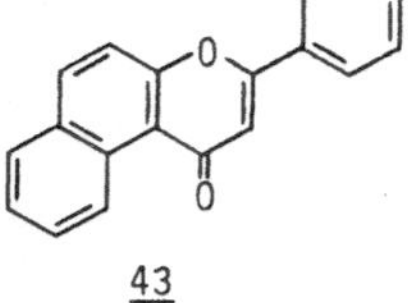

43

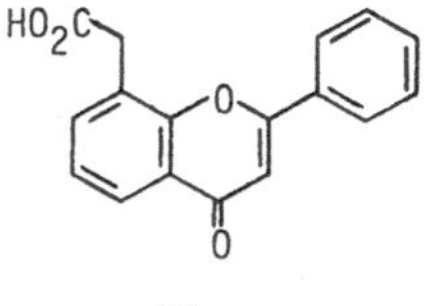

44

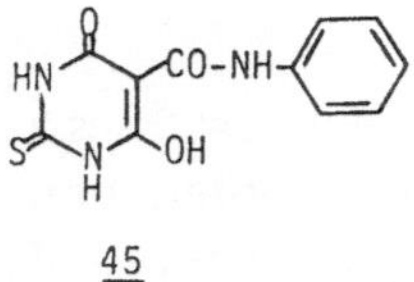

45

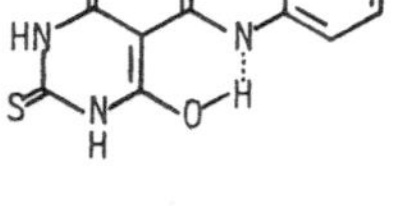

45a

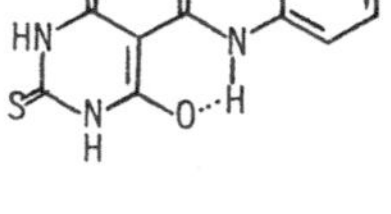

45b

A synthetic compound DuP-785 (brequinar sodium) [34] (structure **33**), possessing antitumor activity against a number of experimental tumors and currently undergoing clinical trials, is 2-phenyl-substituted quinoline derivative. DuP-785 is a potent inhibitor of dihydroorotate dehydrogenase in the *de novo* pyrimidine biosynthesis pathway [35]. An enzyme lactate dehydrogenase inhibitor, gossypol [36] (structure **34**), is a well known antioxidant which plays a crucial role in both the anaerobic and aerobic metabolism of sperm and sperm-generating cells [37]. Gossypol is also a specific inhibitor of DNA polymerase-α [38], a major enzyme involved in DNA replication. The compound may also interfere with the DNA repair process [38]. Here the characteristic "2-phenylnaphthalene" pattern is constituted of all carbon atoms.

There are also many alkaloids with antineoplastic and other types of biological activity which possess the proposed structural pattern: ellipticine [39] (structure **35a**) as well as related 9-methoxyellipticine [39] (structure **35b**) and 9-hydroxyellipticine [40] (structure **35c**) are all clinically active anticancer agents. A synthetic compound of related structure, BD-40 [41] (structure **36**), has demonstrated potent inhibitory activity against a number of experimental tumors and its clinical trials are being conducted. It is of interest to note that the two methyl groups substituted on the ellipticine pyrido[4,3-*b*]carbazole ring (**35**) and those of DMBA (**2**) occupy equivalent positions.

Camptothecin [42] (structure **37a**) is a pentacyclic antitumor alkaloid which displays potent antineoplastic activity against a variety of experimental tumor systems and inhibits DNA synthesis and certain species of RNA synthesis [42, 43]. Some ring-substituted (hydroxy or methoxy) camptothecin derivatives were either isolated in nature or obtained synthetically. The 10-hydroxy derivative of camptothecin, **37b**, and the parent alkaloid are being used for the treatment of liver carcinoma and tumors of the head and neck [42]. The structure of camptothecin not only possesses the characteristic structural pattern but, because of the condensed ring structure, the entire ring system is nearly coplanar.

Acronycine [44] (structure **38a**) is another wide-spectrum antitumor alkaloid which is effective subcutaneously, intraperitoneally or orally. It is of interest to note that the chemically rearranged linear isoacronycine [45] (structure **38b**), which does not contain the characteristic structural pattern, is not active biologically [46]. The strict requirement for, and the dependence on, a specific structure was suggested as being more compatible with a drug receptor-type mechanism rather than with a less precise type of action based on physical characteristics, such as lipophilicity or inadequate solubility [47].

Aside from the plant alkaloids, the flavonoids certainly have offered much to the medicinal use. Flavonoids are among the most abundant groups of relatively low molecule weight compounds in the plant kingdom [48]. They are present in every part of higher plants (many as flavonoid glycosides) including the leaves, flowers, fruits, seeds, nuts, roots, stems and the bark, and are responsible for much of the coloring in nature [48]. In general, the toxicity of these flavonoids is low or even non-existent, as the average human's daily food intake contains about one gram of various flavonoids [49].

The parent nuclei of the flavonoids are flavone (**39**), isoflavone (**40**) and chalcone (**41**). The latter may be regarded as open-chain flavonoids which can be precursors of both the flavonoids and the isoflavonoids [48, 50].

A variety of pharmacological activities of herbal medicine used since the ancient time can be contributed to the flavonoids including the antiinflammatory, antibacterial, antiviral, antiallergic, antimutagenic, anticarcinogenic and antineoplastic, among many others [48–51]. The potent antioxidant activity and inhibitory activity against a number of enzymes by the flavonoids have been reflected in their diversified usefulness such as diuretics, muscle relaxants, human estrogen synthetase inhibitors and treatment of diabetic cataracts [48, 49, 52]. The corresponding isoflavones have shown estrogenic, insecticidal, piscicidal, and antifungal activity [49, 53]. The fact that flavones inhibit estrogen synthetase and that isoflavones act as pro-estrogens is worth noting.

Two benzoflavones possess different actions toward the polycyclic aromatic hydrocarbons. 7,8-Benzoflavone (**42**) inhibits the metabolism (hence activition) of benzo[a]pyrene and DMBA[54], whereas 5,6-benzoflavone (**43**), on the other hand, is a strong inducer of benzo[a]pyrene hydrolase activity [55] and promotes the excretion of the resulting hydroxylated derivatives through mechanisms such as glucuronidation. Therefore, in spite of their different biological actions, both benzoflavones may be considered as anticarcinogen agents. In the area of antineoplastic agents, flavone-8-acetic acid (**44**) was found to exhibit significant activity against a number of experimental tumor systems, particularly it is active against the highly refractory colon 36 tumor [56].

Even certain uncyclized compounds may assume the "2-phenylnaphthalene" type arrangement through hydrogen bond formation. A simple thiobarbituric acid merbarone (**45**) showed exceptional antineoplastic activity against a number of experimental tumor systems by either the ip or oral route [57]. The structure of **45** could be postulated as assuming the proposed structural pattern with hydrogen bonding, shown as either **45a** or **45b**. The coplanar configuration of these rings has later been verified [58].

There are many other biologically active compounds which contain similar pattern. The repeated occurrences of the "2-phenylnaphthalene" type structural pattern among such a huge number of otherwise structurally unrelated natural and synthetic compounds with diverse biological activities suggest that this characteristic molecular arrangement could well be intimately related to some similar structures of biologically pertinent molecules which are of importance to the processes of life. Molecules containing such a pattern, together with the contribution of proper substituents and functional groups, may either mimic, modify, or interfere with the regular biological events, thus exhibit the observed desired or undesired biological and pharmacological properties.

A search of the literature revealed that two groups of biopertinent molecules fulfill the proposed requirements: The purine-pyrimidine (e.g., cytosine-guanine (46) and adenine-thymine (47)) base pairs of the nucleic acids and the steriod hormones (represented by estradiol (48), an estrogen; progesterone (49), a "pregnancy hormone"; and testosterone (50), an androgen). The steroid hormones are depicted in an upside-down manner with respect to the conventional drawings.

The similarity in molecular shapes among cytosine-guanine, progesterone, DMBA, and benzo[a]pyrene was illustrated by means of molecular models some 30 years ago [59]. The molecular planarity of carcinogenic hydrocarbons and nucleotide purine-pyrimidine bonded base pairs was noted even earlier [60]. It is therefore suggested that structural resemblance between the aforementioned biologically active compounds and the nucleic acid base pairs may indicate that enzymes controlling the topological state of DNA (DNA topoisomerases [61]), which are important to the process of DNA replication and genetic recombination, could be reasonable targets for these molecules [62]. A number of reports [63] have already noted that many antineoplastic agents, including ellipticine (**35a**) and camptothecin (**37a**), indeed interfere with the function of DNA topoisomerases.

As early as 1932, it was noted that steroids and some carcinogenic polycyclic hydrocarbons had similar shapes and dimensions [64]. Although the steroid molecules are not altogether planar, the overall conformation of the steroid molecules is closer to the planar structure [65]. Besides, many biomedically useful hormonal therapeutic agents, such as diethylstilbestrol (DES, **51**), the antiestrogenic antineoplastic drugs, such as tamoxifen (**52**), or the previously mentioned flavones (which inhibit the aromatization of androstenedione and testosterone to the estrogens) and isoflavones (many are estrogenic) could assume planar conformation and also possess the uncyclized or cyclized "2-phenylnaphthalene" structural pattern.

In a study of the relationship of polynuclear aromatic hydrocarbons, steroids and carcinogens, it was noticed that there is usually a direct increase in carcinogenicity when the hydrocarbons become structurally more similar to steroids [65]. These investigators thus postulated that the polynuclear aromatic hydrocarbons may act at the same biological sites as steroid hormones [65]. The target tissue for estrogens (estrogen receptors) in rats was found to be located in the uterus, the ovaries and the mammary glands, but not in the small intestine or blood serum [66]. It is therefore conceivable to find that administration of carcinogenic polynuclear hydrocarbons, such as benzo[a]pyrene, 2-hydroxybenzo[a]pyrene, methocho-lanthrene and DMBA, resulted in the tumor formation not at the site of local application but in the breast, the ovaries or the uterus in the rat [67]. Tumor induction by estrogens in female genital organs [68] and detection of high incidences of cancer of the scrotum and testicles among chimney sweepers [69] (which infer to tumor induction in male genital organs by carcinogenic polynuclear hydrocarbons) have also been reported.

The preceding information clearly suggests that compounds containing the "2-phenylnaphthalene" type ring systems as the nucleus with proper substituents attached to both ring units should be explored with synthesis and biological evaluation. This may be one of the rational approaches toward a total design of anticancer drugs with desired biological activity.

References

1. (a) Schneider W and Schroeder K, Aceto-papaverin und Coralyn (Hexadehydro coralydin). *Ber.* **53**: 1459–1469, 1920;
    (b) Zee-Cheng KY and Cheng CC, Practical preparation of coralyne chloride. *J. Pharm. Sci.* **61**: 969–971, 1972; Interaction between DNA and coralyne acetosulfate, an antileukemic compound. *ibid* **62**: 1572–1573, 1973.

2. (a) Bachmann WE and Chemerda JM, The synthesis of 9,10-dimethyl-1,2-benzanthreacene, 9,10-diethyl-1,2-benzanthracene and 5,9,10-trimethyl-1,2-benzanthracene. *J. Am. Chem. Soc.* **60**: 1023–1026, 1938;
(b) Pullman A and Pullman B, Electronic structure and carcinogenic activity of aromatic molecules. New developments. *Adv. Cancer Res.* **3**: 117–169, 1955.

3. (a) Wall ME, Wani MC and Taylor HL, Plant antitumor agents 27. Isolation, structure, and structure-activity relationships of alkaloids from *Fagara macrophylla. J. Nat. Prod.* **50**: 1095–1099, 1987;
(b) Zee-Cheng KY and Cheng CC, Synthesis of 5,6-dihydro-6-methoxynitidine and a practical preparation of nitidine chloride. *J. Heterocycl. Chem.* **10**: 85–88, 1973.

4. (a) Dunlap CE and Warren S, The carcinogenic activity of some new derivatives of aromatic hydrocarbons. I. Compounds related to chrysene. *Cancer Res.* **3**: 606–607, 1943;
(b) Amin S, Huie K, Melikian AA, Leszczynska JM and Hecht SS, Comparative metabolic activation in mouse skin of the weak carcinogen 6-methylchrysene and the strong carcinogen 5-methylchrysene. *Cancer Res.* **45**: 6406–6412, 1985.

5. Cheng CC, Engle RR, Hudson JR, Ing RB, Wood HB, Yan SJ and Zee-Cheng RKY, Absence of mutagenicity of coralyne and related antileukemic agents: structural comparison with the potent carcinogen 7,12-dimethylbenz[a]anthracene. *J. Pharm. Sci.* **66**: 1781–1783, 1977.

6. US Department of Health and Human Services, Survey of compounds which have been tested for carcinogenic activity. Washington, DC, 1–16, 1951–1986.

7. Berger J, Sternbach LH, Isolation of antibiotic X-465A and its identification with chartreusin. *J. Am. Chem. Soc.* **80**: 1636–1638, 1958.

8. Simonitsch E, Eisenhuth W, Stamm OA and Schmidt H, Über die Struktur des chartreusins. I und II. *Helv. Chem. Acta,* **47**: 1459–1484, 1964.

9. Beisler JA, Chartreusin, a glycosidic antitumor antibiotic from *Streptomyces. Prog. Med. Chem.* **19**: 247–268, 1982.

10. Konishi M, Sugawara K, Kofu F, Nishiyama Y, Tomita K, Miyaki T and Kawaguchi H, Elsamicins, new antitumor antibiotics related to chartreusin I. Production, isolation, characterization and antitumor activity. *J. Antibiot.* **39**: 784–791, 1986.

11. Sagawara K, Tsunakawa M, Konishi M, Kawaguchi H, Krishnan B, He CH and Clardy J, Elsamicins A and B, new antitumor antibiotics related to chartreusin. 2. Structures of elsamicins A and B. *J. Org. Chem.* **52**: 996–1101, 1987.

12. Elespuru RK and Gonda SK, Activation of antitumor agent gilvocarcins by visible light. *Science,* **223**: 69–71, 1983.

13. Takahashi K, Yoshida M, Tomita F and Shirahata K, Gilvocarcins, new antitumor antibiotics 2. Structural elucidation. *J. Antibiot.* **34**: 271–275, 1981.

14. Morimoto M, Okubo S, Tomita F and Marumo H, Gilvocarcins, new antibiotics 3. Antitumor activity. *J. Antibiot.* **34**: 701–707, 1981.

15. Sehgal SN, Czerkawski H, Kudelski A, Pandev K, Saucier R and Vezina C, Ravidomycin (AY-25,545), a new antitumor antibiotic. *J. Antibiot.* **36**: 355–361, 1983.

16. Rakhit S, Eng C, Baker H and Singh K, Chemical modification of ravidomycin and evaluation of biological activities of its derivatives. *J. Antibiot.* **36**: 1490–1494, 1983.

17. (a) Okami Y, Antibiotics from marine microorganisms with reference to plasmid involvement. *J. Nat. Prod.* **42**: 583–595, 1979;
(b) Imamura N, Kakinuma K, Ikekawa N, Tanaka H and Ōmura S, Biosynthesis of vineomycins A_1 and B_2. *J. Antibiot.* **35**: 602–608, 1982.

18. (a) Hayakawa Y, Iwakiri T, Imamura L, Seto H and Ōtake N, Studies on the isotetraceneone antibiotics. I. Capomycin, a new antitumor antibiotic. *J. Antibiot.* **38**: 957–959, 1985;
(b) Hayakawa Y, Furahata K, Seto H and Ōtake N, The structure of a new isotetracenone antibiotic, capomycin. *Tetrahedron Lett.* **26**: 3471–3474, 1985.

19. Sezaki M, Kondo S, Maeda K, Umezawa H and Ohno M, The structure of aquaymycin. *Tetrahedron Lett.* **26**: 5171–5190, 1970.

20. Ohta K, Okazaki H and Kishi T, The absolute configuration of P-1894B (vineomycin A₁), a potent prolyl hydroxylase inhibitor. *Chem. Pharm. Bull.* **30**: 762–765, 1982.

21. Irie H, Mizuno Y, Kouno I, Nagasawa T, Tani Y, Yamada H, Taga T and Osagi K, Structures of new antibiotic substances, sakyomicin A, B, C, and D; X-ray crystal and molecular structure of sakyomicin A. *J. Chem. Soc. Chem. Commun.* 174–175, 1983.

22. Hayakawa Y, Iwakiri T, Imamura K, Seto H and Ōtake N, Studies on the isotetracenone antibiotics II. Kerriamycins A, B and C, new antitumor antibiotics. *J. Antibiot.* **38**: 960–963, 1985.

23. Uchida T, Imota M, Watanabe Y, Miura K, Dobashi T, Matsuda N, Sawa T, Naganawa H, Hamada M, Takeuchi T and Umezawa H, Saquayamycins, new aquaymaycin-group antibiotics. *J. Antibiot.* **38**: 1171–1181, 1985.

24. Drantz H, Zähner H, Rohr J and Zeeck A, Metabolic products of microorganisms. 234. Urdamycins, new angucycline antibiotics from *Streptomyces fradiae* I. Isolation, characterization and biological properties. *J. Antibiot.* **39**: 1657–1669, 1986.

25. (a) Kuntsmann MP and Mitscher LA, The structural characterization of tetrangomycin and tetrangulol. *J. Org. Chem.* **31**: 2920–2925, 1966;
    (b) Brown PM and Thomson RH, Naturally occurring quinones, Part XXVI. A synthesis of tetrangulol (1,8-dihydroxy-3-methylbenz[*a*]anthracene-7,12-quinone). *J. Chem. Soc. Perkin. Trans.* **1**: 997–1000, 1976.

26. Liu WC, Parker WL, Slusarchyk DS, Greenwood GL, Graham SF and Meyers E, Isolation, characterization, and structure of rabelomycin, a new antibiotic. *J. Antibiot.* **23**: 437–441, 1970.

27. Rasmussen RR, Nuss ME, Scherr MH, Mueller SL, McAlpine JB and Mitscher LA, Benzanthrins A and B, a new class of quinone antibiotics II. Isolation, elucidation of structure and potential antitumor activity. *J. Antibiot.* **39**: 1515–1526, 1986.

28. Rickards RW and Wu JP, Fujianmycins A and B, new benz[*a*]anthraquinone antibiotics from a *Streptomyces* species. *J. Antibiot.* **38**: 513–515, 1985.

29. Wilton JH, Cheney DC, Hokanson GC, French JC, He CH and Clardy J, A new dihydrobenz[*a*]anthraquinone antitumor antibiotic (PD-116740). *J. Org. Chem.* **50**: 3936–3940, 1985.

30. (a) Rao KV, Biemann K and Woodward RB, The structure of streptonigrin. *J. Am. Chem. Soc.* **85**: 2532–2533, 1963;
    (b) Rivers SL, Whittington RM and Medrek TJ, Methyl ester of streptonigrin (NSC-45384) in treatment of malignant lymphoma. *Cancer Chemother. Rep.* **46**: 17–21, 1965; 1965.

31. Chiu YYH and Lipscomb WN, Molecular and crystal structure of streptonigrin. *J. Am. Chem. Soc.* **97**: 2525–2530, 1975.

32. (a) Liao TK, Nyberg WH and Cheng CC, Synthetic studies of the antitumor antibiotic streptonigrin. I. Synthesis of the A-B ring portion of streptonigrin. *J. Heterocycl. Chem.* **13**: 1063–1065, 1976;
    (b) Wittek PJ, Liao TK and Cheng CC, Synthetic studies of the antitumor antibiotic streptonigrin. 3. Synthesis of the C-D ring of streptonigrin by an unsymmetrical Ullmann reaction. *J. Org. Chem.* **44**: 870–872, 1979.

33. Balitz DM, Bush JA, Brandner WT, Doyle TW, O'Herron FA, Isolation of levendamycin. A new antibiotic from *Streptomyces lavendulae*. *J. Antibiot.* **35**: 259–265, 1982.

34. Dexter DL, Hesson DP, Ardecky RJ, Rao GV, Tippett DL, Dusak BA, Paull KD, Plowman J, DeLarco BM, Narayanan VL and Farbes M, Activity of a novel 4-quinolinecarboxylic acid, NSC-368390 [6-Fluoro-2-(2′-fluoro-1,1′-biphenyl-4-yl)-3-methyl-4-quinolinecarboxylic acid sodium salt] against experimental tumors. *Cancer Res.* **45**: 5563–5568, 1985.

35. Chen SF, Ruben RL and Dexter DL, Mechanism of action of the novel anticancer agent 6-fluoro-2-(2′-fluoro-1,1′-biphenyl-4-yl)-3-methyl-4-quinolinecarboxylic   acid sodium salt (NSC 368390): Inhibition of *de novo* pyrimidine nucleotide biosynthesis. *Cancer Res.* **46**: 5014–5019, 1986.

36. (a) Campbell KN, Morris RC and Adams R, The structure of gossypol. I. *J. Am. Chem. Soc.* **59**: 1723–1728, 1937;

    (b) Adams R, Geissman TA and Edwards JD, Gossypol, a pigment of cottonseed. *Chem. Rev.* **60**: 555–631, 1960.

37. Lee CYG, Moon YS, Yuan JH and Chen AF, Enzyme inactivation and inhibition by gossypol. *Mol. Cell. Biochem.* **47**: 65–70, 1982.

38. Rosenberg LJ, Adlakha RC, Dasai DM and Rao PN, Inhibition of DNA polymerase α by gossypol. *Biochim. Biophys. Acta* **866**: 258–267, 1986.

39. (a) Goodwin S, Smith AF and Horning EC, Alkaloids of *Ochrosia elliptica* Labill. *J. Am. Chem. Soc.* **81**: 1903–1908, 1959;

    (b) Svoboda GH, Poore GA and Montfort ML, Alkaloids of *Ochrosia maculata* Jacq. (*Ochrosia borbonica* Gmel.). *J. Pharm. Sci.* **57**: 1720–1725, 1968.

40. Lesca P, Lecointe P, Paoletti C and Mansuy D, Ellipticines as potent inhibitors of drug metabolism. Protective effect against chemical mutagenesis and carcinogenesis. *Biochimie* **60**: 1011–1018, 1978.

41. Ducrocq C, Bisagni É, Rivalle C and Lhosta JM, Synthesis of 10-substituted 5*H*-pyrido [3′,4′:4,5]pyrrolo[2,3-*g*]isoquinolines. *J. Chem. Soc. Perkin. Trans.* **I**: 142–145, 1979.

42. (a) Wall ME, Wani MC, Cook CE, Palmer KH, McPhail AT and Sim GA, Plant antitumor agents. I. The isolation and structure of camptothecin, a novel alkaloidal leukemia and tumor inhibitor from *Camptotheca acuminata*. *J. Am. Chem. Soc.* **88**: 3888–3890, 1966;

    (b) Perdue RE, Wall ME, Hartwell JL and Abbott BJ, Comparison of the activity of crude *Campotheca acuminata* ethanolic extracts against lymphoid leukemia L-1210. *Lloydia* **31**: 229–236, 1968;

    (c) Hutchinson CR, Camptothecin: Chemistry biogenesis and medicinal chemistry. *Tetrahedron* **37**: 1047–1065, 1981.

43. Kessel D, Bosmann HB and Lohr K, Camptothecin effects on DNA synthesis in murine leukemia cells. *Biochim. Biophys. Acta.* **269**: 210–216, 1972.

44. (a) Svoboda GH, Alkaloids of *Acronychia baueri (Bauerrella anstraliana)*. II. Extraction of the alkaloids and studies of structure-activity relationships. *Lloydia* **29**: 206–224, 1966;

    (b) Brannon DR, Horton DR and Svoboda GH, Microbial hydroxylation of acronycine. *J. Med. Chem.* **17**: 653–654, 1974.

45. Funayama S and Cordell GA, Chemistry of acronycine, XI. Rearrangement of dihydronoracronycine to dihydroisonoracronycine. Mechanistic studies. *J. Nat. Prod.* **48**: 938–943, 1985.

46. Cordell GA, Personal Communication, December 3, 1985.

47. Garzon K and Svoboda GH, Acridone alkaloids: Experimental antitumor activity of acronycine. *The Alkaloids* **21**: 1–28, 1983.

48. (a) Harborne JB, Mabry TJ and Malory H (eds), *The Flavonoids,* Academy Press, New York, NY 1975;

    (b) Harborne JB and Malory TJ (eds), *The Flavonoids. Advances in Research,* Chapman and Hall, London, 1982.

49. (a) Havsteen B, Flavonoids, a class of natural products of high pharmacological potency. *Biochem. Pharmacol.* **32**: 1141–1148, 1983;

    (b) Middleton E, The flavonoids. *Trends in Pharmacological Res.* 335–338, 1984.

50. Hahlbrock K, Flavonoids. In: *The Biochemistry of Plants* (ed Conn EE) Vol 7, Ch. 14, Academy Press, New York, NY, pp. 425–456, 1981.

51. (a) Wattenberg LW and Leong JL, Inhibition of the carcinogenic action of 7,12-dimethyl-benz[a]anthracene by beta-naphthoflavone. *Proc. Soc. Exptl. Biol. Med.* **128**: 940–943, 1968;

   (b) Mitscher LA, Rao GSR, Khanna I, Vegsoglu T and Drake S, Antimicrobial agents from higher plants: prenylated flavonoids and other phenols from *Glycyrrhiza lepidota*. *Phytochemistry* **22**: 573–576, 1983;

   (c) Huang MT, Wood AW, Newmark HL, Sayer JM, Yagi H, Jerina DM and Conney AM, Inhibition of the mutagenicity of bay-region diol-epoxides of polycyclic aromatic hydrocarbons by phenolic plant flavonoids. *Carcinogenesis* **4**: 1631–1637, 1983.

52. (a) Simpson TH and Uri N, Hydroxyflavones as inhibitors of the aerobic oxidation of unsaturated fatty acids. *Chem. Ind.* 956–957, 1956;

   (b) Mehta AC and Seshadri TR, Flavonoids as antioxidants. *J. Sci. Indian Res.* **18B**: 24–28, 1959;

   (c) Borchardt RT and Huber JA, Catechol *O*-methyltransferase 5. Structure-activity relationships for inhibition by flavonoids. *J. Med. Chem.* **18**: 120–122, 1975;

   (d) Varma SD and Kinoshita JH, Inhibition of lens aldose reductase by flavonoids — their possible role in the prevention of diabetic cataracts. *Biochem. Pharmacol.* **25**: 2505–2513, 1976;

   (e) Kellis JT and Vickery LE, Inhibition of human estrogen synthetase (aromatase) by flavones. *Science* **225**: 1032–1034, 1984.

53. Ollis WD, The isoflavonoids. In: *The Chemistry of Flavonoids Compounds* (ed Geissman TA) pp. 353–405, MacMillan, New York, NY, 1962.

54. (a) Diamond L and Gelboin HV, Alpha-naphthoflavone: An inhibitor of hydrocarbon cytotoxicity and microsomal hydroxylase. *Science* **166**: 1023–1025, 1969;

   (b) Diamond L, Miller J and Gelboin HV, The effects of two isomeric benzoflavones on aryl hydrocarbon hydroxylase and the toxicity and carcinogenicity of polycyclic hydrocarbones. *Cancer Res.* **32**: 731–736, 1972;

   (c) Schwartz AG, Protective effect of benzoflavone and estrogen against 7,12-dimethyl-benz(a)anthracene-induced cytotoxicity in cultured liver cells. *Cancer Res.* **34**: 10–15, 1974.

55. (a) Wattenberg LW, Page MA and Leong JL, Induction of increased benzopyrene hydroxylase activity by flavones and related compounds. *Cancer Res.* **28**: 934–937, 1968;

   (b) Wattenberg LW and Leong JL, Inhibition of the carcinogenic action of benzo(a)pyrene by flavones. *Cancer Res.* **30**: 1922–1925, 1970.

56. (a) Double JA, Bibby MC and Loadman PM, Pharmacokinetics and antitumor activity of LM595 in mice bearing transplantable adenocarcinomas of the colon. *Brit. J. Cancer* **54**: 595–600, 1986;

   (b) Plowman J, Narayanan VL, Dykes D, Szarvasi E, Briet P, Yoder OC and Paull KD, Flavoneacetic acid: A novel agent with preclinical antitumor activity against colon adenocarcinoma 38 in mice. *Cancer Treat. Rep.* **70**: 631–635, 1986.

57. (a) Brewer AD, Minatelli JA, Plowman J, Paull KD and Narayanan VL, 5-(N-Phenylcarboxamido)-2-thiobarbituric acid (NSC 336628), a novel potential antitumor agent. *Biochem. Pharmacol.* **34**: 2047–2050, 1985;

   (b) Cooney DA, Covey JM, Kang GJ, Dalal M, McMahon JB and Johns DJ, Initial mechanistic studies with merbarone (NSC 336628). *Biochem. Pharmacol.* **34**: 3395–3398, 1985.

58. Brewer AD, Ferguson G and Parvez M, Triethylammonium salt of 1,2,3,4-tetrahydro-6-hydroxy-4-oxo-N-phenyl-2-thio-5-pyrimidinecarboxamide, ethanol solvate. *Acta Crystallogr.* **C43**: 144–147, 1987.

59. Huggins C and Yang NC, Induction and extinction of mammary cancer. *Science* **137**: 257–262, 1962.

60. Haddow A, New facts and concepts: A general survey. *Canadian Cancer Conference* **2**: 361–374, 1957.

61. (a) Gellert M, DNA topoisomerases. *Ann. Rev. Biochem.* **50**: 879–910, 1981;

   (b) Liu LF, DNA topoisomerases enzymes that catalyze the breaking and rejoining of DNA. *CRC Critical Rev. Biochem.* **5**: 1–24, 1983;

(c) Chen GL and Liu LF, DNA topoisomerases as therapeutic targets in cancer chemotherapy. *Ann. Rep. Med. Chem.* **21**: 257–262, 1986.

62. (a) Tewey KM, Chen GL, Nelson EM and Liu LF, Intercalative antitumor drugs with the breakage-reunion reaction of mammalian DNA topoisomerase II. *J. Biol. Chem.* **259**: 9182–9187, 1984;

    (b) Ross WE, DNA topoisomerases as targets for cancer chemotherapy. *Biochem. Pharmacol.* **34**: 4191–4195, 1985;

    (c) Alexander RB, Nelson WG and Coffey DS, Synergistic enhancement by tumor necrosis factors of *in vitro* cytotoxicity from chemotherapeutic drugs targeted at DNA topoisomerase II. *Cancer Res.* **47**: 2403–2406, 1987.

63. (a) Pommier Y, Schwartz RE, Swelling LA and Kohn KW, Effects of DNA intercalating agents on topoisomerase II induced DNA strand cleavage in isolated mammalian cell nuclei. *Biochemistry* **24**: 6406–6410, 1985;

    (b) Mattern MR, Mong SM, Bartus HF, Mirabelli CK, Crooke ST and Johnson RK, Relationship between the intercellular effects of camptothecin and the inhibition of DNA topoisomerase I in cultural L 1210 cells. *Cancer Res.* **47**: 1793–1798, 1987.

64. Glusker JP, Structure aspects of steroid hormones and carcinogenic polycyclic aromatic hydrocarbons. In: *Biochemical Actions of Hormones* (ed Litwack G), Vol. VI, Ch. 3, pp. 121–204, Academy Press, New York, NY 1979.

65. Yang NC, Castro AJ, Lewis M and Wong TW, Polynuclear aromatic hydrocarbons, steroids and carcinogenesis. *Science* **134**: 386–387, 1961.

66. (a) Glascock RF and Hoekstra WG, Selective accumulation of tritium-labelled hexosterol by the reproductive organs of immature female goats and sheep. *Biochem. J.* **72**: 673–682, 1959;

    (b) Toft D and Gorski J, A receptor molecule for estrogens: Isolation from the rat uterus and preliminary characterization. *Proc. Natl. Acad. Sci. USA* **55**: 1574–1581, 1966.

67. Ebright RH, Wong JR and Chen LB, Binding of 2-hydroxybenzo(*a*)pyrene to estrogen receptors in rat cytosol. *Cancer Res.* **46**: 2349–2351, 1986.

68. (a) Nissen ED and Kent DR, Liver tumors and oral contraceptives. *Obstet. Gynecol.* **46**: 460–467, 1975;

    (b) Siiteri PK, Steroid hormones and endometrial cancer. *Cancer Res.* **38**: 4360–4366, 1978.

69. Pott P, Cancer scroti. *Chirugical Observations Relative to the Cancer of the Scrotum.* 63–68, 1775.

Cyclic Nucleotide Metabolism as a Target in Chemotherapy

B. Jastorff, E. Maronde, M. X. P. van Bemmelen, M. Zorn
and R. Störmann

Department of Bio-Organic Chemistry, University Bremen, D-2800 Bremen, Germany

Abstract. Adenosine-3′,5′-monophosphate (cyclic AMP, cAMP) acts as a second messenger in a multitude of cellular events, including regulation of cell proliferation and differentiation. Abnormal concentrations of cAMP have been correlated with several regulatory diseases and cancer. Hundreds of "first-generation" cAMP analogues have been described aiming at a therapeutical agent. Reasons for lack of success are discussed.

Regulation of cell proliferation and differentiation in tumor cell lines and triggering programmed cell death (apoptosis) by means of cyclic nucleotide analogues yielded promising results. Several mechanisms of action are discussed. So far, 8-chloro-cAMP is found to be the most active tumor growth inhibitor. The role of metabolites of cyclic nucleotide analogues in growth inhibition and apoptosis is currently debated and discussed.

Gene regulation and differentiation in *Dictyostelium discoideum* offers a well-studied model for the complex biochemical mechanisms of cAMP actions in mammalian cells.

We review a more systematic approach to the development of cAMP analogues which led to the "test-kit"-concept. This set of second-generation cAMP derivatives has been systematically used to compare essential interactions between cAMP and its different receptor proteins. Distinct fingerprints of cAMP-binding proteins could be identified. Models for protein kinase binding sites are proposed using sequence alignment and analogue mapping.

We describe a new generation of cAMP analogues which are resistant to enzymatic hydrolysis, highly lipophilic and protein kinase site-selective. Side effects and misleading cell biological results may be minimized using these compounds.

1. Introduction

Adenosine-3′,5′-monophosphate (cyclic AMP, cAMP, compound 1), shown in Fig. 1 in its two possible conformations about the glycosidic bond, *syn* and *anti*, is involved in a multitude of cellular events. Its regulatory signal is ubiquitous throughout all organisms and affects diverse physiological and biochemical responses [1]. It is involved in hormone-mediated responses of cells, in the control of cell proliferation and differentiation [2–4], and in the onset of programmed cell death (apoptosis) [5]. As a second messenger it serves to control enzymatic activities, ion concentrations and gene expression and it is also known to be a chemotactic first messenger in some organisms like *Dictyostelium discoideum* [6]. The cAMP signal is perceived by different kinds of receptor proteins, e.g. catabolite gene activator-protein (CAP, CRP respectively) in bacteria [7], cell

surface receptors (CSR) in *Dictyostelium discoideum* and two general types of
cAMP-dependent protein kinases (cAKs) as well as specific ion channels [8] in
eukaryotes.

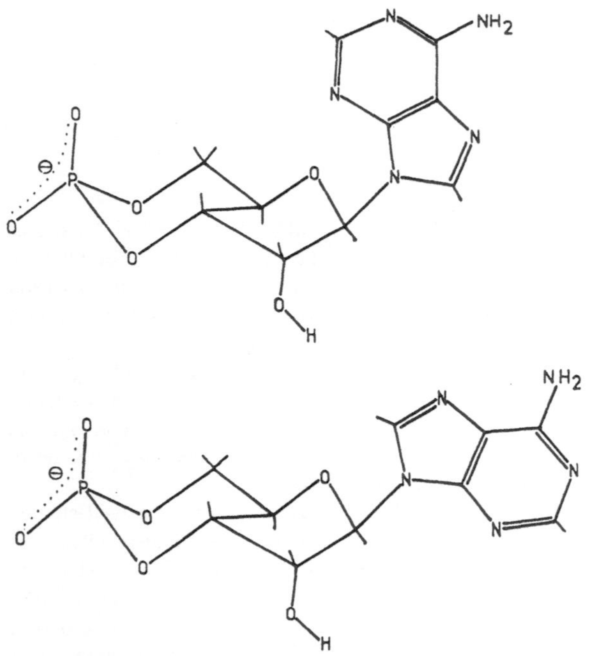

Fig. 1. Structure of cyclic AMP (cAMP, 1) in its two possible conformations about the glycosidic
bond, *syn* (upper) and *anti* (lower)

Those physiological activities which depend on intracellular cAMP concentration are strongly affected by proteins involved in the metabolism of cAMP, the
most prominent being adenylate cyclase and the enzyme family of cyclic phosphodiesterases (cPDEs) [9].

The complex network of cAMP metabolism and different cAMP receptors and
mediators is uniquely characterized in the model system of *Dictyostelium discoideum* [10–12]. Numerous similarities between the regulatory properties of this
system and that of mammalian cells have been pointed out by van Haastert et al.
[13].

When in the late sixties it became obvious that cyclic AMP was involved in
nearly all regulatory processes in man, and that several diseases were correlated
with cAMP levels below or above the standard values for healthy tissues (see
Table I) [1], pharmaceutical companies world-wide became interested in cyclic

nucleotide analogues as therapeutical agents. Attempts were made to develop cAMP analogues which could specifically revert abnormal cAMP metabolism in metabolic disorders to normal without affecting healthy tissue.

Table I. Regulatory diseases correlated with a pathological cAMP concentration

cAMP-regulated cellular processes	Pathological changes in cAMP levels	Diseases
Carbohydrate metabolism	Increased	Hyperglycemia
		Diabetes
Hormone synthesis	Increased	Hyperthyroidism
and secretion	Decreased	Hypothyroidism
Water and electrolyte metabolism	Increased	Cholera
Autoimmune response	Decreased	Bronchial asthma
Cell cycle regulation	Decreased	Cell transformation
Proliferation (e.g. in epithelial cells)		Psoriasis

Since the first report on chemical modifications of cAMP [14], several hundreds of cAMP analogues have been synthesized and screened for pharmacological activity in a first attempt to develop therapeutically useful derivatives [14–16]. The majority of these analogues comprises 2-, 6-, and 8-substituted derivatives, which are easily accessible from a synthetic point of view. The excessive costs of synthezising large amounts of the analogues needed for screening in animal models, and lacking specificity leading to side effects prevented the pharmacological study from being continued. Research was limited to non-systematic gathering of kinetic data for protein kinases and cPDEs [17]. Thus, only very few of these derivatives were investigated on a broader scale. All these compounds exhibited agonistic behaviour with protein kinases type I and II alike. They also were susceptible to hydrolysis by cPDEs, and many of them interfered with cyclic guanosine-3′,5′-monophosphate-dependent protein kinases. Nevertheless, some of these first-generation analogues remained as important tools in biochemical research (Fig. 2, compounds 2–8). The most prominent member of this group is N6.2′-dibutyryl cAMP (Bt$_2$-cAMP, Fig. 2, compound 2).

A second generation of cAMP analogues was developed during the early eighties as the pharmaceutical industry became interested in finding a cAMP antagonist useful for treatment of, e.g. diabetes. Resistance to degradation by cPDEs was also required for the newly synthezised derivatives. The first antagonist of cAMP, Rp-cAMPS, derived by substitution of sulfur for the equatorial exocyclic oxygen (Fig. 2, compound 9), was discovered during a close collaboration between industry and basic research [18–20]. The diastereomeric Sp-cAMPS with the sulfur substitution for the axial exocyclic oxygen (Fig. 2, compound 10)

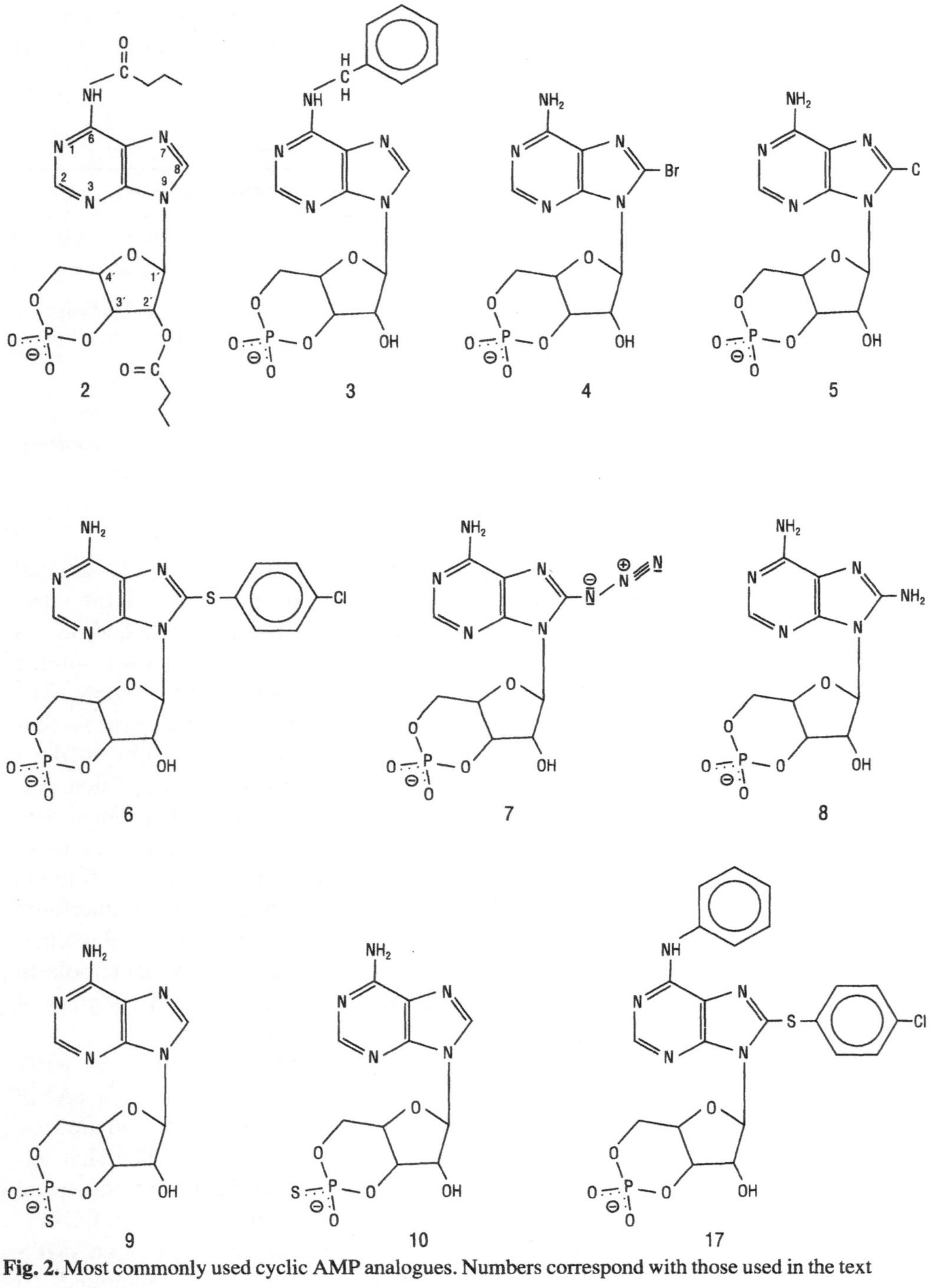

Fig. 2. Most commonly used cyclic AMP analogues. Numbers correspond with those used in the text

is a weak agonist of cAMP action *in vitro* and *in vivo* [18–20]. Lengthy synthesis and purification procedures, leading to high costs, again prevented these compounds from being thoroughly evaluated for pharmacological usefulness. All analogues of the first and second generation tested so far have shown too many side effects to become therapeutical agents for metabolic diseases, the main reason being interference with target enzymes in the healthy tissue. Still they constitute powerful tools for biochemical research.

2. Use of cAMP Analogues in Growth Regulation of Tumor Cells

The primary mediator for cAMP effects in mammalian cells is the cAMP-dependent protein kinase (EC 2.7.1.37., cAK) [21], existing as two isozymes, type I and II [22], which are tetrameric holoenzymes consisting of two regulatory (R) and two catalytic (C) subunits. The isozymes differ primarily in the sequence of their regulatory subunits [23]. The catalytic (C) subunits show extensive sequence homologies with all protein kinases [24].

In the absence of cAMP the dimer of regulatory subunits (R_2) acts as an inhibitor of the phosphotransferase activity of the C-subunits [25]. Upon elevation of the intracellular cAMP level up to four molecules of cAMP bind with high affinity to C_2R_2 and promote dissociation of the holeonzyme into R_2 and two monomeric C-subunits, which phosphorylate many different proteins inside the cell:

$$R_2C_2 + 4cAMP + 2MgATP = [R_2 * cAMP_4] + 2[C * MgATP]$$

The R subunits have a well-defined domain structure which includes two different cAMP binding sites: a kinetically stable site, termed Site B (Site 1) and a kinetically labile Site A (Site 2) [26–28]. These two sites exhibit different affinity for cAMP analogues. Generally, analogues with hydrophobic *N*6-substituents were preferentially bound to Site A, while modifications in C2- or C8-positions of the cAMP molecule led to Site B specificity [26, 29]. Furthermore, it was established that binding of analogues at either site stimulates binding at the other site in a cooperative manner [30].

Although there have been numerous attempts to relate cAMP concentration to the regulation of cell growth, its precise role has not been clearly defined. Early studies on the regulation of cell growth by cAMP employed first-generation analogues like Bt_2-cAMP and 8-bromo-cAMP (Fig. 2, compounds 2 and 4), which required unphysiologically high millimolar concentrations, or agents like forskolin and choleratoxin, that raised intracellular cAMP to abnormal and continuously high levels [4, 31, 32].

Because the involvement of cAMP in cellular proliferation and growth control [33, 34] holds the promise of developing a new antitumor agent on the basis of a

Table II. Growth inhibition of human cancer cell lines by cyclic AMP analogues

Cell line		Growth inhibition IC$_{50}$ (μM)		
		8-Cl	N^6-Benzyl	N^6-Phenyl-8-p-chlorophenylthio
Breast cancer	MCF-7	10	20	17
	MCF-7 *ras*	5	15	20
	MDA-MB-231	18	20	30
	T-47D	20	20	19
	ZR-75-1	20	20	19
	BT-20	19	27	20
Colon cancer	LS-174T	1	18	27
	WiDr	10	15	27
	HT-29	10	25	25
Lung cancer	A549	10	18	20
Fibrosarcoma	GT-1080	20	20	25
Ewing's sarcoma	A4573	18	15	20
Glioma	FOG	18	10	15
	U251	25	18	18

Note. 2–3 × 10^5 cells/60 mm dish were seeded, and 24 h later (Day 0), the medium was removed, and fresh medium and the analogues 8-chloro-cAMP (8-Cl), N6-benzyl-cAMP (N^6-Benzyl), N6-phenyl-8-p-chlorophenylthio-cAMP (N^6-Phenyl-8-p-chlorophenylthio) were added then and every 48 h thereafter. Cell growth was measured by counting the cell number at Day 3 and Day 4 after treatment of cells with each analogue at several concentrations at Day 0 and Day 2. Media used for cell culture contained 10% fetal bovine serum [35]. IC$_{50}$, concentration inhibiting 50% of cell proliferation. Data taken from [34].

modified cAMP molecule, these early studies were taken up again in pioneering biochemical and cell biological studies by Cho-Chung and coworkers. 8-Chloro-cAMP (Fig. 2, compound 5) N6-benzyl-cAMP (Fig. 2, compound 3) and the doubly modified analogue N6-phenyl-8-(p-chlorophenyl-thio)-cAMP (Fig. 2, compound 17) emerged as the most potent analogues. They produced growth inhibition at micromolar concentrations in 14 different human cancer cell lines (Table II), while showing no signs of toxic effects [35]. The growth inhibitory effect is correlated with a change in cell morphology and, in the case of many leukemia cell lines, also with the expression of differentiation markers [34]. When cultured cells were treated with micromolar concentrations of N6-benzyl-cAMP or 8-chloro-cAMP for a few days, intracellular concentrations of cAK I decreased about 60–80%, with a concomitant increase of cAK II [34]. Combination of these two cAMP analogues produced synergistic effects [36]. Also the regulatory subunit of cAK II is translocated to the nucleus in response to 8-chloro-cAMP treatment of cancer cells [37].

These findings led Cho-Chung and coworkers to propose the following hypothesis: cAK I and cAK II have different functions in regulation of cellular processes. While cAK I is necessary for proliferation, a high level of cAK II is required for differentiation. Any cAMP analogue that preferentially binds to Site B, but not to Site A of cAK II, produces a holoenzyme of cAK II with two bound molecules of the cAMP analogue ($[R[cAMP^*]]_2C_2$) which still inhibits the activity of the catalytic subunit C, but exposes a region of the protein that serves as a nuclear translocation site [38]. Thus cAK II is translocated to the nucleus where it boosts the expression of its regulatory subunit by a still unclear mechanism. This ultimately leads to a halt in proliferation and the start of differentiation.

As a result of these lines of research 8-chloro-cAMP (Fig. 2, compound 5) has already become the most prominent candidate for use in cancer therapy. This has been made possible by the scaling-up of several steps in the synthesis of this analogue by Robins and coworkers [39]. 8-chloro-cAMP is presently in preclinical testing at the NIH.

The biochemical mechanism leading to the unique properties of 8-chloro-cAMP has not been firmly established. Although disputed by Cho-Chung and coworkers, recent results suggest that metabolites of 8-chloro-cAMP, especially 8-chlo-roadenosine, are responsible for some of the effects on cell growth regulation [40] or may exert at least a synergistic effect with 8-chloro-cAMP. cAMP analogues with binding properties similar to 8-chloro-cAMP, but resistant to biological degradation, can help to clarify the points of doubt (see below).

Due to the long and strict trial which a potential drug has to go through, it is far from being sure that 8-chloro-cAMP will soon become a therapeutic agent. Especially tests of both its acute and chronic toxicity will be critical hurdles on its way towards clinical application.

3. Triggering Programmed Cell Death with cAMP Analogues

Another line of research with respect to potential tumor therapy points to the possibility of triggering programmed cell death with cAMP analogues. In the last decade it has become clear that a number of cell types can undergo programmed self-destruction (apoptosis), that is essential for embryogenesis, metamorphosis and related processes under normal conditions [41–44]. A key step in apoptosis is a protein synthesis-dependent DNA fragmentation leading ultimately to cell lysis. It has been shown that this process can be triggered in leukemia cells by agents which elevate intracellular cAMP levels such as cholera toxin, and also by a number cAMP analogues at micromolar concentrations (Table III) [5]. Lanotte and coworkers found a correlation between the potency of cAMP derivatives in triggering apoptosis and their affinity for protein kinase I [5]. This interpretation has been strengthened by the finding that type I Site A-selective and Site B-selective cAMP analogues in combination exhibit synergism in triggering apoptosis [5]. Recent experiments with more than 50 cAMP analogues and some of their metabolic products point to an involvement of metabolically produced adenosine analogues in apoptosis or in other mechanisms of cell death (i.e. necrosis) [45].

Table III. Potency of cAMP analogues as inducers of cytolysis

Compound (mM)	EC$_{50}$ (µM)
N^6-monobutyryl-cAMP	305
$N^6,O2'$-dibutyryl-cAMP	280
8-Piperidino-cAMP	1,000
N^6-benzoyl-cAMP	295
8-p-Chlorophenylthio-cAMP	30
8-Aminohexylcarbamoyl-cAMP	825
8-Chloro-cAMP	50
8-Bromo-cAMP	860
8-S-ethyl-cAMP	215
8-S-methyl-cAMP	375
8-Aminohexylamino-cAMP	1,100
8-Amino-cAMP	10
8-Methylamino-cAMP	2,100

Note. The table gives the concentration of cAMP analogues required for 50% decrease of the viability of the leukemia cells as EC$_{50}$. 10^5 Cells/ml were incubated for 48 h in the presence of the analogue, and the remaining mitochondrial dehydrogenase activity was determined. Media contained 5% fetal calf serum. Data taken from [5].

Adenosine receptor agonists have been shown to induce apoptosis in thymocytes, suggesting involvement of the adenosine receptor [46].

These results suggest a potential for developing an alternative treatment for leukemia, which does not use generally cytotoxic substances and can be performed as an extracorporal treatment of bone marrow. This would markedly reduce the potential for side effects of cyclic nucleotide analogues on healthy tissues and particularly diminish e.g. the risk of genotoxicity of current therapeutic drugs.

Detailed *in vitro* studies of the mechanism of cAMP analogue-induced apoptosis are under way to design a suitable compound for pharmacological testing.

4. Possible Role of Metabolism in Growth-Inhibiting Properties of Cyclic Nucleotide Analogues

The assessment of the pharmacological and therapeutical value of any new drug is essentially impossible without a firm knowledge of its pharmaco-kinetics and metabolic transformations. Nucleotide metabolism was extensively studied during the sixties and early seventies, leading to the elucidation of the main pathways of degradation and the enzymes involved (for an overview see Scheme 1) [47].

Under physiological *in vivo* conditions, and in long term cell biological experiments, slow degradation processes can lead to accumulation of analogue metabolites, which may produce deleterious side effects or, in the case of biochemical investigations, misleading results. Evidence suggests that, in the case of the most

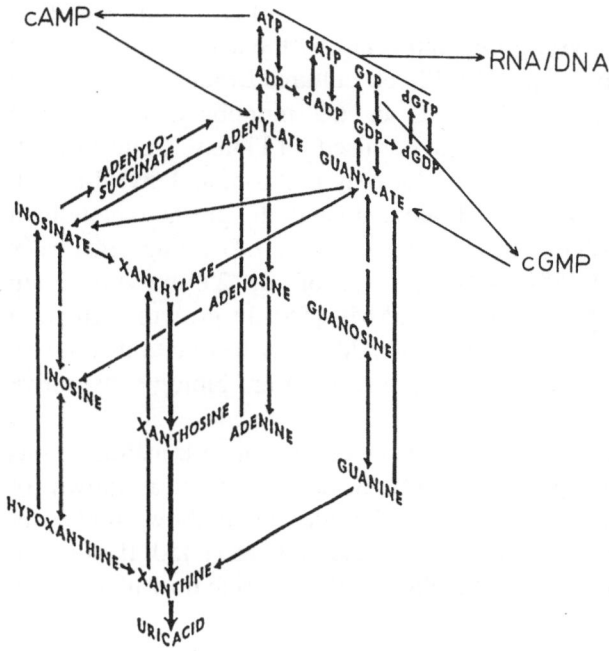

Scheme 1. Metabolic pathways which cyclic nucleotides and their analogues may follow after initial hydrolysis by cPDEs

promising cAMP analogues for tumor inhibition (e.g. 8-chloro-cAMP and 8-amino-cAMP), metabolites take part in exerting the observed tumor inhibitory properties [40, 48]. For a more cautious interpretation of cell biological results with cAMP analogues, their metabolites have to be considered as substrates for RNA/DNA synthesis and as ligands for all proteins regulated by purines, purine nucleosides, purine nucleotides or purine-containing cofactors.

Studies on the metabolism of many cyclic nucleotide analogues were performed using purified enzymes of the common nucleotide degradation pathway (cyclic phosphodiesterases, phosphatases, adenosine deaminases). The key step in cAMP analogue metabolism is hydrolysis by cPDEs, which has been studied extensively [15, 49–51]. More complex systems such as serum-containing cell culture medium [40], cell cultures and whole tissue or perfused organs [52–54] have only rarely been used.

4.1. Metabolism of cAMP Analogues by Esterases and Monooxygenases

Transformation of cAMP analogues by enzymes not involved in the nucleotide degradation pathway must also be taken into account, as illustrated e.g. by the metabolism of the most widely used analogue, Bt_2-cAMP (Fig. 2, compound 2). Hydrolysis of the labile $O2'$-butyryl group can occur by the action of unspecific esterases, as well as chemically [55]. In PC12 cell cultures Braumann et al. [56] showed the conversion of Bt_2-cAMP to both $N6$- and $O2'$-monobutyryl-cAMP and even to cAMP in the course of 3 days. In the perfused rat kidney, Coulson et al. [53] demonstrated conversion of Bt_2-cAMP not only to $N6$-monobutyryl-cAMP and butyrate, but also integration into anabolic and catabolic purine metabolism via cAMP, AMP, ADP and ATP. Since a substituted 2'-OH-group drastically decreases cAK activation [57, 58], only removal of the $2'O$-butyryl group produces the cAMP agonist $N6$-monobutyryl-cAMP [55]. In growth regulation and differentiation the biological activity of Bt_2-cAMP may be due to its conversion to $N6$-monobutyryl-cAMP and butyrate which both are biologically active [59–62].

For cyclic nucleotides resistant to break-down by cPDEs, other transformations must be considered. In the case of e.g. Rp-cAMPS, desulfuration (as shown for phosphorothioate xenobiotics [63]), or aromatic N-oxidation (as shown for 9-substituted adenine derivatives [64]) by monooxygenase systems (P450, flavin-containing monooxygenase) are possible metabolic reactions. Up to now no research has been done in this area.

4.2. Metabolism by Cyclic Phosphodiesterases

In most cases the first step of cAMP analogue metabolism is hydrolysis of the 3',5'-cyclic phosphate by cyclic phosphodiesterases. In vertebrates this reaction usually yields the correspondent nucleoside-5'-monophosphate. Cyclic nucleotide analogs which are degradable by cPDEs can enter the adenine and/or guanine anabolic or catabolic pathways. The resulting complications for investigation of the molecular mechanisms of cyclic nucleotide effects can only be prevented by the use of nonhydrolysable analogues.

So far, only substitution of the equatorial exocyclic oxygen atom by sulfur, as in Rp-cAMPS (Fig. 2, compound 9, or analogs of 9), abolishes hydrolysis of the cyclic phosphate moiety by cPDEs [49, 65]. The corresponding Sp-derivatives

Table IV. Multiple isoenzymes of cyclic nucleotide phosphodiesterases [9]

cPDE families	Number of characterized subfamilies
I. Ca^{2+}-calmodulin-dependent cPDEs	6
II. cGMP-stimulated cPDEs	3
III. cGMP-inhibited cPDEs	4
IV. cAMP-specific cPDEs	4
V. cGMP-specific cPDEs	3

have also been found to be resistant to cPDEs, provided a second substitution is introduced at the base moiety [49], i.e. compounds 13–16 in Fig. 6.

In most, if not all, tissues several cPDE isoenzymes are present which show different substrate specificity and regulatory properties [66–68]. Analogues of cyclic nucleotides may not only act as substrates [49, 51, 69] or inhibitors, but also as regulatory effectors of cPDE activity, namely on cGMP-activated and cGMP-inhibited cPDEs (Table IV) [50, 69–71]. Not only cyclic nucleotide analogues are able to inhibit cPDEs, but also the well known methylxanthines [72], acylpeptides [73] and other heterocyclic compounds [74].

4.3. *Effects of the Anabolic Metabolites of cAMP Analogues*

After the first metabolic step of forming a nucleoside-5′-monophosphate, different metabolic pathways may be followed depending on the specific nucleotide analogue and the metabolizing tissue. The anabolic pathway leads to the correspondent 5′-diphosphates, which may act on ADP-reductase, and 5′-triphosphates that may interfere with adenylate cyclase, G-proteins, DNA and RNA synthesis. Thus anabolic processes provide another possible mechanism for the observed growth inhibition. Recent studies in our laboratory have shown that, in the case of 8-chloro-cAMP, micromolar concentrations of its metabolite 8-chloro-5′-AMP effectively inhibit proliferation of glioma cells [75]. The amounts of this metabolite formed by cPDE hydrolysis in cell culture medium are large enough to make a significant contribution to the proliferation inhibition seen with the cyclophosphate. However, 8-chloro-5′-AMP did not induce differentiation in glioma cells and thus cannot constitute the sole agent for proliferation blocking [75]. Interference with 5′-nucleotidase, AMP-kinase and AMP-deaminase are possible mechanisms for 8-chloro-5′-AMP growth inhibiting effects as proposed for 8-bromo-5′-AMP [76]. Modified polyadenylate may also affect cellular growth and proliferation by inhibiting adenylate cyclase [77].

4.4. *Effects of the Catabolic Metabolites of cAMP Analogues*

At least in part, biological effects of cAMP analogues may also result from conversion to the corresponding modified adenosine, which can be accumulated if subsequent degradation steps are slow. This situation has been found, e.g. with 8-chloro-cAMP (Fig. 2, compound 5) in CHO cells [40]. Results from our laboratory also suggest a prominent role for the dephosphorylated metabolite of N6-benzyl-cAMP in proliferation control of C6 glioma cells. In this case the metabolite N6-benzyladenosine is more active than its parent cyclic nucleotide and is present in the cell culture medium. Addition of cPDE to the culture medium potentiates the growth inhibitory effect of N6-benzyl-cAMP (Fig. 3) [78].

Adenosine has established regulatory roles in a wide range of tissues as diverse

(A)

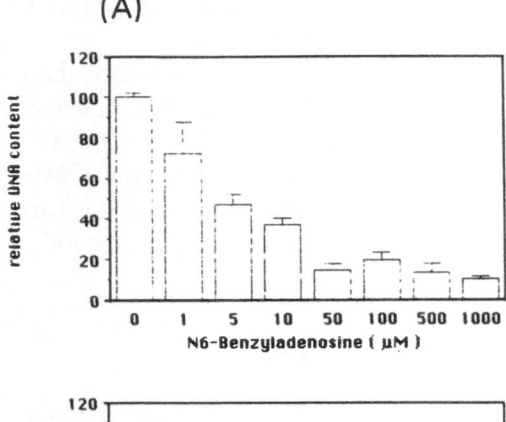

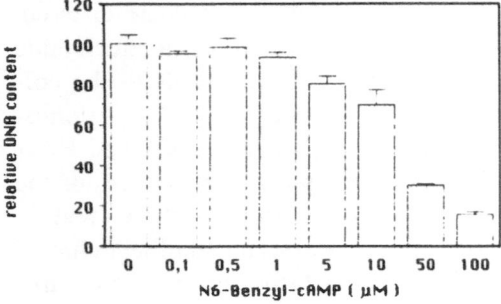

(B)

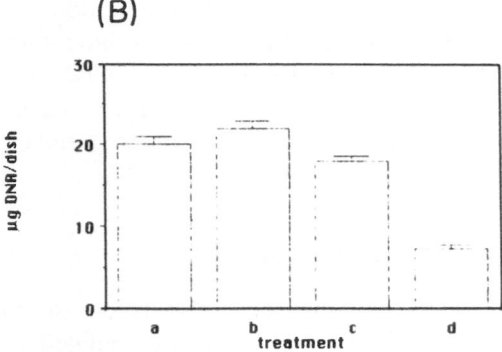

Fig. 3. (A) Effect on DNA content in cultured C6 glioma cells by either *N*6-benzyl-cAMP or *N*6-benzyladenosine (control = 100%). **(B)** DNA contents of C6 glioma cells cultured under different conditions: a—untreated cells; b—in the presence of 0.02 U exogenous cPDE; c—in the presence of 10 μM *N*6-benzyl-cAMP, and d—10 μM *N*6-benzyl-cAMP and 0.02 U exogenous cPDE. 3.8×10^4 cells per cm^2 were seeded in multiwell-dishes, and 24 h later (Day 0), the medium was removed and fresh medium and the analogues *N*6-benzyladenosine or *N*6-benzyl-cAMP were added then and at Day 3. DNA-content was determined on Day 5. All cultures were performed in the presence of 0.5% heat-inactivated fetal calf serum

as the heart and vascular tissue, the central and peripheral nervous system, the kidney, adipose tissue, blood platelets and macrophages, including a first messenger function in intercellular communications (for an overview on adenosine-dependent regulatory pathways, see Scheme 2) [79].

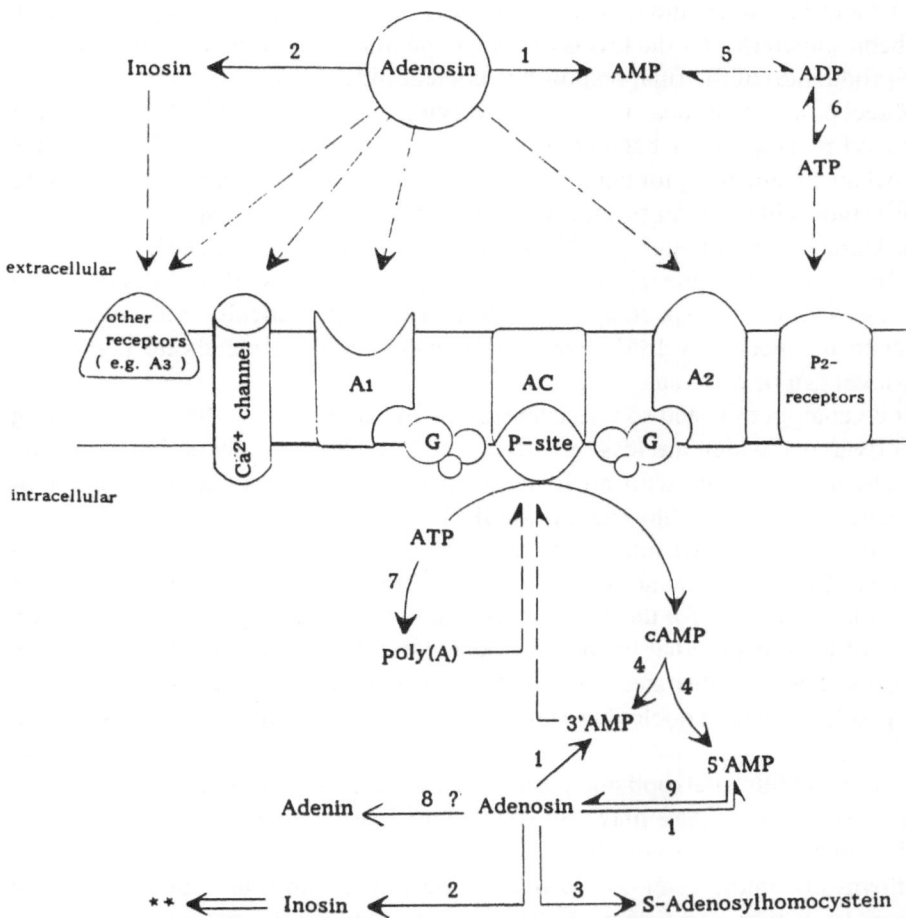

Scheme 2. Regulatory and metabolic processes with the involvement of adenine nucleotides or nucleosides in cellular control mechanisms. Numbers denote the following enzymes: 1: adenosine kinase; 2: adenosine deaminase; 3: S-adenosylhomocysteine transferase; 4: phosphodiesterase; 5: adenylate kinase; 6: phosphokinase; 7: poly(A) polymerase; 8: phosphorylase; 9: 5'-nucleotidase

With respect to regulation of cyclic nucleotide metabolism and growth control, the following functions of adenosine need to be considered.

(i) Adenosine can regulate adenylate cyclase activity through two G-protein-linked transmembrane receptors (the stimulatory R_a or A2 receptor and the inhibitory R_i or A1 receptor) [80, 81].

(ii) An intracellular receptor site of low potency, mediating adenylate cyclase inhibition, the P-site, seems to be located on the catalytic subunit of the cyclase itself [80, 81].

(iii) cAMP has been claimed to be the primary second messenger associated with A1 and A2 adenosine receptors. In many cases, adenosine receptor agonists have been shown to alter the levels of cAMP, but the involvement of such changes in the production of the final response is unclear [82].

A2 receptors are almost ubiquitous, varying only in their levels in different tissues. A1 receptors have been preferentially found in brain, heart and in lipocytes [83]. All adenosine receptor agonists are modified adenosine analogues. Structure activity studies have so far produced no analogue with absolute specificity for the A1 or A2 receptor, although preferential binding is observed [83, 84].

AMP– and adenosine signals are switched off by AMP– and adenosine deaminases (ADA). Especially ADA has been systematically investigated in respect to its specificity [85]. For cAMP analogues their metabolic fate at the ADA-level can be estimated or tested *in vitro*.

For all analogues tested so far, deamination by ADA leads to the corresponding oxo derivatives, which are less cytotoxic [86, 87]. Drastic reduction of deamination velocity correlates with an increase in cytotoxicity, as exemplified for 8-bromoadenosine and 8-chloroadenosine [87].

Adenosine analogues may also interfere with the purine nucleoside phosphorylase (PNP), which catalyzes the removal of the ribose moiety from purine nucleosides. Although for the natural substrates of PNP, the equilibrium lies far on the side of the reaction that forms the riboside, it may be shifted, when adenosine analogues serve as substrates. The facile conversion of 8-chloroadenosine to the corresponding adenine nucleobase and ribose-1-phosphate has been demonstrated [86].

In the catabolic metabolism, purine bases are further oxidized by xanthine oxidase. Xanthine oxidase may also oxidize nucleoside analogs directly, because its substrate specificity is very low.

Pathways of interconversion of nucleosides, nucleotides and nucleobases are different from tissue to tissue. All these diverse metabolic pathways and their products have to be evaluated for every new analogue with a therapeutical perspective.

5. Protein Kinase as a Target for the Design of Tumor Inhibitory cAMP Analogues: Approaches for SAR Studies

Taking into account the numerous unresolved issues regarding the role of metabolites, none of the proposed antitumor drugs seems ideally suited for pharmacological development. The analogues in question were discovered by random screening.

Since the targets of cyclic nucleotides in proliferation control are the well-studied cyclic AMP-dependent protein kinases, current knowledge of molecular interactions determining affinity of cAMP analogues for cAKs can be exploited for a more rational design of new candidates for tumor inhibition.

Different methods have been used to identify the specific molecular interactions (and concomitantly the type of amino acid involved) between cyclic nucleotides and their target proteins:

(i) Mapping the binding site by means of cyclic nucleotide analogues.

(ii) Affinity labeling.

(iii) Isolation and crystallisation of cyclic nucleotide binding or metabolizing proteins.

(iv) Sequencing of proteins or their genes, sequence alignment, site directed mutagenesis within motifs predicted to be in contact with the ligand.

(v) Computer modelling of a homologus sequence based on the tertiary structure obtained by X-ray analysis.

Major advances in cyclic nucleotide research, such as the elucidation of the principal molecular interactions between cAMP and the binding sites of the regulatory subunits of cAKs, have been possible using the mapping approach [29, 57, 88–92].

Mapping studies have been performed on purified cAK holoenzymes and isolated R-subunits [88–90], leading to a model of molecular interactions proposed by Jastorff et al. [57]. The types of amino acids involved in recognition and binding of cAMP to the R subunits were deduced from these studies. Subsequent systematic binding studies on highly purified isolated regulatory subunits of cAKs showed the presence of two kinetically different intrasubunit binding sites for cAMP, designated Site A (Site 2) and Site B (Site 1) [26–28]. Binding of cAMP to both sites is neccessary to activate the enzyme [93]. Cyclic AMP analogues have been particularly useful in characterizing these sites. First, certain cAMP analogues selectively bind to one or the other of the two types of the regulatory subunits; second, binding of a cyclic nucleotide at one site stimulates binding at the other site [94]. Cyclic nucleotide analogues were classified as site-selective, isoenzyme-selective or non-selective ligands. The kinetics of each individual site in cAK I and II was studied using more than 100 cAMP derivatives [29, 92].

The selection of analogues tested was largely based on availability and was therefore more random than systematic. The approach followed in our laboratory used a limited set of cyclic nucleotide derivatives with systematic variation of molecular interactions donated by the cAMP molecule (Fig. 4) [95, 96]. All compounds of this testkit have been characterized according to lipophilicity [97, 98], cPDE-stability [50, 51] and chemical stability under physiological conditions.

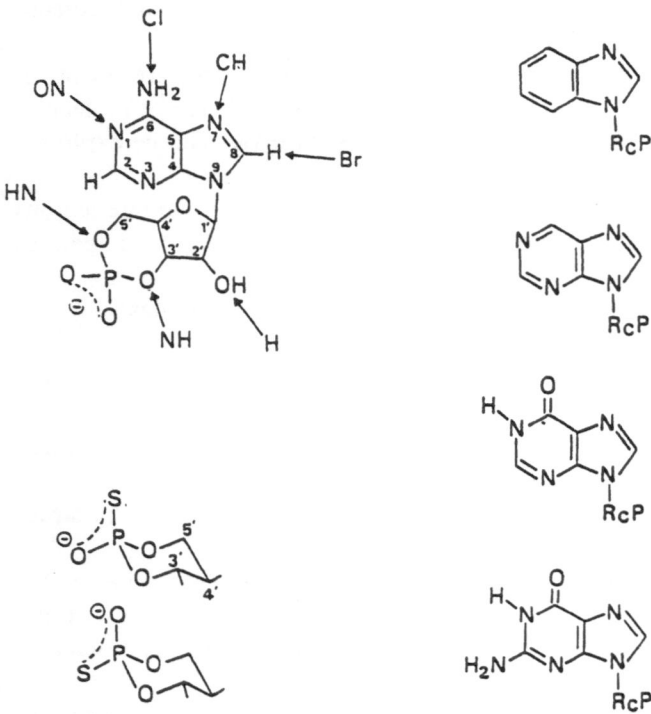

Fig. 4. Systematic modifications at positions of cyclic AMP which may directly interact with a protein and aromatic heterocyclic bases with increasing polarity and different potential for charge-transfer interactions (RcP denotes ribose-3′,5′-monophosphate). This set of compounds (testkit) serves to define all essential molecular interactions between cyclic AMP and its binding proteins

In addition, the theoretical electronic properties of the base moieties have been calculated by MNDO [51, 99].

Using this testkit, the contribution to cAMP binding of each potentially interacting atomic group can be evaluated. Several different proteins of cAMP metabolism in *E. coli* [100], *Dictyostelium discoideum* [101], yeast [49] and mammals [51, 71, 88, 89, 99] have been mapped with the testkit analogues. Four different types of cyclic AMP binding sites were thus identified. Cyclic AMP is bound by a characteristic network of hydrogen bonds, ionic interactions and stacking, or pure hydrophobic interactions of the base moiety, which is present either in *syn* or *anti* conformation (Fig. 5). The fingerprint of essential interactions obtained for cAMP with cAK is identical for all cAKs mapped so far, from protista to mammals. The first antagonist of cAK activation was found using this concept [18–20, 102, 103]. It also led to a third generation of cAMP analogues, which constitute cPDE-resistant, site-selective and isoenzyme-selective compounds (Fig. 6) [104]. Compounds 13–16 are highly lipophilic [99]. Promising initial results with some of

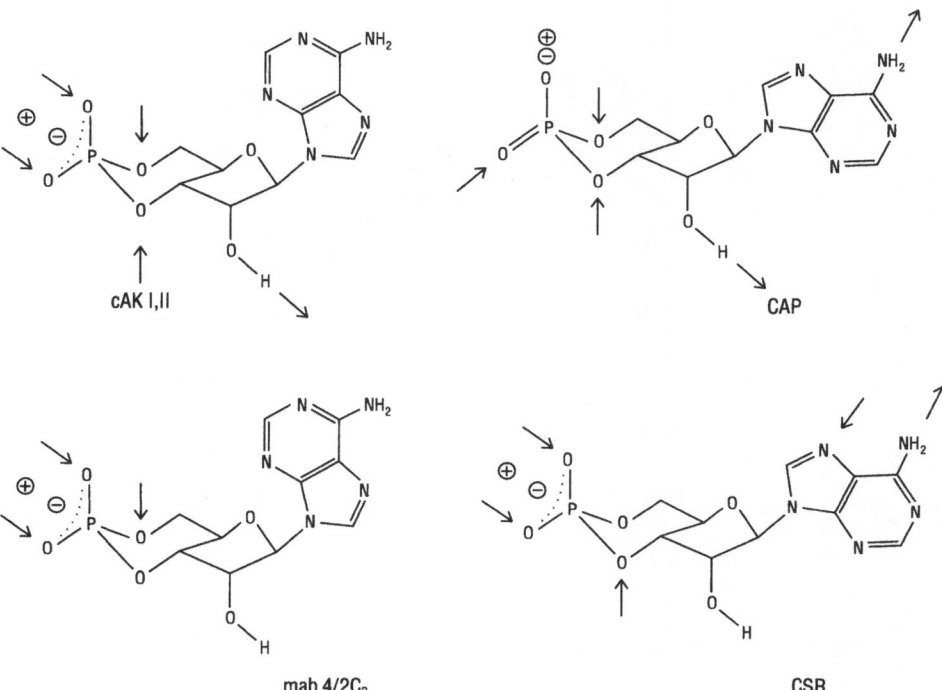

Fig. 5. Fingerprint of molecular interactions between cAMP and its protein receptors. cAK I.II: cyclic AMP-dependent protein kinases type I and II; CAP: catabolite activator protein; CSR: cyclic AMP cell surface receptor of *Dictyostelium discoideum;* mab 4/2 C₂: monoclonal antibody raised against cyclic AMP bound to bovine serum albumin via a succinyl ester linkage at the 2′-O-position [99]. The cAKs and mab 4/2 C₂ bind cAMP in *syn*-conformation, CAP and CSR bind cAMP in *anti*-conformation. Ionic interactions and hydrogen bonding are shown for each model; base stacking, dipole interactions and hydrophobic interactions, which are also essential for binding to some of these proteins, are omitted for clarity

these analogues in growth inhibition and differentiation in human cancer cells [105], and in prespore gene induction in *Dictyostelium discoideum* have been obtained [106].

In close correspondence to this approach are more direct methods of determining the amino acids involved in cAMP binding by cAK. Affinity labeling of the R subunits was performed by Taylor et al. using 8-azido-cAMP (Fig. 2, compound 7) as a labeling analogue [107]. In conjunction with the establishment of the primary sequence of R-I and R-II [108, 109], this method allowed to define amino acids which lie near or within the binding site. Thus, Tyr-371 of cAK R-I at Site B has been shown to interact with cAMP [107].

A break-through in cyclic nucleotide research was the first crystal structure of the catabolite gene activator protein CAP from *E. coli* by Weber and Steitz [110].

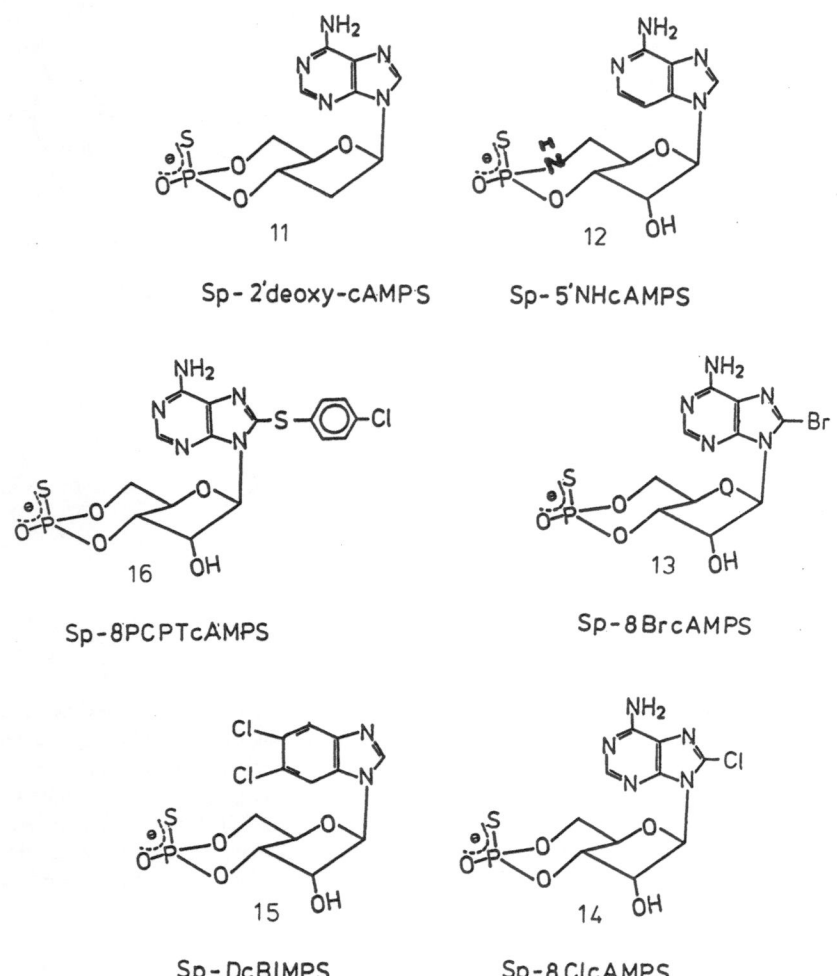

Fig. 6. The new generation of process-selective cAMP analogues, all showing agonistic properties with cAMP-dependent protein kinase, resistance to hydrolysis by cPDE and increased hydrophobicity in the case of compounds 13–16

Alignment of the amino acid sequences of cAK R-I and R-II [108, 109] with that of CAP established a relatedness between the bacterial cAMP binding site and the two cAMP binding domains in the cAKs [111]. A model of each binding domain was constructed based on the crystal structure of CAP by building the R-subunit sequences into the α-carbon backbone of the cAMP binding domain of CAP [111]. Two residues interacting directly with cAMP in CAP were found invariant in cAKs: In CAP, Arg-82 interacts with the negative charge at the phosphate moiety, and Glu-72 forms a hydrogen bond to the 2′-OH of the ribose ring [110]. The corresponding Arg-209/Glu-200 in Site A, and Arg-333/Glu-324 in Site B, respectively, were proposed to bind the cyclophosphate region in cAK R-I similarily [111, 112]. Kinetic changes in several mutant cAKs, which have been obtained by either site-directed mutagenesis or induced point mutations are in accord with these assignments, and so are the photolabeling experiments [112, 113]. These data also agree with the model based on mapping with the testkit analogues.

In addition to the interactions proposed by Weber et al. [111], our testkit-based mapping of cAMP binding sites on cAKs gives evidence for other essential interactions, particularly to 3′-O and 5′-O of the cyclophosphate ring (Fig. 5) [29, 88, 92].

Up to now, primary sequences have been determined for the different types of regulatory subunits R-I$_\alpha$, R-I$_\beta$, R-II$_\alpha$, R-II$_\beta$ from several mammals [108, 109, 114–121], R-I from the insect *Drosophila* [122], the nematode *Caenorhabditis elegans* [123], R-D from the cellular slime mould *Dictyostelium discoideum* [124] and the baker's yeast regulatory subunit [125, 126].

In an attempt to find further amino acid residues which may be involved in cAMP binding, we aligned all the sequences of all putative cAMP binding domains of these proteins, including the two published cAMP-gated olfactory channels [127, 128] (Table V) [129]. (A slightly different and limited alignment of cAMP binding sites in the channels with cyclic nucleotide-dependent protein kinases has also been published recently [8]; for a review on alignment of cAKs with cGMP-dependent kinases see [130].) In our model a short concensus sequence Phe–Gly–Glu or Tyr–Phe–Gly–Glu, which includes the invariant Glu proposed by Weber et al. [111], could be identified tentatively, suggesting an interaction of a second aromatic residue (e.g. Tyr-321 in R-I Site B and Phe-198 in R-I Site A) with the adenine base. Also, in a position corresponding to Tyr-371 in Site B of R-I and R-II, we found Phe to be present in Site A. A second putative consensus sequence involves the invariant Arg. The sequence X–Pro–Arg, with X denoting a hydrogen bond donor residue or positively charged ionic residue, is proposed to be bound to both exocyclic oxygens of the cyclophosphate ring. Different affinity for cAMP at Sites A and B may arise form the variance in X (Arg in R-I Site B, Lys in R-II Site B, and Thr in R-I and R-II Sites A) and the differing aromatic side chains interacting with the adenine base (two Tyr in both Sites B and two Phe in both Sites A) (see Table V and Fig. 7). We also propose an Asp residue (Asp-288 in R-I), identified by modelling studies, which was conserved in all mammalian cAKs, to interact with the 5′-O of cAMP (Fig. 7). The identification of the amino acids responsible for site-selectivity or isozyme-selectivity has not yet been possible. Site-directed mutagenesis experiments are under way, to test our extended cAMP binding model for the cAKs.

Table V. Consensus sequences in cyclic nucleotide binding proteins

Amino acid binding to	Purine 2'-OH	eq., ax. phosphate-O
Olfactory channel (rat)[127]	(519) S C*F G E*I S I L N I	K*G S*K M G (535)
Olfactory channel (bovine)[128]	(517) S C F G E I S I L N I	K G S K M G (533)
R-II B-site (mammalian α, β)[114-121]	(330) Q Y*F G E L A L V T N K P R A A S (346)	
R-I B-site (mammalian α)[114-121]	(320) D Y F G E I A L L M N R P R A A T (336)	
R-I B-site (mammalian β)[114-121]	(320) D Y F G E I A L L L N R P R A A T (336)	
R-I B-site (insect)[122]	(318) D Y F G E I A L L L D R P R A A T (334)	
R-I B-site (nematode)[123]	(316) D Y F G E I A L L L D R P R A A T (332)	
R-I B-site (slime mould)[124]	(258) D Y F G E I A L L T D R P R A A T (274)	
R-I A-site (mammalian α, β)[114-121]	(196) G S F*G E L A L I Y G T P R A A T (212)	
R-I A-site (insect)[122]	(192) G S F G E L A L I Y G T P R A A T (208)	
R-I A-site (nematode)[123]	(192) G S F G E L A L I Y G T P R A A T (208)	
R-I A-site (slime mould)[124]	(132) G S F G E L A L I Y G S P R A A T (148)	
R-II A-site (mammalian α, β)[114-121]	(200) G S F G E L A L M Y N T P R A A T (216)	
R A-site (yeast)[125,126]	(245) S S F G E L A L M Y N S P R A A T (261)	
R "B"-site (yeast)[125,126]	(363) D Y F G E V A L L N D L P R Q A T (380)	
CAP (E. coli)[110]	(68) D F I G E I G L F E E G Q E R S A (84)	

Note. Putative cAMP binding domains of regulatory subunits of protein kinases type I (R-I$_\alpha$ and R-I$_\beta$) and type II (R-II$_\alpha$ and R-II$_\beta$), the CAP of *E. coli* and the cAMP-gated olfactory channels were aligned and amino acids designated by their one-letter code. Bracketed numbers in front and at the end of the cited sequences denote the position of the first and last amino acid shown according to the individual numbering of each sequence. Superscript numbers refer to the publications, from which the sequences were taken.

Amino acids proposed to directly interact with a particular part of the cyclic nucleotide are indicated (∗); eq.: equatorial, ax.: axial exocyclic phosphate oxygen atoms. The proposed interaction of 5'-O with an Asp-residue and the known interaction of the base with a second aromatic amino acid in cAMP-dependent protein kinases (Tyr in Site B and Phe in Site A) is not included in this table, because the residues in question lie outside the regions shown.

6. *Dictyostelium Discoideum:* A Model System for Gene Regulation by Cyclic Nucleotide Analogues

Several members of the slime mould subclass *Dictyostelia* are the only organisms known to use cAMP as first messenger. Thus, *Dictyostelium discoideum,* *D. rosarium, D. mucuroides* and *D. purpureum* use cAMP as a chemotactic signal [6]. *D. discoideum* is the most intensively studied of all these species by far. *D. discoideum* lives as single amoebae in soil where it feeds on bacteria. When food is depleted, cells aggregate by means of chemotaxis and form a multicellular pseudoplasmodium. In that stage they differentiate into either spore or stalk cells.

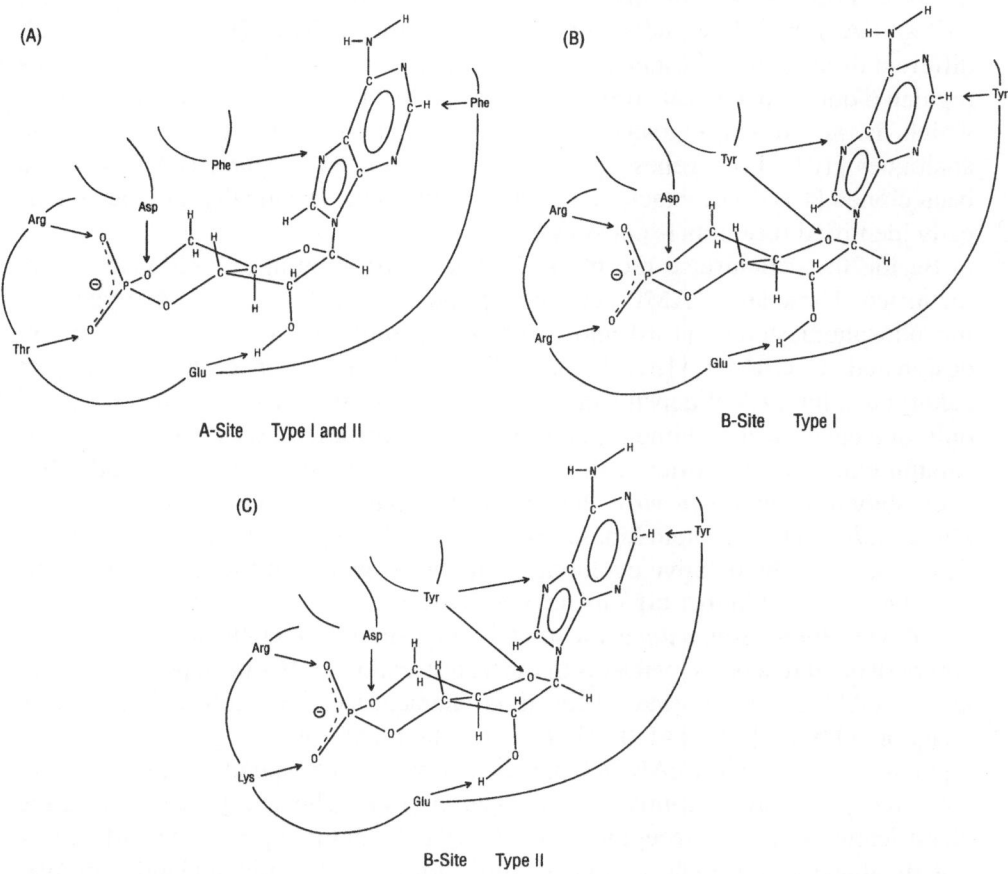

Fig. 7. Model for the binding of the exocyclic oxygen atoms, the ribose and the purine base of cAMP to the regulatory subunits of cAMP-dependent protein kinases. (**A**) Site A of mammalian R-Iα, R-Iβ, R-IIα and R-IIβ, *Drosophila* R-I and *Caenorhabditis elegans* R-I; in yeast and *Dictyostelium discoideum* only the Thr residue is changed to Ser. (**B**) Site B of mammalian R-Iα and R-Iβ, *Drosophila* R-I, *Caenorhabditis elegans* R-I and *Dictyostelium discoideum* R-D. (**C**) Site B of mammalian R-IIα and R-IIβ

cAMP constitutes both the chemotactic extracellular hormone [131] and the first messenger for differentiation. Although intracellular receptors for cAMP have been found, their role in chemotaxis and development is still controversial. *D. discoideum* constitutes a good example of the diversity of cAMP receptors which can be found in a single organism. Receptors for cAMP are found both at the cell surface and intracellularly. Both groups of receptors are involved in the processes by which cAMP acts as a chemoattractant and as a morphogenetic signal.

Kinetic studies have characterized four presumably interconvertible cAMP receptor classes on the cell surface: high- and low-affinity fast dissociating A Sites (A^H and A^L) and slow and superslow dissociating B Sites (B^S and B^{SS}), with different dissociation constants and dissociation rates [132, 133]. They are down-regulated during persistent stimulation with cAMP [10]. Recently a fifth receptor which is not down-regulated has been characterized by cyclic nucleotide analogues [101]. Four genes of the chemotactic cAMP receptor (cAR1-4) have been cloned [134]. The exact relationship between the kinetically and the genetically identified receptors is not yet known.

Beside the cell-surface receptors, further cAMP-binding proteins have been identified. These are a cAMP-dependent protein kinase [135], a cAMP hydrolyzing phosphodiesterase [136] and a recently identified cAMP binding protein, designated as CABP1 [137, 138]. Unlike its homologous proteins in higher eukaryotes, the cAMP-dependent protein kinase from *D. discoideum* consists of only one catalytical and one regulatory subunit (for a review, see [139]). The R subunit exhibits similarities with its homologues in higher organisms, and it has been shown to inhibit *in vitro* the catalytic activities of both bovine [135] and *Dictyostelium* [140] protein kinase. The amino acid sequences show almost identical motifs for the putative cyclophosphate binding sites of the R subunits of *D. discoideum* and of higher organisms (see Table V).

Dictyostelium discoideum is the model organism for which the most systematic comparison of receptors by means of the testkit analogues has been performed. All known cAMP receptors have been thus characterized: the cell surface cAMP receptor subtypes [101, 141–143], the cytosolic cAMP binding protein [144], the regulatory subunit of the cAK [88] and the cell surface phosphodiesterase [51]. All these receptors have recently been compared [144]. This study has allowed the classification of cAMP receptors of *D. discoideum* in three groups according to a statistical analysis of their relative affinities for a set of cyclic nucleotide derivatives: (i) the cell surface receptors, (ii) the cAK and CAPB1, and (iii) the cell surface cPDE.

From these data it is obvious that *D. discoideum* constitutes an excellent model for the design of selectively acting cyclic nucleotide derivatives. The characterization of the different cAMP receptors within the cell and at its surface has allowed the selection of sets of compounds with high selectivity. Thus neither the 5'-NH-substitution nor the removal of the 2'-OH group (2'-deoxy-cAMP) disturbs the binding to the cell surface receptor, but these modifications drastically reduce interaction with the cAK. In contrast, hydrophobic substitutions at C6 or C8 of the base moiety affect the affinity for the cAK only marginally, while leading to strongly decreased binding to the cell surface receptor.

The cPDEs of *D. discoideum* are in many respects similar to their homologues in higher organisms. Substitution of one of the exocyclic oxygens by a sulfur atom in either axial or equatorial positions either prevents or reduces the hydrolysis by cPDEs of *D. discoideum*. Therefore, it is possible to develop derivatives with receptor selectivity which are at the same time resistant to degradation. Rp- and Sp-5'NHcAMPS (see Fig. 6, compound 12) and Rp- and Sp-2'deoxy-cAMPS (see Fig. 6, compound 11) are powerful tools for selectively acting on the cell surface receptor. On the other hand, Sp-8-bromo-cAMPS (Fig. 6, compound 13) and Sp-5,6-dichloro-benzimidazoleriboside-3',5'-monophosphate (Fig. 6, compound 15), combine resistance to hydrolysis with high lipophilicity and a selective activation of the cAK. Thus, selective regulation of single components of the cAMP signal response system is possible in *Dictyostelium*. If a *Dictyostelium*-like cell surface receptor is ever found in higher organisms, these derivatives will probably constitute a powerful tool to investigate its role.

7. Conclusions and Perspectives

The mechanisms of cAMP analogue induced growth inhibition or apoptosis will be further elucidated if cAMP analogues become available which are capable of selectively acting on a single binding site of a single protein target. The model system of *Dictyostelium discoideum* not only demonstrates that this is possible, but it also points out the way to develop these new cAMP analogues.

Two important concepts have evolved from the study of cAKs with cAMP analogues: "site-selectivity" and "isoenzyme-selectivity". A site-selective compound displays higher affinity for one of the two cAMP-binding sites on the regulatory subunit of cAK, but is not neccessarily also isoenzyme-selective. An isoenzyme-selective analogue binds with higher affinity to one of the two cAK types I or II, but is not always targeted to only one binding site.

In the last decade, there has been a considerable effort to develop site-selective and isoenzyme-selective compounds by introducing modifications at the base moiety of cAMP [26–28, 89, 91, 145] and the phosphate group [92]. In spite of the fact that both isoforms of the regulatory subunit show extensive sequence homologies (Table V), it has been possible to find isoenzyme-selective analogues.

For pharmacological studies it is desirable to obtain an analogue which is not only selective, but specific for the binding site and the isoenzyme that is to be triggered. Site-selective or isoenzyme-selective compounds may still act on the "non-selected" site or isoenzyme, if micromolar concentrations are reached in cells. The search for specificity instead of only selectivity may prove difficult but, using our systematic testkit-based approach, we have recently been able to synthezise one cAMP analogue (8-piperidino-cAMPS-Rp) that is Site A-specific with cAK-R-I and Site B-specific with cAK-R-II. It does not bind to Site B of cAK-R-I or Site A of cAK-R-II below the millimolar range [146]. One of the hardest problems still to solve may be the development of a cell-type or tissue-specific cyclic nucleotide analogue, which would keep the level of interference with normal metabolic processes low.

Since a broad spectrum of cyclic nucleotide binding proteins is found in a great variety of cell types and organisms, a thorough knowledge of the isoenzymes present in a cellular system, their mechanism of action, their specific function and their binding properties is needed. This knowledge is crucial in the development of a new cyclic nucleotide drug for cancer therapy.

A second problem which should be kept in mind is the possible formation of metabolites and their action on cell regulation. If a selectively acting derivative is metabolized, it always has to be taken into account that one or more of the resulting metabolites may either counteract or amplify the effect of the cyclic nucleotide. Although a synergistic effect of cyclic nucleotides and some of their degradation products would allow lower doses of a potential combination drug, it also raises the probability of producing unwanted side effects. For some cell types that express mainly one type of cAK, it may be possible to use a synergistic pair of analogues at lower doses than one analogue at a high dose.

If metabolites are undesirable, it will always be possible to develop degradation-resistant analogues which conserve their selective action. For this purpose, thio-derivatives are the tool of choice.

Although general therapeutical applications of cyclic nucleotide analogues in cancer treatment may still be a distant goal, some specific techniques, such as the extracorporal treatment of blood cells from blood cancer patients with cyclic nucleotide analogues, may constitute a real chance in a near future.

For basic cancer research, the use of cAMP analogues will undoubtedly be one of the most promising methods to unravel the precise mechanisms by which cyclic nucleotides exert their actions.

References

1. *Handbook of Experimental Physiology* **58**, 1982.
2. Ryan WL and Heidrick ML, Role of cyclic nucleotides in cancer. *Adv. Cyclic Nucleotide Res.* **4**: 81–115, 1974.
3. Sheppard JR, The role of cyclic AMP in the control of cell division. In: *Cyclic AMP, Cell Growth, and the Immune Response*, pp. 290–301. Braun W, Lichtenstein LM and Parker CW (eds), Springer, Berlin–Heidelberg–New York–London–Paris–Tokyo, 1974.
4. Boynton AL and Whitfield JF, The role of cyclic AMP in cell proliferation: A critical assessment of the evidence. *Adv. Cyclic Nucleotide Res.* **15**: 193–294, 1983.
5. Lanotte M, Riviere JB, Hermouet S, Houge G, Vintermyr OK, Gjersten BT and Døskeland SO, Programmed cell death (apoptosis) is induced rapidly and with positive cooperativity by activation of cAMP-kinase I in a myeloid leukemia cell line. *J. Cell. Physiol.* **146**: 73–88, 1991.
6. Konijn TM, Cyclic AMP as a first messenger. *Adv. Cyclic Nucleotide Res.* **1**: 17–31, 1972.
7. Peterkowsky P, Cyclic nucleotides in bacteria. *Adv. Cyclic Nucleotide Res.* **7**: 1–48, 1976.
8. Kaupp UB, The cyclic nucleotide-gated channels of vertebrate photoreceptors and olfactory epithelium, *Trends in Neurol. Sci.* **14**: 150–157, 1991.
9. Beavo J and Houslay MD (eds), *Cyclic Nucleotide Phospodiesterases: Structure, Regulation and Drug Action.* J. Wiley and Sons, London–New York–Sidney, 1990.

10. Janssens PMW and van Haastert PJM, Molecular basis of transmembrane signal transduction in *Dictyostelium discoideum. Microbiol. Rev.* **51**: 396–418, 1987.
11. Gerisch G, Cyclic AMP and other signals controlling cell development and differentiation in *Dictyostelium. Annu. Rev. Biochem.* **56**: 853–879, 1987.
12. Peters DJM, Cammans M, Smit S, Spek W, Van Lookeren Campagne MM and Schaap P, Control of cAMP-induced gene expression by divergent signal transduction pathways. *Dev. Gen.* **12**: 25–34, 1991.
13. van Haastert PJM, Janssens PMW and Erneux C, Sensory transduction in eukaryotes. A comparison between *Dictyostelium* and vertebrate cells. *Eur. J. Biochem.* **195**: 289–303, 1991.
14. Posternack T, Sutherland EW and Henion WF, Derivates of cyclic-3',5'-adenosine monophosphate. *Biochim. Biphys. Acta* **65**: 558–560, 1962.
15. Miller JP, Cyclic nucleotide analogs. In: *Cyclic-3',5'-Nucleotides: Mechanism of Action,* pp. 77–104, J. Schulz (ed), Wiley and Sons, London–New York–Sidney, 1977.
16. Miller JP and Robins RK, The chemical modification of cyclic AMP and cyclic GMP. *Annu. Rep. Med. Chem.* **11**: 291–300, 1976.
17. Revankar GR and Robins RK, Chemistry of cyclic nucleotides and cyclic nucleotide analogs. *Handbook Exp. Pharm.* **58**: 17, 1982.
18. Rothermel JD, Stec WJ, Baraniak J, Jastorff B and Parker-Botelho LH, Inhibition of glycogenolysis in isolated rat hepatocytes by the RP diastereomer of adenosine cyclic 3',5'-phosphorothioate. *J. Biol. Chem.* **258**: 12125–12128, 1983.
19. Rothermel JD, Jastorff B and Parker-Botelho LH, Inhibition of glucagon-induced glycogenolysis in isolated rat hepatocytes by the RP-diastereomer of adenosine cyclic 3',5'-phosphorothioate. *J. Biol. Chem.* **259**: 8151–8155, 1984.
20. Parker-Botelho LH, Rothermel JD, Coombs RV and Jastorff B, cAMP analog antagonists of cAMP action. *Meth. Enzymol.* **159**: 159–172, 1988.
21. Krebs EG and Beavo JA, Phosphorylation-dephosphorylation of enzymes. *Annu. Rev. Biochem.* **48**: 923–959, 1979.
22. Corbin JD, Keely FL and Park CR, The distribution and dissociation of cyclic adenosine 3':5'-monophosphate-dependent protein kinases in adipose, cardiac and other tissues. *J. Biol. Chem.* **250**: 218–225, 1975.
23. Hofmann F, Beavo JA, Bechtel PT and Krebs EG, Comparison of adenosine 3':5'-monophosphate-dependent protein kinases from rabbit skeletal and bovine heart muscle. *J. Biol. Chem.* **250**: 7795–7801, 1975.
24. Hanks SK, Quinn AM and Hunter T, The protein kinase family: conserved features and deduced phylogeny of the catalytic domains. *Science* **241**: 42–52, 1988.
25. Scot JD, Fisher EH, Takio K, Demaille JG and Krebs EG, Amino acid sequence of the heat-stable inhibitor of the cAMP-dependent protein kinase from rabbit skeletal muscle. *Proc. Natl. Acad. Sci. USA* **82**: 5732–5736, 1985.
26. Døskeland SO, Øgreid D, Ekanger R, Sturm PA, Miller JP and Suva RH, Mapping of the two intrachain cyclic nucleotide binding sites of adenosine cyclic 3',5'-phosphate dependent protein kinase I. *Biochemistry* **22**: 1094–1101, 1983.
27. Rannels SR and Corbin JD, Two different intrachain cAMP binding sites of cAMP-dependent protein kinases. *J. Biol. Chem.* **255**: 7085–7088, 1980.
28. Døskeland SO, Evidence that rabbit muscle protein kinase has two kinetically distinct binding sites for adenosine 3',5'-cyclic monophosphate. *Biochem. Biophys. Res. Commun.* **83**: 542–549, 1978.
29. Øgreid D, Ekanger R, Suva RH, Miller JP and Døskeland SO, Comparison of the two classes of binding sites (A and B) of type I and type II cyclic-AMP-dependent protein kinases by using cyclic nucleotide analogs. *Eur. J. Biochem.* **181**: 19–31, 1989.

30. Rannels SR and Corbin JD, Studies on the function of the two intrachain cAMP binding sites of protein kinase. *J. Biol. Chem.* **256**: 7871–7876, 1981.

31. Pastan I, Johnson GS and Anderson WB, Role of cyclic nucleotides in growth control. *Annu. Rev. Biochem.* **44**: 491–522, 1975.

32. Friedman DL, Role of cyclic nucleotides in cell growth and differentiation. *Physiol. Rev.* **56**: 652–708, 1976.

33. Friedman DL, Regulation of cell cycle and cellular proliferation by cyclic nucleotides. *Handbook of Exper. Pharmacol.* **58**: 151–188, 1982.

34. Cho-Chung YS, Clair T, Tagliaferri P, Ally S, Katsaros D, Tortora G, Neckers L, Avery TL, Crabtree GW and Robins RK, Site-selective cyclic AMP analogs as new biological tools in growth control, differentiation, and proto-oncogene regulation, *Cancer Invest.* **7**: 161–177, 1989.

35. Katsaros D, Tortora G, Tagliaferri P, Clair T, Ally S, Neckers L, Robins RK and Cho-Chung YS, Site-selective cAMP analogs provide a new approach in the control of cancer cell growth. *FEBS Lett.* **223**: 97–103, 1987.

36. Tagliaferri P, Katsaros D, Clair T, Ally S, Tortora G, Neckers L, Rubalcava B, Parandoosh Z, Chang Y, Revankar GR, Crabtree GW, Robins RK and Cho-Chung YS, Synergistic inhibition of growth of breast and colon human cancer cell lines by site-selective cyclic AMP analogues. *Cancer Res.* **48**: 1642–1650, 1988.

37. Ally S, Clair T, Katsaros D, Tortora G, Yokozaki H, Finch RA, Avery TL and Cho-Chung YS, Inhibition of growth and modulation of gene expression in human lung carcinoma in athymic mice by site-selective 8-Cl-cyclic adenosine monophosphate. *Cancer Res.* **49**: 5650–5655, 1989.

38. Cho-Chung YS, Tortora G, Clair T, Ally S and Yokozaki H, Nuclear location signal and nuclear translocation of type II protein kinase regulatory subunit in site-selective cyclic AMP analog-induced growth control and differentiation. *J. Cell Biol.* **107**: 492a, 1988.

39. Robins RK, Revankar GR and Chang Y, Treatment of malignant tumors with 8-chloroadenosine 3′,5′-cyclic phosphate, 8-aminoadenosine 3′,5′-cyclic phosphate and preparations thereof. PCT/US88/01217 European Patent, 1989.

40. Van Lookeren-Campagne MM, Vilalba Diaz F, Jastorff B and Kessin RH, 8-Chloroadenosine 3′,5′-monophosphate inhibits the growth of chinese hamster ovary and Molt-4 cells through its adenosine metabolite. *Cancer Res.* **51**: 1600–1605, 1991.

41. Wyllie AH, Kerr JF and Currie AR, Cell death: The significance of apoptosis. *Int. Rev. Cytol.* **68**: 251–305, 1980.

42. Budzik GP, Huston JM, Ikawa H and Donahoe PK, The role of zinc in Müllerian duct regression. *Endocrinol.* **110**: 1521–1525, 1982.

43. Snow A. In: *Perspectives on Mammalian Cell Death,* Potten CS (ed), p. 322, University Press, Oxford, 1987.

44. Koury MJ and Bondurant MC, Erythropoietin retards DNA break-down and prevents programmed cell death in erythroid progenitor cells. *Science* **248**: 378–381, 1990.

45. Lanotte M, Døskeland SO and Jastorff B, unpublished results, 1991.

46. Kizaki H, Suzuki K, Tadakuma T and Ishimura Y, Adenosine receptor-mediated accumulation of cyclic AMP-induced T-lymphocyte death through internucleosomal DNA cleavage. *J. Biol. Chem.* **265**: 5280–5284, 1990.

47. Henderson JF and Paterson RP, Catabolism of purine nucleotides. In: *Nucleotide metabolism: An introduction,* Academic Press, 1973.

48. Saito M, Ayako N, Kataoka S, Ohshita K and Yamaji N, Cytotoxic effect of 8-aminoadenosine 3′,5′-cyclic monophosphate on FM3A mouse mammary tumor cells. *J. Pharmacobio-Dyn.* **12**: 357–362, 1989.

49. Van Lookeren-Campagne MM, Vilalba Diaz F, Jastorff B, Winkler E, Genieser HG and Kessin RH, Characterization of the low Km cAMP-phosphodiesterase with cAMP analogues. Applica-

tions in mammalian cells that express the yeast PDE2 gene. *J. Biol. Chem.* **265**: 5847–5854, 1990.

50. Braumann T, Erneux C, Petridis G, Stohrer WD and Jastorff B, Hydrolysis of cyclic nucleotides by purified cGMP-stimulated phosphodiesterase: structural requirements for hydrolysis. *Biochim. Biophys. Acta* **871**: 199–206, 1986.

51. van Haastert PJM, Dijkgraaf PAM, Konijn TM, Abbad EG, Petridis G and Jastorff B, Substrate specificity of cyclic nucleotide phosphodiesterase from beef heart and from *Dictyostelium discoideum. Eur. J. Biochem.* **131**: 659–666, 1983.

52. Coulson R and Hartington WW, Renal metabolism of *N*6, *O2'*-dibutyryl adenosine 3',5'-monophosphate. *Am. J. Physiol.* **273**: F75–F84, 1979.

53. Coulson R, Baraniak J, Stec WJ and Jastorff B, Transport and metabolism of N6- and C8-substituted analogues of adenosine 3',5'-cyclic monophosphate and adenosine 3',5'-cyclic phosphorothioate by the isolated perfused rat kidney. *Life Sci.* **32**: 1489–1498, 1983.

54. Scheinman SJ, Stec WJ and Coulson R, Effects of (Sp)- and (Rp)-adenosine cyclic 3',5'-phosphorothioates on electrolyte excretion by the isolated perfused rat kidney. *Miner. Electrolyte Metab.* **11**: 85–90, 1985.

55. Kaukel E and Hilz H, Permeation of dibutyryl cAMP in HeLa cells and its conversion to monobutyryl cAMP. *Biochem. Biophys. Res. Commun.* **46**: 1011–1018, 1972.

56. Braumann T, Jastorff B and Richter-Landsberg C, Fate of cyclic nucleotides in PC 12 cell cultures: Uptake, metabolism, and effects of metabolites on nerve growth factor-induced neurite outgrowth. *J. Neurochem.* **47**: 912–919, 1986.

57. Jastorff B, Hoppe J and Morr M, A model for the chemical interactions of adenosine 3':5'-monophosphate with the R subunit of protein kinase type I. Refinement of the cyclic phosphate binding moiety of protein kinase type I. *Eur. J. Biochem.* **101**: 555–561, 1979.

58. Richter-Landsberg C and Jastorff B, *In vitro* phosphorylation of microtubule-associated protein 2: Differential effects of cyclic AMP analogues. *J. Neurochem.* **45**: 1218–1222, 1985.

59. Kruh J, Effects of sodium butyrate, a new pharmacological agent, on cells in culture. *Mol. Cell. Biochem.* **42**: 65–82, 1982.

60. Yusta B, Ortiz-Caro J, Pascual A and Aranda A, Comparison of the effects of forskolin and dibutyryl cyclic AMP in neuroblastoma cells: Evidence that some of the actions of dibutyryl cyclic AMP are mediated by butyrate. *J. Neurochem.* **51**: 1808–1818, 1988.

61. Kumar S, and de Vellis J, Glucocorticoid-mediated functions in glial cells. In: *Glial Cell Receptors,* pp. 243–264, Kimelberg KH (ed), Reven Press, New York, 1988.

62. Graham KA and Buick RN, Sodium butyrate induces differentiation in breast cancer cell lines expressing the estrogen receptor. *J. Cell. Physiol.* **136**: 63–71, 1988.

63. Guengerich FP, Separation and purification of multiple forms of microsomal cytochrome P-450. Activities of different forms of cytochrome P-450 towards several compounds of environmental interest. *J. Biol. Chem.* **252**: 3970–3979, 1977.

64. Gorrod JW and Lam SP, Microsomal cytochrome P-450 mediated N-oxygenation of amino azaheterocycles. In: *Biological Oxidation Systems,* Vol. I. Reddy CC, Hamilton GA and Madyastha KM (eds), pp. 147–161, Academic Press, Orlando–London, 1990.

65. Burgers PMJ, Eckstein F, Hunneman DH, Baraniak J, Kinas RW, Lesiak K and Stec WJ, Stereochemistry of hydrolysis of adenosine 3',5' cyclic phosphorothioate by the cyclic phosphodiesterase from beef heart. *J. Biol. Chem.* **254**: 9959–9961, 1979.

66. Beavo J, Multiple phosphodiesterase isoenzymes: background, nomenclature and implications. In: *Cyclic Nucleotide Phosphodiesterases: Structure, Regulation and Drug Action.* Beavo J and Houslay MD (eds), pp. 3–15, J. Wiley and Sons, London–New Yrok–Sydney, 1990.

67. Beavo JA, Multiple isoenzymes of nucleotide phosphodiesterase. *Adv. Second Messenger Phos. Prot. Res.* **22**: 12–54, 1989.

68. Beebe SJ, Beasley-Leach A and Corbin JD, Cyclic AMP analogs used to study low Km hormone-sensitive phosphodiesterase. *Meth. Enzymol.* **159**: 531–540, 1988.

69. Couchie D, Petridis G, Jastorff B and Erneux C, Characterization of phosphodiesterase catalytic sites by means of cyclic nucleotide derivatives. *Eur. J. Biochem.* **136**: 571–575, 1983.

70. Erneux C and Miot F, Cyclic nucleotide analogs used to study phosphodiesterase catalytic and allosteric sites. *Meth. Enzymol.* **159**: 520–530, 1988.

71. Erneux C, Couchine D, Dumont JE and Jastorff B, Cyclic nucleotide derivatives as probes of phosphodiesterase catalytic and regulatory sites. *Adv. Cyclic Nucleotide and Prot. Phos. Res.* **16**: 107–118, 1984.

72. Wells JN and Miller JR, Methylxanthine inhibitors of phosphodiesterases. *Meth. Enzymol.* **159**: 489–496, 1988.

73. Hosono K, Acyl peptide inhibitors of phosphodiesterase produced by *Bacillus subtilis*. *Meth. Enzymol.* **159**: 497–504, 1988.

74. Manganiello V, Degerman E and Elks M, Selective inhibitors of specific phosphodiesterases in intact adipocytes. *Meth. Enzymol.* **159**: 504–520, 1988.

75. Maronde E, Richter-Landsberg C and Jastorff B, unpublished results, 1991.

76. Martin TFJ and Kowalchyk JA, Growth inhibition by adenosine 3′,5′-monophosphate derivatives does not require 3′,5′ phosphodiester linkage. *Science* **213**: 1120–1122, 1981.

77. Bushfield M, Shoshani I, Cifuentes M, Stübner F and Johnson RA, Inhibition of adenylate cyclase by polyadenylate. *Archs. Biochem. Biophys.* **278**: 88–98, 1990.

78. Zorn M and Jastorff B, unpublished results, 1991.

79. Gerlach E and Becker BF, *Topics and Perspectives in Adenosine Research*. Springer, Berlin–Heidelberg–New York–London–Paris–Tokyo, 1987.

80. Bushfield M and Johnson RA, Regulation of adenylate cyclase by adenosine: Characterization of the P-site. *Biochem. Soc. Trans.* **18**: 150–151, 1990.

81. Daly JW, Adenosine receptors. *Adv. Cyclic Nucleotide and Prot. Phos. Res.* **19**: 29–46, 1985.

82. Burnstock G, Purine Receptors. In: *Adenosine Receptors in the Nervous System*. Ribeiro JA (ed), pp. 1–14, Taylor & Francis Ltd., London, 1989.

83. Daly JW, Adenosine receptors: Targets for future drugs. *J. Med. Chem.* **25**: 197–206, 1982.

84. Kusachi S, Thompson RD, Yamada N, Daly TD and Olsson RA, Dog coronary artery adenosine receptor: Structure of the N^6-aryl subregion. *J. Med. Chem.* **29**: 989–996, 1986.

85. Niedzwicki JG and Abernethy DR, Structure-activity relationship of ligands of human plasma deaminase 2. *Biochem. Pharm.* **41**: 1615–1624, 1991.

86. Bennett Jr. LL, Chang CH, Allan PW, Adamson DJ, Rose LM, Brockman RW, Secrist III JA, Shortancy A and Montgomery JA, Metabolism and metabolic effects of halopurine nucleosides in tumor cells in culture. *Nucleosides Nucleotides* **4**: 107–116, 1985.

87. Koontz JW and Wicks WD, Cytotoxic effects of two novel 8-substituted cyclic nucleotide derivatives in cultured rat hepatoma cells. *Mol. Pharmacol.* **18**: 65–71, 1980.

88. De Wit RJW, Hoppe J, Stec WJ, Baraniak J and Jastorff B, Interaction of cAMP derivatives with the "stable" cAMP-binding site in the cAMP-dependent protein kinase type I. *Eur. J. Biochem.* **122**: 95–99, 1982.

89. De Wit RJW, Hekstra D, Jastorff B, Stec WJ, Baraniak J, Van Driel R and van Haastert PJM, Inhibitory action of certain cyclophosphate derivatives of cAMP on cAMP-dependent protein kinases. *Eur. J. Biochem.* **142**: 255–260, 1984.

90. Yagura TS and Miller JP, Mapping adenosine cyclic 3′,5′-phosphate binding sites on type I and type II adenosine cyclic 3′,5′-phosphate-dependent protein kinases using ribose ring and cyclic phosphate ring analogues of cyclic 3′,5′-phosphate. *Biochemistry* **20**: 879–887, 1981.

91. Øgreid D, Ekanger R, Suva RH, Miller JP, Sturm P, Corbin JD and Døskeland SO, Activation of protein kinase isozymes by cyclic nucleotide analogs used single or in combination. Principles

for optimizing the isozyme specificity of analog combinations. *Eur. J. Biochem.* **150**: 219–227, 1985.

92. Dostmann WRG, Taylor SS, Genieser HG, Jastorff B, Døskeland SO and Øgreid D, Analogs of adenosine cyclic-3',5'-phosphorothioates: Probing the cyclic nucleotide binding sites of cAMP-dependent protein kinases I and II. *J. Biol. Chem.* **265**: 10484–10491, 1990.

93. Houge G, Steinberg RA, Øgreid D and Døskeland SO, The rate of recombination of subunits (RI and C) of cyclic AMP-dependent protein kinase depends on whether one or two cyclic AMP molecules are bound per RI monomer. *J. Biol. Chem.* **265**: 19507–19516, 1990.

94. Beebe SJ and Corbin JD, Cyclic nucleotide-dependent protein kinases. In: *The Enzymes,* Krebs EG and Boyer PD (eds), Vol. 17A, pp. 43–111, Academic Press, Orlando–London, 1986.

95. Jastorff B, Abbad EG, Petridis G, Tegge W, de Wit RJW, Erneux C, Stec WJ and Morr M. Systematic use of cyclic nucleotide analogues—Mapping of essential interactions between nucleotides and proteins. *Nucleic Acids Res. Symp. Ser.* **9**: 219–223, 1981.

96. Jastorff B, The effects of cyclic nucleotide derivatives on cell metabolism. In: *Cell Regulation by Intracellular Signals.* Dumont E and Swillens S (eds), pp. 195–207, Plenum Press, New York–London, 1982.

97. Braumann T and Jastorff B, Physico-chemical characterization of cyclic nucleotides by reversed-phase high-performance liquid chromatography. I. Cation binding in the mobile phase. *J. Chromatogr.* **329**: 321–330, 1985.

98. Braumann T and Jastorff B, Physico-chemical characterization of cyclic nucleotides by reversedphase high-performance liquid chromatography. II. Quantitative determination of hydrophobicity. *J. Chromatogr.* **350**: 105–118, 1985.

99. Naß N, Colling C, Cramer M, Genieser HG, Butt E, Winkler E, Jaenicke L and Jastorff B, Mapping of the epitope/paratope interactions of a monoclonal antibody directed against adenosine-3',5'-monophosphate. *Biochemical J.* (submitted).

100. Scholübbers HG, van Knippenberg PH, Baraniak J, Stec WJ, Morr M and Jastorff B, Investigations on stimulation of lac transcription *in vivo* in *Escherichia coli* by cAMP analogues. *Eur. J. Biochem.* **138**: 101–109, 1984.

101. van Ments-Cohen M, Genieser HG, Jastorff B, van Haastert PJM and Schaap P, Kinetics and nucleotide specificity of a surface cAMP binding site in *Dictyostelium discoideum,* which is not down-regulated by cAMP. *Microbiol. Lett.* **82**: 9–14, 1991.

102. Schmid E, Schmid W, Jantzen M, Mayer D, Jastorff B and Schütz G, Transcription activation of the tyrosine aminotransferase gene by glucocorticoids and cAMP in primary hepatocytes. *Eur. J. Biochem.* **165**: 499–506, 1987.

103. Adashi EY, Resnick CE and Jastorff B, Blockade of granulosa cell differentiation by an antagonistic analog of adenosine 3',5'-cyclic phosphate (cAMP): central but non-exclusive intermediary role of cAMP in follicle-stimulating hormone action. *Mol. Cell. Endocrinol.* **72**: 1–11, 1990.

104. Øgreid D, Genieser HG, Døskeland SO and Jastorff B, unpublished results, 1991.

105. Yokozaki H, Tortora G, Pepe S, Maronde E, Genieser HG, Jastorff B and Cho-Chung YS, Unhydrolyzable analogues of adenosine 3',5'-monophosphate demonstrating growth inhibition and differentiation in human cancer cells. *Cancer Res.,* submitted for publication.

106. Schaap P, Genieser HG, Bottin U and Jastorff B, unpublished results, 1991.

107. Bubis J and Taylor SS, Correlation of photolabeling with occupancy of cAMP-binding sites in the regulatory subunit of cAMP-dependent protein kinase I. *Biochemistry* **26**: 3478–3486, 1987.

108. Titani K, Sasagawa G, Ericsson LH, Kumar S, Smith SB, Krebs EG and Walsh KA, Amino acid sequence of the regulatory subunit of bovine type I adenosine cyclic 3',5'-phosphate dependent protein kinase. *Biochemistry* **23**: 4193–4199, 1984.

109. Takio K, Smith SB, Krebs EG, Walsh KA and Titani K, Amino acid sequence of the regulatory subunit of bovine type II adenosine cyclic 3′,5′-phosphate dependent protein kinase. *Biochemistry* **23**: 4200–4206, 1984.

110. Weber IT and Steitz TA, Structure of a complex of catabolite gene activator protein and cyclic AMP refined at 2.5 Å resolution. *J. Mol. Biol.* **198**: 311–326, 1987.

111. Weber IT, Steitz TA, Bubis J and Taylor SS, Predicted structures of cAMP binding domains of type I and II regulatory subunits of cAMP-dependent protein kinase. *Biochemistry* **26**: 343–351, 1987.

112. Taylor SS, cAMP-dependent protein kinase. Model for an enzyme family. *J. Biol. Chem.* **264**: 8443–8446, 1989.

113. Steinberg RA, Russel JL, Murphy CS and Yphantis DA, Activation of type I cyclic AMP-dependent protein kinases with defective cyclic AMP-binding sites. *J. Biol. Chem.* **262**: 2664–2671, 1987.

114. Scott JD, Glaccum MB, Zoller MJ, Uhler MD, Helfman DM, McKnight GS and Krebs EG, The molecular cloning of a type II regulatory subunit of the cAMP-dependent protein kinase from rat skeletal muscle and mouse brain. *Proc. Natl. Acad. Sci. USA* **84**: 5192–5196, 1987.

115. Kuno T, Ono Y, Hirai M, Hashimoto S, Shuntoh H and Tanaka C, Molecular cloning and cDNA structure of the regulatory subunit of type I cAMP-dependent protein kinase from rat brain. *Biochem. Biophys. Res. Commun.* **146**: 878–883, 1987.

116. Sandberg M, Tasken K, Øyen O, Hansson V and Jahnsen T, Molecular cloning, cDNA structure and deduced amino acid sequence for a type I regulatory subunit of cAMP-dependent protein kinase from human testis. *Biochem. Biophys. Res. Commun.* **149**: 939–945, 1987.

117. Nowak I, Seipel K, Schwarz M, Jans DA and Hemmings BA, Isolation of a cDNA and characterization of the 5′ flanking region of the gene encoding the type I regulatory subunit of the cAMP-dependent protein kinase. *Eur. J. Biochem.* **167**: 27–33, 1987.

118. Clegg CH, Cadd GG and McKnight GS, Genetic characterization of a brain-specific form of the type I regulatory subunit of cAMP-dependent protein kinase. *Proc. Natl. Acad. Sci. USA* **85**: 3703–3707, 1988.

119. Sandberg M, Levy FO, Øyen O, Hansson V and Jahnsen T, Molecular cloning, cDNA structure and deduced amino acid sequence for the hormone-induced regulatory subunit (RIIβ) of cAMP-dependent protein kinase from rat ovarian granulosa cells. *Biochem. Biophys. Res. Commun* **154**: 705–711, 1988.

120. Øyen O, Myklebust F, Scott JD, Hansson V and Jahnsen T, Human testis cDNA for the regulatory subunit RIIα of cAMP-dependent protein kinase encodes an alternate amino-terminal region. *FEBS Lett.* **246**: 57–64, 1989.

121. Levy FO, Øyen O, Sandberg M, Tasken K, Eskild W, Hansson V and Jahnsen T, Molecular cloning, complementary deoxyribonucleic acid structure and predicted full-length amino acid sequence of the hormone-inducible regulatory subunit of 3′-5′-cyclic adenosine monophosphate-dependent protein kinase from human testis. *Mol. Endocrinol.* **2**: 1364–1373, 1988.

122. Kalderon D and Rubin GM, Isolation and characterization of *Drosophila* cAMP-dependent protein kinase genes. *Genes Dev.* **2**: 1539–1556, 1988.

123. Lu X, Gross RE, Bagchi S and Rubin CS, Cloning, structure, and expression of the gene for a novel regulatory subunit of cAMP-dependent protein kinase in *Caenorhabditis elegans. J. Biol. Chem.* **265**: 3293–3303, 1990.

124. Mutzel R, Lacombe ML, Simon MN, de Gunzburg J and Veron M, Cloning and cDNA sequence of the regulatory subunit of cAMP-dependent protein kinase from *Dictyostelium discoideum. Proc. Natl. Acad. Sci. USA* **84**: 6–10, 1987.

125. Toda T, Cameron S, Sass P, Zoller M, Scott JD, McMullen B, Hurwitz M, Krebs EG and Wigler M, Cloning and characterization of *BCY1*, a locus encoding a regulatory subunit of the

cyclic AMP-dependent protein kinase in *Saccharomyces cerevisiae, Mol. Cell. Biol.* **7**: 1371–1377, 1987.

126. Kunisawa R, Davis TN, Urdea MS and Thorner J, Complete nucleotide sequence of the gene encoding the regulatory subunit of 3′,5′-cyclic AMP-dependent protein kinase from the yeast *Saccharomyces cerevisiae. Nucleic Acids Res.* **15**: 368–369, 1987.

127. Dhallan RS, Yau KW, Schrader KA and Reed RR, Primary structure and functional expression of a cyclic nucleotide-activated channel from olfactory neurons. *Nature* **347**: 184–187, 1990.

128. Ludwig J, Margalit T, Eismann E, Lancet D and Kaupp UB, Primary structure of cAMP-gated channel from bovine olfactory epithelium. *FEBS Lett.* **270**: 24–29, 1990.

129. Jastorff B, Störmann R, van Bemmelen M and Dostmann W, A stroll along the channels. *Nature,* submitted for publication.

130. Weber IT, Shabb JB and Corbin JD, Predicted structures of the cGMP binding domains of the cGMP-dependent protein kinase: A key alanine/threonine difference in evolutionary divergence of cAMP and cGMP sites. *Biochemistry* **28**: 6122–6127, 1989.

131. Konijn TM, van de Meene JGC, Bonner JT and Barkley DS, The acrasin activity of adenosine 3′,5′-cyclic phosphate. *Proc. Natl. Acad. Sci. USA* **58**: 1152–1154, 1967.

132. van Haastert PJM and De Wit RJW, Demonstration of receptor heterogeneity and negative cooperativity by nonequilibrium binding experiments. The cell surface cAMP receptor of *Dictyostelium discoideum. J. Biol. Chem.* **259**: 13321–13328, 1984.

133. van Haastert PJM, De Wit RJW, Janssens PMW, Kesbeke F and De Goede J, G-protein mediated interconversions of cell surface cAMP receptors and their involvement in excitation and desensitization of guanylate cyclase in *Dictyostelium discoideum. J. Biol. Chem.* **261**: 6904–6911, 1986.

134. Klein PS, Sun TJ, Saxe CL, Kimmel AR, Johnson RL and Devreotes PN, A chemoattractant receptor controls development in *Dictyostelium discoideum. Science* **241**: 1467–1472, 1988.

135. Leichtling BH, Majerfeld LH, Coffman DJ and Rickenberg HV, Identification of the regulatory subunit of a cAMP-dependent protein kinase in *Dictyostelium discoideum. Biochem. Biophys. Res. Commun.* **105**: 949–955, 1982.

136. Malchow D, Nagele B, Schwarz H and Gerisch G, Membrane bound cAMP phosphodiesterase in chemotactically responding cells of Dictyostelium discoideum. Eur. J. Biochem. 28: 136–142, 1972.

137. Tsang AS and Tasaka M, Indentification of multiple cAMP binding proteins in developing *Dictyostelium discoideum* cells. *J. Biol. Chem.* **262**: 7700–7704, 1986.

138. Kay CA, Noce T and Tsang AS, Translocation of an unusual cAMP receptor to the nucleus during development of Dictyostelium discoideum. Proc. Natl. Acad. Sci. USA 84: 2322–2326, 1987.

139. Veron M, Mutzel R, Lacombe ML, Simon MN and Wallet V, cAMP-dependent protein kinase from *Dictyostelium discoideum. Dev. Genet.* **9**: 247–258, 1987.

140. de Gunzburg J and Veron MA, cAMP-dependent protein kinase is present in differentiating *Dictyostelium discoideum* cells. *EMBO J.* **1**: 1063–1068, 1982.

141. Mato JM, Jastorff B, Morr M and Konijn TM, A model for cAMP chemoreceptor interaction in *Dictyostelium discoideum. Biochim. Biophys. Acta* **544**: 309–314, 1978.

142. van Haastert PJM and Kien E, Binding of cAMP derivatives to *Dictyostelium discoideum* cells. Activation mechanism of the cell surface cAMP receptor. *J. Biol. Chem.* **258**: 9636–9642, 1983.

143. van Haastert PJM, Kesbeke F, Konijn TM, Baraniak J, Stec WJ and Jastorff B, (Rp)-cAMPS, an antagonist of cAMP in *Dictyostelium discoideum.* In: *Biophosphates and their Analogues — Synthesis, Structure, Metabolism and Activity,* pp. 469–483. Bruzik KS and Stec WJ (eds), Elsevier, Amsterdam, 1987.

144. van Ments-Cohen M and van Haastert PJM, The cyclic nucleotide specificity for eight cAMP-binding proteins in *Dictyostelium discoideum* is correlated into three groups. *J. Biol. Chem.* **264**: 8717–8722, 1989.

145. Robinson-Steiner AM and Corbin JD, Probable involvement of both intrachain cAMP binding sites in activation of protein kinase. *J. Biol. Chem.* **258**: 1032–1040, 1983.
146. Øgreid D, Genieser HG, Jastorff B, Dostmann W and Døskeland SO, unpublished results, 1991.

Resistance to Anthracyclines in Multidrug Resistant Cells. Role of the Membrane Transport

A. Garnier-Suillerot, F. Frezard and J. Tarasiuk

Laboratoire de Chimie Bioinorganique (LPCB URA CNRS 198), UFR de Medecine et Biologie Humaine, Universite Paris Nord, 74 rue Marcel Cachin, 93012 Bobigny, France

Abstract. One of the phenotypes of multidrug resistance is characterized by a decrease in the intracellular concentration of drug in resistant cells as compared to sensitive cells. This is correlated with the presence, in the membrane of resistant cells, of a 150–180 kDa glycoprotein, P-glycoprotein, responsible for an active efflux of the drug. The fluorescence emission spectra from anthracycline-treated cells suspended in buffer have been used to measure the uptake of anthracycline derivatives in drug-sensitive and drug-resistant K562 cells. The initial rate of uptake and the kinetics of active efflux under the effect of P-glycoprotein have been measured as a function of the extracellular pH, pH_e, and as a function of temperature. Our data show that anthracycline derivatives gain access to cells by free permeation of the neutral form of the drug through the lipid domain, under the action of a driven force provided by DNA in the nucleus. At the steady state, in sensitive cells, the equilibrium transmembrane concentrations verified the Henderson–Hasselbach relation $[DH^+]_i/[DH^+]_e = [H^+]_i/[H^+]_e$, where $[DH^+]_i$ and $[DH^+]_e$ stand for the concentration of protonated form of the free drug inside and outside the cells respectively. This relation is not verified for resistant cells. The mechanisms of influx and efflux are the same in sensitive cells. In the case of resistant cells, we propose the coexistence of two types of efflux: (a) an efflux which is the same as seen in sensitive cells and involves a free permeation of the neutral form of the drug, (b) an efflux which is energy-dependent and involves the P-glycoprotein. Our data strongly suggest that it is the neutral form of the drug which participates in the active transport.

1. Introduction

Tumor cell resistance to cytotoxic drugs is considered to be one of the major causes of failure of clinical chemotherapy [1, 2]. Operationally one observes that some tumor cell populations are either inherently non-responsive to a spectrum of antitumor agents, or initially respond to chemotherapy, but unfortunately, after repeated exposition to drug, cellular resistance to antineoplastic agents appears. The problem is that the tumors cells become resistant not only to the drug which has been used during the treatment but also to other drugs which are structurally and functionally unrelated [3]. For instance, tumor cells selected for resistance to adriamycin are also resistant to actinomycin, colchicine and vincaalcaloids such as vincristine and vinblastine. This phenomenon of broad resistance is termed "pleiotropic drug resistance" or "multidrug resistance" (MDR). The characteristic features of multidrug resistant cells are: (i) decreased drug accumulation due to increased drug efflux [4–7], (ii) partial reversibility of MDR by calcium channel blockers such as verapamil [8–10], (iii) increased expression of the *mdr1* gene which encodes a 150- to 180-kDa plasma membrane glycoprotein, termed *P*-glycoprotein [7, 11–13]. MDR can be transferred to drug-sensitive cells by *mdr* gene sequences. This results indicate that the presence of *P*-glycoprotein confers multidrug resistance. Moreover, in general there is a relationship between the degree of resistance and the amount of *P*-glycoprotein [14–16].

Cells exhibiting the MDR phenotype can be made more sensitive to the cytotoxic effects of drugs by treatment with calcium channels blockers such as verapamil [8–10]. However, clear potentiation effects of verapamil *in vitro* in MDR were obtained at concentrations which cannot readily be achieved in patients [17]. The search for agents which can potentiate drug sensitivity in resistant tumor cells at clinically achievable concentration is a very important point. This requires the determination of the precise role of *P*-glycoprotein and involves the determination and the comparison of the mechanisms of drug uptake and release by resistant and sensitive cells. For this, one needs to distinguish between parameters which depend on the nature of the drug and those which depend on membrane properties. Numerous papers dealing with this problem have been written [18–25]. However, analysis of drug transport proves to be still technically demanding. For example, influx measurements are complicated by rapid and substantial adsorption of hydrophobic drugs to the cell surface. Moreover, due to the absence of methods for the determination of intracellular free drug concentration, informations about efflux mechanisms are still lacking.

Recently we have developed a simple spectrofluorometric method which allows the rapid and accurate determination of the kinetics of influx and efflux of anthracycline derivatives in the cells [26–30]. These measurements are made as incubation of the drug with the cells proceeds without compromising cell viability. We have thus obtained kinetic and thermodynamic parameters which are absolutely necessary to get a systematic insight into the mechanism of drug transport.

In this paper we will present: (i) the fluorometric method which has been used, (ii) the data obtained when the uptake and release of anthracycline derivatives by sensitive and resistant cells were followed as a function of pH and as a function of temperature.

2. The Fluorometric Method

The anthracycline derivatives used were adriamycin (ADR), daunorubicin (DNR), 4′-*O*-tetrahydropyranyl-adriamycin (THP-ADR), carminomycin (CAR) and aclacinomycin (ACM). Their structures are shown in Figs. 1 and 2. At 37 °C, in the presence of 0.1 M NaCl, the pK_a of deprotonation of these drugs are equal to 8.4 for ADR and DNR, 7.7 for THP-ADR, 7.5 for CAR and 7.0 for ACM. One can

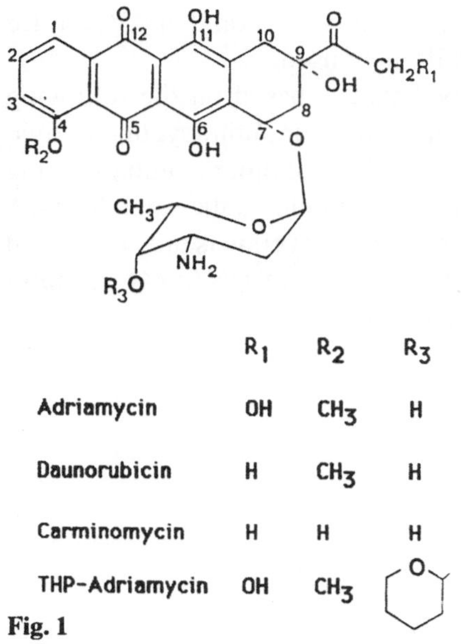

	R_1	R_2	R_3
Adriamycin	OH	CH_3	H
Daunorubicin	H	CH_3	H
Carminomycin	H	H	H
THP-Adriamycin	OH	CH_3	

Fig. 1

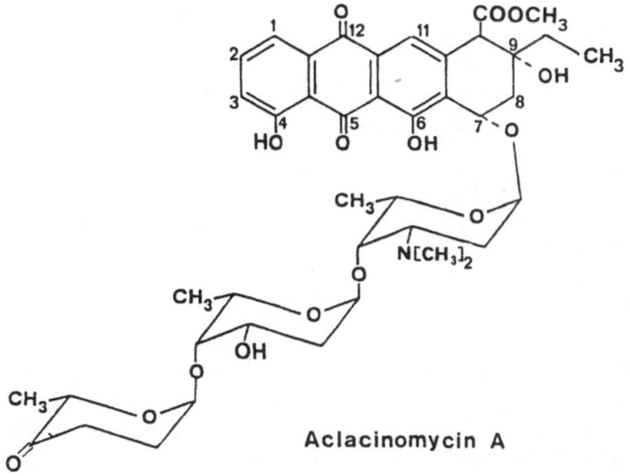

Aclacinomycin A

Fig. 2

calculate that in these conditions and at pH 7.2, the percentage of drug in the neutral form equals 5 for ADR and DNR, 30 for THP-ADR, 38 for CAR and 63 for ACM [27].

The visible spectra of ADR, DNR, CAR and THP-ADR are similar and exhibit a maximum absorbance at 480 nm. ACM lacks a hydroxyl group in position C11 and, therefore, its absorption is maximal at 436 nm. The fluorescence spectra are obtained by excitation at 480 nm for the first four derivatives and at 436 nm for aclacinomycin. K562 sensitive and resistant cells were used.

The fluorometric method that we have developed is based on the following observations: (1) in the cells the strongest binding sites for anthracycline derivatives are provided by DNA and in a first approximation the other cellular binding sites can be neglected; (2) the binding capacity of the nuclei is the same for both sensitive and resistant cells; (3) the fluorescence of anthracycline is only quenched when the molecule is intercalated between the base pairs of DNA; (4) transport across the cell membrane is the rate-limiting step [26].

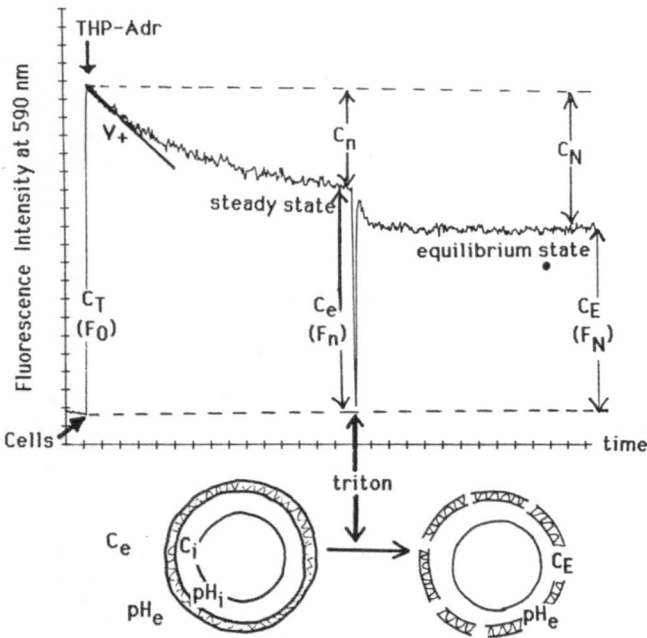

Fig. 3. Fluorometric determination of the kinetics of uptake of anthracycline derivatives by cells. The fluorescence intensity at $\lambda_{em} = 590$ nm in the case of THP-ADR, ADR, DNR and CAR ($\lambda_{ex} = 480$ nm) and at $\lambda_{em} = 585$ nm in the case of ACM ($\lambda_{ex} = 436$ nm) is recorded as a function of time. Cells are suspended in a cuvette filled with 2 ml of buffer at pH_e under vigorous stirring (n cells/cm^3). At $t = 0$, small aliquots of a stock anthracycline solution is added to the cells yielding an anthracycline concentration of C_T; the fluorescence intensity is then F_0. The slope of the tangent to the curve $F = f(t)$ at $t = 0$ is $(dF/dt)_{t=0}$ and the initial rate of uptake $V_+ = (dF/dt)_{t=0} \cdot C_T/F_0$ (M · s^{-1}). Once the steady state is reached, the fluorescence intensity is F_n and the overall concentration of drug intercalated between the base pairs in the nucleus is $C_n = C_T (F_0 - F_n)/F_0$. The extracellular free drug concentration is $C_e = C_T \cdot F_n/F_0$. The addition of 0.05% triton X-100 yields the equilibrium state. The fluorescence intensity is then F_N and the concentration of drug intercalated between the base pairs is $C_N = C_T \cdot (F_0 - F_N)/F_0$. The cytoplasmic pH_i is equal to the extracellular pH_e and the cytoplasmic free drug concentration $C_E = C_T \cdot F_N/F_0$

Figure 3 illustrates the principle of the method. In a typical experiment, 2×10^6 cells are suspended in 2 ml glucose-containing Hepes buffer under continuous stirring; 20 μl of the stock anthracycline solution is added quickly to this suspension yielding a micromolar concentration referred to as C_T. The decrease of fluorescence intensity at 590 nm, in the case of ADR, DNR, THP-ADR and CAR and at 585 nm, in the case of ACM, was followed as a function of time. The initial rate of uptake $V_+ = (dF/dt)_{t=0} \cdot C_T/F_0$, where $(dF/dt)_{t=0}$ is the slope of the tangent to the curve $F = f(t)$ and F_0 is the fluorescence intensity at $t = 0$ of an anthracycline solution of concentration C_T. After some time, depending on the type of drug used, on the extracellular pH (pH_e) outside the cells and on the temperature, the curve $F = f(t)$ reaches a plateau and the fluorescence intensity is equal to F_n. Outside the cells, pH_e has a value imposed by the buffer and the free drug concentration is C_e; inside the cell, pH_i equals 7.2 ± 0.1 [28] and the free drug concentration is C_i in thermodynamic equilibrium with the drug bound to the nucleus (overall molar concentration equals C_n). Once the steady state is reached, cell membranes are permeabilized with 0.05% w/v Triton X-100 yielding the equilibrium state which is characterized by a new value of the fluorescence intensity, F_N. Outside the permeabilized cells, pH_e value is the same as at the steady state, but the free drug concentration is different and equals C_E; inside the cell, pH_i now equals pH_e and C_i equals C_E in equilibrium with the drug bound to the nucleus (overall molar concentration C_N). Thus this method allows the accurate determination of the initial rate of uptake, V_+, the concentration of free drug in the cytoplasm at the steady state, the overall concentration of drug bound to the nucleus at the steady state and at the equilibrium state [28].

3. Uptake and Release of Anthracyclines by Drug-Resistant and Drug-Sensitive Cells as a Function of pH_e

The kinetics of uptake, V_+, as well as the amount of drug bound to the nucleus, strongly depends on the pH_e outside the cells. For experiments performed in the pH_e range 6.5–8.2, one observes that the initial rate of uptake, $(V_+)_{t=0}$, and the concentration of drug bound to the nucleus at the steady state, C_n, increases as pH_e increases. On the other hand, as can be seen in Table 1, the kinetics of uptake determined at pH_e 7.2 follows the order $V_+(ADR) < V_+(DNR) < V_+(THP\text{-}ADR) < V_+(CAR) < V_+(ACM)$. This order is the same as the pK_a values. Both these results strongly suggest that the neutral form of the drug selectively permeates the membrane [27–29].

We have thoroughly studied the uptake of THP-ADR by sensitive and resistant K562 cells as a function of pH_e. This drug enters the cells very rapidly [23]; for instance, when 10^6 cells/cm^3 are incubated with 1 μM drug at 37°C, the steady state is reached within about 10 minutes at $pH_e = 7.0$ and less than 5 minutes at $pH_e = 7.8$. We have calculated, for different pH_e values, the concentration of free drug outside the cells, C_e, and inside the cells, C_i, at the steady state. Using the Henderson–Hasselback equation, it is possible to calculate the concentration of free drug inside the cell in the neutral form $[D^0]_i$ and protonated form $[DH^+]_i$, with $[D^0]_i + [DH^+]_i = C_i$ and $pH_i = pK_a + \log[D^0]_i/[DH^+]_i$ at $pH_i = 7.2$ and the concentration of free drug outside the cell in the neutral form $[D^0]_e$ and protonated form $[DH^+]_e$, with $[D^0]_e + [DH^+]_e = C_e$ and $pH_e = pK_a + \log[D^0]_e/[DH^+]_e$ at pH_e.

Table 1. Kinetics of uptake and release of anthracycline derivatives by sensitive and resistant K562 cells at 37°C

	Cells	ADR	DNR	THP-ARD	CAR	ACM
V_+	S	0.037±0.005	0.47±0.06	6.8±0.9	8.8±1.2	15.4±2
(nm · s⁻¹)	R	0.007±0.001	0.36±0.05	4.9±0.7	8.8±1.2	15.4±2
pK_a		8.4	8.4	7.7	7.5	7.0
$k_+^0 \times 10^9$	S	0.61±0.08	8±1	25±3	26±3	25±3
s⁻¹(cells/cm³)⁻¹	R	0.13±0.02	6±1	16±2	26±3	25±3
$k_+^{7.2} \times 10^9$	S		0.47±0.07	6.0±0.8	8.5±1.0	
s⁻¹(cells/cm³)⁻¹	R		0.36±0.05	3.8±0.5	8.5±1.0	
$k_-^{7.2} \times 10^9$	S		0.70±0.15	7.4±1.5	6.0±1.5	
s⁻¹(cells/cm³)⁻¹	R		3.2±0.6	23±4	18±3	
$(k_-^{7.2})_a \times 10^9$	R		2.5	15	13	
s⁻¹(cells/cm³)⁻¹						

Note. V_+ is the initial rate of uptake when 10^6 cells/cm³ are incubated with 1 μM drug at $pH_e = 7.2$, pK_a is the pK_a of deprotonation of the drug at 37°C in the pH range 0–9, $k_+^{7.2}$ and $k_-^{7.2}$ are the influx and efflux coefficients, respectively, for the drugs at pH 7.2, $(k_-^{7.2})_a$ is the active efflux coefficient for the drugs in the case of resistant cells, S and R stand for sensitive and resistant cells, respectively.

Figure 4 shows that, within the limits of experimental errors, the relation $[DH^+]_i/[DH^+]_e = [H^+]_i/[H^+]_e$ is verified in the case of sensitive cells. In other words, at the steady state, the concentration of drug in the neutral form is the same inside the cells and outside the cells. This is not true for resistant cells [28].

Recently, Mayer et al. have characterized the uptake of lipophilic amine containing compound into large unilamellar vesicles in response to proton gradients [30]. They have shown that dibucaine uptake in response to ΔpH proceeds rapidly in a manner consistent with permeation of the neutral form of the drug reaching a Henderson–Hasselbach equilibrium. Our data show that such observations hold for the lipophilic amine anthracycline-cells system [28].

The permeation of the neutral form of the drug is corroborated by the study of

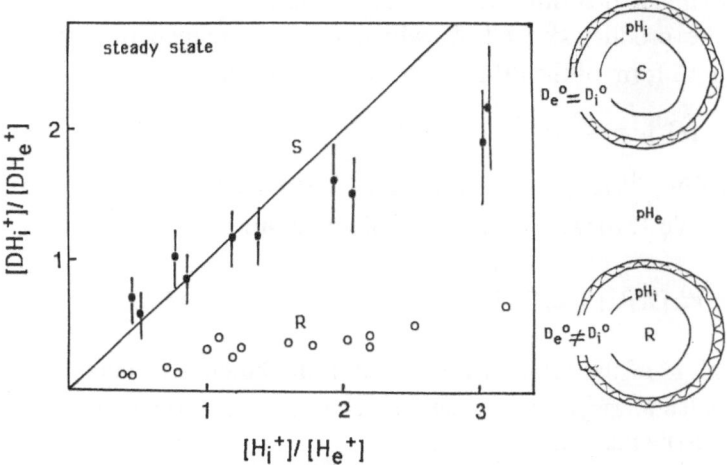

Fig. 4. Plots of $[DH^+]_i/[DH^+]_e$ as a function of $[H^+]_i/[H^+]_e$ for drug-sensitive (•) and drug-resistant cells (○). $[DH^+]_i$ and $[DH^+]_e$ are the concentrations of protonated form of free THP-ADR in the cytoplasm at pH_i 7.2 and outside the cell at pH_e. These concentrations have been calculated for different values of pH_e ranging from 6.8 to 7.8, using the Henderson-Hasselbach equation

The permeation of the neutral form of the drug is corroborated by the study of the variation of the initial rate of uptake, V_+, as a function of pH_e and as a function of the concentration. Two types of experiments were performed: (i) V_+ was measured at a constant initial drug concentration C_T equal to 1 µM and at various pH_e values ranging over 6.7–8.0; (ii) pH_e was constant and equal to 7.2 and the concentration of drug was varied over 0–6 µM. In both cases one observes a linear relation between V_+ and $[D^0]_e$ with $V_+ = 0$ when $[D^0]_e = 0$.

V_+ has the same value (a) for sensitive cells, (b) for resistant cells in the presence of verapamil, (c) for resistant cells and sensitive cells incubated in the presence of azide and absence of glucose, i.e. energy deprived. V_+ is slightly lower for resistant cells (Table 1). This clearly shows that the drug in the neutral form is the predominant species which crosses the membrane.

The uptake of anthracyclines by cells occurs by passive diffusion of the neutral form, the following equations stand for the flux and the initial rate V_+:

$$J_+ = P_+^0 \cdot [D^0] \qquad \text{and} \qquad (V_+)_{t=0} = J_+ \cdot n \cdot S$$

where P_+^0 is the permeability constant for the neutral form of the drug, n the number of cells/cm^3 and S the membrane exchange area per cell, i.e.

$$(V_+)_{t=0} = P_+^0 \cdot S \cdot n \cdot [D^0]_e$$

Up to now it is not possible to determine the membrane exchange area S, therefore we have calculated the product $k_+^0 = P_+^0 \cdot S$, which we have named the influx coefficient for the neutral form of the drug. V_+ can thus be written as

$$V_+ = k_+^0 \cdot n \cdot [D^0]_e = k_+^0 \cdot n \cdot C_e / (1 + 10^{pK_a - pH_e}) = k_+^{pH_e} \cdot n \cdot C_e$$

with $k_+^{pH_e} = k_+^0 / (1 + 10^{pK_a - pH_e})$; $k_+^{pH_e}$ is the influx coefficient for the drug at pH_e.

As can be seen in Table 1, for k_+^0 we have the following order:

$$k_+^0(ADR) < k_+^0(DNR) < k_+^0(THP\text{-}ADR) \approx k_+^0(CAR) \approx k_+^0(ACM)$$

Indeed, if only the charge of the drug was involved in the kinetics of uptake, k_+^0 would be the same for all anthracycline derivatives. This means that other physico-chemical aspects have to be taken into account.

One important point of our method is that it is possible to calculate the kinetics of efflux of the drug outside the cells at the steady state. This requires to take into account the following points: (i) at the steady state, the kinetics of influx of the drug inside the cell $(V_+^{pH_e})_s$ is equal to the kinetics of efflux $(V_-^{pH_i})_s$; (ii) $(V_+^{pH_e})_s = k_+^{pH_e} \cdot n \cdot (C_e)_s$ and $(V_-^{pH_i})_s = k_-^{pH_i} \cdot n \cdot (C_i)_s$, where $(C_e)_s$ is the extracellular drug concentration in buffer at pH_e and $(C_i)_s$ the free drug concentration in the cytoplasm at pH_i at the steady state. As $(V_+^{pH_e})_s = (V_-^{pH_i})_s$, it follows that

$$k_-^{pH_i} = k_+^{pH_e} \cdot (C_e)_s / (C_i)_s$$

We have determined the parameters of efflux $k_-^{7.2}$ at $pH_e = 7.2$ for DNR, THP-ADR and CAR. The values obtained for sensitive and resistant cells as well as the values of $k_+^{7.2}$ are shown in Table 1. As can be seen, $k_+^{7.2}$ and $k_-^{7.2}$ are of the same order of magnitude for sensitive cells. In the case of resistant cells, the influx and efflux coefficients are very different. However, in the presence of verapamil or alternatively in the absence of glucose and presence of azide, $k_+^{7.2}$ and $k_-^{7.2}$ are equal.

It is reasonable to propose in resistant cells the coexistence of two types of efflux: (a) an efflux which is the same as in sensitive cells and involves a free permeation of the neutral form of the drug, and (b) an efflux which is energy-de-pendent and involves the P-glycoprotein.

At the steady state, the kinetics of influx equals that of efflux. At $pH_e = 7.2$,

$$k_+^{7.2} \cdot (C_e)_s = (k_-^{7.2})_p \cdot (C_i)_s + (k_-^{7.2})_a \cdot (C_i)_s$$

where $(k_-^{7.2})_p$ and $(k_-^{7.2})_a$ stand for the passive and active efflux coefficients respectively.

As we have shown, $k_+^{7.2} = (k_-^{7.2})_p$. It follows that

$$(k_-^{7.2})_a = (k_+^{7.2}) \cdot [(C_e)_s / (C_i)_s - 1]$$

As can be seen in Table 1,

$$(k_-)_a(\text{DNR}) < (k_-)_a(\text{THP-ADR}) \approx (k_-)_a(\text{CAR})$$

4. Temperature Dependence of the Uptake and Release of Anthracycline by Drug-Resistant and Drug-Sensitive Cells

The kinetics of uptake V_+ and the amount of drug bound to the nucleus strongly depend on the temperature. For experiments performed in the temperature range 5°–40°C, one observes that the kinetics of uptake decreases as the temperature decreases, whereas the amount of drug bound to the nucleus at the steady state increases as the temperature decreases. For instance, when 1 μM THP-ADR is incubated with 10^6 cells/cm^3, the initial rate of uptake is 4 nM · s^{-1} at 37°C and 0.5 nM · s^{-1} at 6°C, whereas the overall concentrations of drug bound to the nucleus at the steady state is 0.6 μM for sensitive cells and 0.25 μM for resistant cells at 37°C and 0.78 μM for sensitive cells and 0.5 μM for resistant cells at 10°C [31].

Using the data obtained at the steady state, we have estimated the enthalpy of the binding of anthracycline derivatives to cell nuclei. For this purpose we have made the simple hypothesis that the drug-nucleus interaction can be represented by the equilibrium

nuclei + drug ↔ drug bound

The overall binding constant of the drug to the nuclei may be defined as

$$K_{\text{nucleus}} = C_n/C_i[N]$$

where $[N]$ stands for the free DNA concentration that remains nearly constant throughout the experiments performed with the same number of cells in the same proliferative state. The exact value of $[N]$ is unknown, however, the determination of the enthalpy ΔH_0 from the slope of $\ln K_{\text{nucleus}}$ versus $1/T$ does not requires this knowledge, the slope of C_n/C_i versus $1/T$ being the same since $[N]$ is constant. The values of ΔH_0 (kJ/mol) obtained for ADR, THP-ADR and ACM are -36 ± 3, -32 ± 3, and -39 ± 3, respectively. These values are the same for both sensitive and resistant cells.

For comparison we have determined the enthalpies for the binding of these three drugs to naked DNA. They are -35 ± 3, -35 ± 3 and -30 ± 3 kJ/mol for ADR, THP-ADR and ACM, respectively. Within the limit of experimental errors, the enthalpy of fixation of one drug to DNA is the same, whereas the DNA is free or forms part of the nucleus inside the cells. In both cases the process is exothermic.

This observation gives additional evidence that the drug bound to the nucleus inside the cell is in thermodynamic equilibrium with the free drug in the cytoplasm.

In relation with the kinetics of uptake of anthracycline by cells we have evaluated the outer membrane "fluidity" of sensitive and resistant cells using TMA-DPH [32]. The values of the fluorescence polarization parameter P, that we have measured at different temperatures, are consistently higher for drug-resistant cells than for drug-sensitive cells. This is in agreement with previous observations showing that the membrane fluidity of some resistant cell lines decreases [33–35]. In both type of cells we have obtained a thermotropic transition near 20°C in agreement with literature data [36]. Our first idea was that the kinetics of uptake of anthracycline as a function of temperature should also exhibit a thermotropic transition near 20°C. However, the Arrhenius plot of $\ln V_+$ as a function of $1/T$ is monophasic. The values of the activation energy E_a (kJ/mol) for uptake of THP-ADR is 45 ± 3 and 37 ± 2 for sensitive and resistant-cells, respectivelly; these values are 47 ± 3 and 40 ± 3 for ACM in the case of sensitive and resistant-cells respectively.

5. Conclusion

Our data clearly demonstrates that the mechanisms of efflux and influx are the same in sensitive-cells, which involves passive diffusion of the neutral form of the drug through lipid domains [37–39]. A variety of studies have suggested structural homogeneity in the lipid membrane. The membrane has been schematically pictured as being composed of different lipid domains. The domain structure may be a consequence of the interaction of certain intramembranous proteins with the lipid.

Our data suggest that the passage of the unprotonated form of the anthracycline molecules occurs in lipid domains whose transition temperature is lower than 5°C. This could be in relation with the fact that phospholipids such as phosphatidylcholine, which are the main components of erythrocyte membrane, have a transition temperature lower than 0°C. The values of the activation energy obtained for drug-resistant cells are slightly lower than those for drug-sensitive cells. We can tentatively suggest that this is due to the fact that resistant cells may contain more phospholipids than do sensitive cells [40].

Moreover, our data clearly show that anthracycline derivatives gain access to cells by passive diffusion of the neutral form of the drug under the action of a driven force provided by DNA in the nucleus; a mechanism that does not depend on the presence of a specific carrier proteins. This was confirmed by the demonstration that anthracycline derivatives can rapidly accumulate into DNA-containing LUV (large unilamellar vesicles) systems exhibiting neither a membrane potential nor a pH gradient [41].

Now concerning the active efflux of the drug out of resistant cells, our data lead us to propose that anthracycline derivatives are actively pumped out in the neutral form. It is also interesting to note that, for the various anthracycline derivatives used in this study, the kinetics of uptake, the kinetics of release and the degree of resistance follow the same order [29].

Acknowledgement. This investigation was supported by Université Paris Nord, le Centre National de la Recherche Scienifique, l'Institut Curie, L'Association pour la Recherche sur le Cancer and the Polish Minister of National Education.

References

1. Bradley G, Juranka PF and Ling V, Mechanism of multidrug resistance. *Biochim. Biophys. Acta* **948**: 87–128, 1988.
2. Kessel DH, Mechanisms of anthracycline resistance. In: *Bioactive Molecule: Anthracycline and Anthracenedione-Based Anticancer Agents* (ed Lown JW), Vol. 6, pp. 599–628, Elsevier, Amsterdam, 1988.
3. Fojo A, Akiyama S, Gottesman MM and Pastan I, Reduced drug accumulation in multiple drug-resistant human KB carcinoma cell lines. *Cancer Res.* **45**: 3002–3007, 1985.
4. Dano K, Active outward transport of daunorubicin in resistant Ehrlich ascites tumor cells. *Biochim. Biophys. Acta* **323**: 466–483, 1973.
5. Inaba M, Kobayashi H, Sakurai Y and Johnson RK, Active efflux of daunorubicin and adriamycin in sensitive and resistant sublines of P388 leukemia. *Cancer Res.* **39**: 2200–2203, 1979.
6. Skovsgaard T, Circumvention of resistance to daunorubicin by *N*-acetyldaunorubicin in Ehrlich ascites tumor. *Cancer Res.* **40**: 1077–1083, 1980.
7. Riordan JR and Ling V, Genetic and biochemical characterization of multidrug resistance. *Pharmacol. Ther.* **28**: 207–227, 1985.
8. Tsuruo T, Lida H, Tsukagoshi Y and Sakuari Y, Potentiation of vincristine and adriamycin effects in human hemopoietic tumor cell lines by calcium antagonists and calmodulin inhibitors. *Cancer Res.* **43**: 2267–2272, 1983.
9. Kessel D and Wilberding C, Anthracycline resistance in P388 murine leukemia and its circumvention by calcium antagonists. *Cancer Res.* **45**: 1687–1691, 1985.
10. Kessel D and Wilberding C, Mode of action of calcium antagonists which alter anthracycline resistance. *Biochem. Pharmacol.* **33**: 1157–1160, 1984.
11. Pastan I and Gottesmen MM, Multidrug-resistance in human cancer. *N. Engl. J. Med.* **316**: 1388–1393, 1987.
12. Skovsgaard T, Mechanism of resistance to daunorubicin in Ehrlich ascites tumor cells. *Cancer Res.* **38**: 1785–1791, 1978.
13. Inaba M and Johnson RK, Uptake and retention of adriamycin and daunorubicin by sensitive and anthracycline-resistant sublines of P388 leukemia. *Biochem. Pharmacol.* **27**: 2123–2130, 1978.
14. Juliano RL and Ling V, A surface glycoprotein modulating drug permeability in Chinese hamster ovary cell mutants. *Biochim. Biophys. Acta* **455**: 152–162, 1976.
15. Beck WT, Mueller TJ and Tanzer LR, Altered surface membrane glycoproteins in Vinca alkaloid resistant human leukemic lymphoblasts. *Cancer Res.* **39**: 2070–2076, 1979.

16. Riordan JR and Ling V, Purification of P-glycoprotein from plasma membrane vesicles of Chinese hamster ovary cell mutants with reduced colchicine permeability. *J. Biol. Chem.* **254**: 12701–12705, 1979.
17. Ozols RF, Cunnion RE, Klecker RW, Hamilton TC, Ostchega Y, Parrillo JE and Young RC, Verapamil and adriamycin in the treatment of drug-resistant ovarian cancer patients. *J. Clin. Oncol.* **5**: 641–647, 1987.
18. Burke TG, Morin MJ, Sartorelli AC, Lane PE and Tritton TR, Function of the anthracycline amino group in cellular transport and cytotocicity. *Mol. Pharmacol.* **31**: 552–556, 1987.
19. Skovsgaard T and Nissen MI, Membrane transport of anthracyclines. *Pharmacol. Ther.* **18**: 293–311, 1982.
20. Siegfried JM, Burke TG and Tritton TR, Cellular transport of anthracyclines by passive diffusion. Implications for drug resistance. *Biochem. Pharmacol.* **34**: 593–598, 1985.
21. Skovsgaard T, Transport and binding of daunorubicin, adriamycin and rubidazone in Ehrlich ascites tumour cells. *Biochem. Pharmacol.* **26**: 215–222, 1977.
22. Fourcade A, Farhi J-J, Bennoun M and Tapiero H, Uptake, efflux, and hydrolysis of aclacinomycin A in Friend leukemia cells. *Cancer Res.* **42**: 1950–1954, 1982.
23. Munck J-N, Fourcade A, Bennoun N and Tapiero H, Relationship between the intracellular level and growth inhibition of a new anthracycline 4' O-tetrahydropyranyl-adriamycin in Friend leukemia cell variants. *Leukemia Res.* **9**: 289–296, 1985.
24. Peterson C and Trouet A, Transport and storage of daunorubicin and doxorubicin in cultures fibroblasts. *Cancer Res.* **38**: 4645–4649, 1978.
25. Dalmark M and Storm HH, A Fickian diffusion transport process with features of transport catalysis. *J. Gen. Physiol.* **78**: 349–364, 1981.
26. Tarasiuk J, Frezard F, Garnier-Suillerot A and Gategno L, Anthracycline incorporation in human lymphocytes. Kinetics of uptake and nuclear concentration. *Biochim Biophys. Acta* **1013**: 109–117, 1989.
27. Frezard F and Garnier-Suillerot A, Comparison of the binding of anthracycline derivatives to purified DNA and to cell nuclei. *Biochim. Biophys. Acta* **1036**: 121–127, 1990.
28. Frezard F and Garnier-Suillerot A, Determination of the osmotic active drug concentration in the cytoplasm of anthracycline-resistant and -sensitive K562 cells. *Biochim. Biophys. Acta* **1091**: 29–35, 1991.
29. Frezard F and Garnier-Suillerot A, Comparison of the membrane transport of anthracycline derivatives in drug-resistant and drug-sensitive K562 cells. *Eur. J. Biochem.* **196**: 483–491, 1991.
30. Mayer LD, Bally MB and Cullis PR, Uptake of ariamycin into large unilamellar vesicles in response to a pH gradient. *Biochim. Biophys. Acta* **857**: 123–126, 1986.
31. Tarasiuk J and Garnier-Suillerot A, submitted for publication.
32. Kuhry JG, Fonteneau P, Duportail G, Maechling C and Laustriat G, TMA-DPH: a suitable fluorescence polarization probe for specific plasma membrane fluidity studies in intact living cells. *Cell. Biophys.* **5**: 129–140, 1983.
33. Wheeler C, Rader R and Kessel D, Membrane alterations associated with progressive adriamycin resistance. *Biochem. Pharmacol.* **31**: 2691–2693, 1982.
34. Ramu A, Glaubiger D, Madrath IT and Joshi A, Plasma membrane lipid structural order in doxorubicin-sensitive and -resistant P388 cells. *Cancer Res.* **43**: 5533-5537, 1983.
35. Tapiero H, Mishal Z, Wioland M, Silber A, Fourcade A and Zwingelstein G, Changes in biophysical parameters and in phospholipid composition associated with resistance to doxorubicin. *Anticancer Res.* **6**: 649–652, 1986.
36. Deliconstantinos G, Kopeikina-Tsiboukidou L and Villiotou V, Evaluation of membrane fluidity effects and enzyme activities alterations in adriamycin neurotocicity. *Biochem. Pharmacol.* **36**: 1153–1161, 1987.

37. Klausner RD, Kleinfeld AM, Hoover RL and Karnovsky MJ, Lipid domains in membranes. Evidence derived from structural perturbations induced by free fatty acids and lifetime heterogeneity analysis. *J. Biol. Chem.* **255**: 1286–1295, 1980.
38. Karnovsky MJ, Kleinfeld AM, Hoover RL and Klausner RD, The concept of lipid domains in membranes. *J. Cell Biol.* **94**: 1–6, 1982.
39. Rodgers W and Glaser M, Characterization of lipid domains in erythrocyte membranes. *Proc. Natl. Acad. Sci* **88**: 1364–1368, 1991.
40. Vrignaud P, Mantaudon D, Londos-Gagliardi D and Robert J, Fatty acid composition transport and metabolism in doxorubicin-sensitive and -resistant rat glioblastoma cells. *Cancer Res.* **46**: 3258–3261, 1986.
41. Frezard F and Garnier-Suillerot A, DNA-containing liposomes as a model for the study of cell membrane permeation by anthracycline derivatives. *Biochemistry* **30**: 5038–5043, 1991.

Tumor Necrosis Factor: Mechanism of Action and its Potential for Anticancer Therapy

W. Fiers,   R. Beyaert,   P. Brouckaert,   E. Decoster,   D. De Valck,
B. Everaerdt, J. Grooten, A. Lenaerts, C. Libert, K. Schulze-Osthoff,
N. Takahashi,   S. Van Bladel,   C. Van Dorpe,   B. Vanhaesebroeck,
X. Van Ostade and F. Van Roy*

Laboratory of Molecular Biology, State University, K.L. Ledeganckstraat 35, 9000 Gent, Belgium

Abstract. Tumor Necrosis Factor (TNF) is secreted by appropriately induced monocytes and also by some T-cells. The subunit of TNF is a 17 kDa polypeptide, while the native molecule corresponds to a trimer. The three-dimensional structure has been solved at 2.6 Å resolution. Studies on mutants provided evidence for localization of receptor binding sites (three per trimer) in clefts between subunits (one site for each cleft). TNF (in combination with IFN-γ) is a promising anticancer agent, provided the therapeutic index can be improved. The antitumor activity can be enhanced by the presence of LiCl, but this combination is not without effects on normal cells; when TNF + LiCl was injected in the skin, a rapid, local inflammation was observed.

TNF has two pathways of action, a nucleus-independent cytotoxic effect and transcriptional activation of a defined set of genes leading e.g. to IL6 synthesis. This transcriptional activation of the IL6 gene is also very strongly enhanced by Li^+, suggesting that Li^+ acts early in the common signal transduction pathway. There is an overlap with the signal transduction pathway induced by IL6, as also some IL1 effects are dramatically increased by Li^+.

Expression of the endogenous or an exogenous TNF gene in a sensitive cell renders it resistant to TNF. Are TNF-producing tumor cells more tumorigenic? We compared the tumorigenicity in nude mice of tumor cells varying in their TNF expression level. No cachexia was observed. However, with these subcutaneous tumors, the expression level of TNF correlated with a much reduced tumor growth. In fact, the (locally produced) TNF seems to induce a host response leading to encapsulation of the tumor. This may possibly be due to the mitogenic effect of TNF on fibroblasts.

TNF and IL1 are strong inducers of IL6. The latter peaks at about 2–3 h and then disappears from circulation. However, in the case of highly toxic treatments (high mTNF dose, TNF together with IL1 or TNF together with RU486), the IL6 level in circulation at later times even

* Research supported by the Belgian Fund for Medical Scientific Research (FGWO), the "Algemene Spaar- en Lijfrentekas" (ASLK), the National Incentive Program in Life Sciences, the Interuniversity Attraction Poles (IUAP), "Levenslijn", the "Vlaams Actieprogramma Biotechnologie" and the National Lottery. RB, CL, SVB and BV hold fellowships from the NFWO, and DDV, AL and CVD from the IWONL. NT is a recipient of a grant from the Canon Foundation. FVR is a Senior Research Associate with the NFWO.

increases and this correlates with subsequent death. Possibly small amounts of IL6 are protective, but high concentrations contribute to lethality.

One can tolerize mice towards moderate doses of TNF by daily administration of small concentrations of TNF for 5–6 days. This tolerization also works in mice bearing a B16 melanoma tumor. After tolerization, the anticancer treatment can be started by paralesional administration of moderate doses of TNF + IFN-γ. This leads to complete tumor regression with a survival of 80%. These tumor model systems offer promising indications for therapeutic use of TNF in cancer patients.

1. Introduction

In 1975, Lloyd Old and his colleagues at the Sloan-Kettering Institute in New York [1] could demonstrate that animals, such as mice or rabbits, treated with *Bacillus* Calmette–Guérin for ten days, followed by injection of endotoxin, produce a protein released into the serum, which they called Tumor Necrosis Factor (TNF). This factor was so named because when injected into animals carrying a methylcholanthrene-induced sarcoma, a hemorrhagic necrosis followed by rapid regression of the tumor could be observed. Furthermore, a number of malignant cell lines *in vitro* become selectively killed by TNF, while normal cells are not affected or may even show a proliferative response. It was subsequently shown that there is a strong synergism between TNF and interferon (IFN), such that many more tumors become sensitive when the combination TNF plus IFN is used, while this combination of cytokines is still not harmful to normal cells [2, 3]. These remarkable properties of TNF attracted the attention of molecular biologists; we and others cloned in 1984 the TNF gene from man and from various other species, and expressed these genes with very high efficiency in *E. coli* [4, 5]. From then on, TNF became available in large quantities for chemical and biological research, and for preclinical and clinical evaluations. Also, probes and serological reagents became available, which led to an exponential growth in our understanding and knowledge of TNF.

2. TNF: Structure/Function Relationship

The three-dimensional structure of TNF has been solved at 2.9 Å resolution [6], and subsequently even at 2.6 Å [7]. These results not only confirmed the trimeric structure of TNF [8], but moreover provided many surprises, such as the startling resemblance of TNF to capsid proteins of some viruses. In order to understand the molecular mechanism of action of TNF, the first question which needs to be resolved is where the binding site(s) to the TNF receptor is (are) located. We therefore isolated a series of TNF mutants, which all have lost biological activity on the standard tester cell line L-929 [9]. Many inactive mutants corresponded to proteins which were no longer soluble upon expression in an *E. coli* cell, and these were discarded, as presumably the mutation interfered with proper folding of the polypeptide. But a number of inactive mutants produced a protein with properties similar to wild-type TNF in various physico-chemical criteria, not only solubility, but also unchanged circular dichroism spectrum and maintenance of a trimeric structure as assessed by cross-linking experiments.

It was clear from our results that those inactivating mutations which did not interfere with normal folding, clustered around the cleft between two subunits at the lower half of the TNF molecule. Undoubtedly, this locus corresponds to the receptor interaction site. Obviously, considering the trimeric structure, there are three such receptor-binding sites per molecule, which suggests that TNF acts by clustering of receptor molecules. This conclusion is in complete agreement with various reports which document that some anti-receptor monoclonal antibodies mimic all TNF activities tested [10, 11]. There is, so far, no clear evidence that TNF is fulfilling any function in addition to cross-linking of the receptors; both the clustering of inactivating mutations at a single area, as well as the efficient mimicking by antibodies, strongly argue against an additional activity.

3. TNF and Lithium

We have previously reported that for a number of cell types there is a remarkable synergy between TNF and lithium salts in causing cytotoxicity [12]. This effect is very specific for the lithium ion. Lithium salts have been in clinical use since considerable time, and the pharmacology is well known. The concentrations required for effective synergy are in the same range as those achieved in therapeutic protocols (e.g. 1.25 mM). Indeed, we were able to demonstrate with two types of tumor, a murine fibrosarcoma and a human xenograft in nude mice, that there was a very effective antitumor activity upon treatment with TNF together with lithium chloride, but not, or very much less, with either agent alone. Under these conditions and with the concentrations used, there were no treatment-related deaths.

When we injected TNF together with lithium chloride into normal skin, an erythema was rapidly formed [13]. Histologic examination showed that there was a marked initial dermal and subcutaneous neutrophil infiltration. This was followed somewhat later by monocytes. This inflammatory skin reaction was remarkably specific. Nothing could be seen when either agent was injected alone. Furthermore, when IL-1 was injected together with lithium salt, no skin reaction was observed.

4. TNF Expression and its Role in TNF Resistance

In 1986, Rubin et al. [14] reported that L-cells selected for resistance to TNF, in fact had switched on the endogenous TNF gene and produced small amounts of

TNF. We followed up this observation and selected TNF-resistant L-929 derivatives. Two types were obtained [15]: the first had indeed become a TNF producer; this cell line was stably TNF-resistant, even in the presence of actinomycin D (ActD). The second type of resistant cells was of a different nature; the resistant phenotype was not stably maintained in the absence of selection pressure; also, these cells became sensitive to TNF when treated in the presence of ActD. As many phenomena can happen during a selection procedure, we then transfected sensitive L-929 cells with the TNF gene [16]. It turned out that cells could be transfected at high efficiency using the neomycin gene as a selection marker, and all cells which had taken up the TNF gene became TNF-resistant. The degree of resistance was complete when the murine TNF gene was used for transfection, while an intermediate result was obtained with the human TNF gene, in which case the resistance was higher towards the hTNF than towards the mTNF. Remarkably, such resistance could not be obtained by means of the lymphotoxin gene, suggesting that perhaps a membrane-bound form of the cytokine is required. Furthermore, an active TNF molecule seems to be essential, as transfection with a mutant TNF gene, in which the receptor interaction site has been crippled, failed to render the cells resistant.

The resistant cells no longer have TNF receptors on their surface, neither free nor covered with ligand. A likely mechanism for explaining the resistance seems to be that membrane-bound TNF associates with TNF receptor during biosynthesis, and before reaching the cell surface, and this interaction would cause the down-modulation of the receptor. Obviously, such a hypothesis assumes that the intracellular interaction between TNF and receptor, both newly synthesized, does not trigger a TNF response. It may be, for example, that it is only at the cell surface that the TNF receptor properly associates with other signal-transducing molecules.

It has been reported that some cell lines derived from human tumors constitutively synthesize TNF [17, 18]. Could this occur by selection due to the presence of endogenous TNF? The next important question would then be whether such TNF-producing tumor cells would be more tumorigenic or less. By the transfection procedure we obtained a series of L-929 derivatives which differed in their level of TNF production. When these cells were injected subcutaneously as tumor inoculum, it turned out that there was a clear-cut relationship between the time it takes for a tumor to become palpable, and the level of TNF expression [19]. Furthermore, even when tumor growth was initiated, the rate of increase in size of the tumor was more reduced for cells making more TNF. These results were obtained in nude mice, and hence an involvement of a T-cell-dependent immune reaction can be excluded. We must conclude then that TNF activates host mechanisms which limit the growth of the tumor. The tumors generated by the TNF-producing cells were frequently encapsulated, and the TNF produced by the tumor cells might well be responsible for induction of this host response.

5. Tumor Regression in TNF-tolerant Mice

It is well known that a major problem hampering the successful introduction of TNF in the clinic for the treatment of cancer, is the dose-limiting toxicity. It has long been recognized that mice exposed to small doses of endotoxin become tolerant to a challenge with much higher doses [20]. Similar results were subsequently obtained with TNF. Treatment of mice with small doses of TNF for five to six days provided nearly complete protection when the mice were subsequently challenged with a lethal dose of TNF [21]. The molecular and cellular basis for the development of this tolerance is still not known. But an important practical question is, when the mice have been rendered tolerant to the toxic effect of TNF, have they also become tolerant towards the antitumor activity of TNF? This obiously would have severe limitations for a TNF-based therapy. When we treated mice each day with sublethal doses of rmTNF (e.g. 2 µg), an anorectic effect was seen within a day, due to loss of appetite and no water intake. But by the second day, the mice had recovered and after five to six days they had become tolerant to an injection with an otherwise lethal dose of TNF. We then proceeded to the key question, viz. when this pretreatment is done on tumor-bearing mice, would the tumor cells also have become tolerant towards TNF action? Fortunately, this was not so [22]. After tolerization, we treated B16-melanoma-carrying mice by a protocol involving fairly high concentrations of rmTNF (12.5 µg) together with rmIFN-γ (5,000 units). This aggressive treatment was lethal for most of the untolerized animals. But in the tolerized mice, there was a high survival rate, a rapid elimination of the tumor, and the mice became tumor-free for a long time. Similar results were obtained when the B16-melanoma cells were injected in the course of the tolerization treatment, or even when the tolerization was started a few days after the tumor inoculation. These experiments resulted in complete curing and in a survival of 70 to 80% of the mice.

6. Conclusions

TNF is a remarkably pleiotropic molecule, the study of which has contributed enormously to our understanding of various regulatory networks in the human body, such as the immune system, the endocrine system, the central nervous system and the cross-talk between these networks. TNF as well as its antagonists offer interesting prospects in medicine. So far, progress in the clinic has been slow and timid. But new research on preclinical systems has pointed a way to protocols for effective anticancer therapy with acceptable toxicity. Inevitably, more questions have surfaced by these studies than have been answered. But at least directive signs have been established which may point a way to a completely new and effective anticancer therapy.

References

1. Carswell EA, Old LJ, Kassel RL, Green S, Fiore N and Williamson B, An endotoxin-induced serum factor that causes necrosis of tumors. *Proc. Natl. Acad. Sci. USA* **72**: 3666–3670, 1975.
2. Williamson BD, Carswell EA, Rubin BY, Prendergast JS and Old LJ, Human tumor necrosis factor produced by human B-cell lines: Synergistic cytotoxic interaction with human interferon. *Proc. Natl. Acad. Sci. USA* **80**: 5397–5401, 1983.
3. Fransen L, Van der Heyden J, Ruysschaert R and Fiers W, Recombinant tumor necrosis factor: Its effect and its synergism with interferon-γ on a variety of normal and transformed human cell lines. *Eur. J. Cancer Clin. Oncol.* **22**: 419–426, 1986.
4. Pennica D, Nedwin GE, Hayflick JS, Seeburg PH, Derynck R, Palladino MA, Kohr WJ, Aggarwal BB and Goeddel DV, Human tumour necrosis factor: Precursor structure, expression and homology to lymphotoxin. *Nature* **312**: 724–729, 1984.
5. Marmenout A, Fransen L, Tavernier J, Van der Heyden J, Tizard R, Kawashima E, Shaw A, Johnson MJ, Semon D, Müller R, Ruysschaert MR, Van Vliet A and Fiers W, Molecular cloning and expression of human tumor necrosis factor and comparison with mouse tumor necrosis factor. *Eur. J. Biochem.* **152**: 515–522, 1985.
6. Jones EY, Stuart DI and Walker NP C, Structure of tumor necrosis factor. *Nature* **338**: 225–228, 1989.
7. Eck MJ and Sprang SR, The structure of tumor necrosis factor-α at 2.6 Å resolution. Implications for receptor binding. *J. Biol. Chem.* **264**: 17595–17605, 1989.
8. Lewit-Bentley A, Fourme R, Kahn R, Prangé T, Vachette P, Tavernier J, Hauquier G and Fiers W, Structure of tumour necrosis factor by X-ray solution scattering and preliminary studies by single crystal X-ray diffraction. *J. Mol. Biol.* **199**: 389–392, 1988.
9. Van Ostade X, Tavernier J, Prangé T and Fiers W, Localization of the active site of human tumour necrosis factor (hTNF) by mutational analysis. *EMBO J.* **10**: 827–836, 1991.
10. Espevik T, Brockhaus M, Loetscher H, Nonstad U and Shalaby R, Characterization of binding and biological effects of monoclonal antibodies against a human tumor necrosis factor receptor. *J. Exp. Med.* **171**: 415–426, 1990.
11. Engelmann H, Holtmann H, Brakebusch C, Shemer Avni Y, Sarov I, Nophar Y, Hadas E, Leitner O and Wallach D, Antibodies to a soluble form of a tumor necrosis factor (TNF) receptor have TNF-like activity. *J. Biol. Chem.* **265**: 14497–14504, 1990.
12. Beyaert R, Vanhaesebroeck B, Suffys P, Van Roy F and Fiers W, Lithium chloride potentiates tumor necrosis factor-mediated cytotoxicity *in vitro* and *in vivo*. *Proc. Natl. Acad. Sci. USA* **86**: 9494–9498, 1989.
13. Beyaert R, De Potter C, Vanhaesebroeck B, Van Roy F and Fiers W, Induction of inflammatory cell infiltration and necrosis in normal mouse skin by the combined treatment with tumor necrosis factor and lithium chloride. *Am. J. Pathol.* **138**: 727–739, 1991.
14. Rubin BY, Anderson SI, Sullivan SA, Williamson BD, Carswell EA and Old LJ, Non-hematopoietic cells selected for resistance to tumor necrosis factor produce tumor necrosis factor. *J. Exp. Med.* **164**: 1350–1355, 1986.
15. Vanhaesebroeck B, Van Bladel S, Lenaerts A, Suffys P, Beyaert R, Lucas R, Van Roy F and Fiers W, Two discrete types of tumor necrosis factor-resistant cells derived from the same cell line. *Cancer Res.* **51**: 2469–2477, 1991.
16. Vanhaesebroeck B, Decoster E, Van Bladel S, Lenaerts A, Van Ostade X, Van Roy F and Fiers W, Expression of an exogenous TNF gene in TNF-sensitive cells confers resistance to TNF-mediated cell lysis. Submitted.
17. Krönke M, Hensel G, Schlüter C, Scheurich P, Schütze S and Pfizenmaier K, Tumor necrosis factor and lymphotoxin gene expression in human tumor cell lines. *Cancer Res.* **48**: 5417–5421, 1988.

18. Spriggs D, Imamura K, Rodriguez C, Horiguchi J and Kufe DW, Induction of tumor necrosis factor expression and resistance in a human breast tumor cell line. *Proc. Natl. Acad. Sci. USA* **84**: 6563–6566, 1987.
19. Vanhaesebroeck B, Mareel M, Van Roy F, Grooten J and Fiers W, Expression of the tumor necrosis factor gene in tumor cells correlates with reduced tumorigenicity and reduced invasiveness *in vivo*. *Cancer Res.* **51**: 2229–2238, 1991.
20. Johnston CA and Greisman SE, Mechanisms of endotoxin tolerance. In: *Pathophysiology of Endotoxin* (Hinshaw LB, ed), Amsterdam, Elsevier, Vol 2, pp 359–401, 1985.
21. Fraker DL, Stovroff MC, Merino MJ and Norton JA, Tolerance to tumor necrosis factor in rats and the relationship to endotoxin tolerance and toxicity. *J. Exp. Med.* **168**: 95–105, 1988.
22. Takahashi N, Brouckaert P and Fiers W, Induction of tolerance allows separation of lethal and antitumor activities of tumor necrosis factor in mice. *Cancer Res.* **51**: 2366–2372, 1991.

Exploitation of Peptide Transport Systems in the Design of Antimicrobial Agents

D. R. Tyreman, M. W. Smith, G. M. Payne and J. W. Payne*

School of Biological Sciences, University of Wales, Bangor, Gwynedd, LL57 2UW, UK

Abstract. The occurrence of transport systems for the uptake of peptide nutrients is widespread amongst bacteria and fungi. In general, two or three such systems occur, the overlapping and complementary specificities of which ensure the efficient uptake of all peptides comprising 2–5 residues. The broad structural specificities of these systems also permit uptake of modified natural peptides and varied peptide analogues. This situation is being exploited in the synthesis of peptide smugglins, i.e. compounds in which a normally impermeant moiety is incorporated into a peptide so that it can be actively accumulated; once inside the organism, the impermeant moiety can be released enzymically but will be unable to efflux. In this way, potentially toxic compounds can be directed against normally inaccessible, intracellular targets. Various natural, antibacterial peptide smugglins are known to exist.

Here we describe novel assays, using purified peptide binding proteins, to characterise substrate binding, the results of which can be used in the design of optimal peptide carriers for uptake of smugglins. In addition, we report on the use *in vitro* of antibacterial smugglins together with different antimicrobial agents that offer interesting prospects for combinative drug therapy.

1. Introduction

There is an ever present need for new antimicrobial agents to complement the limited current resource of natural antibiotics, and also to counteract the increasing emergence of antibiotic-resistant strains. To this end, conventional screening procedures for new antibiotic activities, that have proved so fruitful to date, are being pursued as diligently as ever, but this is a diminishing resource. Consequently, increasing effort is being devoted to a fundamental approach, aimed at the design of novel antimicrobial compounds created to inhibit selected microbial targets.

In this rational approach, the following scenario often develops. Target microorganisms will be identified, e.g. Gram-positive and/or Gram-negative bacteria, opportunistic, pathogenic fungi, etc. A target enzyme will be selected, ideally specific to the particular microorganism. Considerable effort will be devoted to the isolation and characterisation of the enzyme, including its catalytic mechanism. Based on this information, efficient and specific inhibitors may be

---

* Corresponding author.

developed that will be categorised on the basis of their inhibitory parameters measured with purified enzyme in appropriate assays *in vitro*. When the putative inhibitors are tested against the intact target microorganisms *in vitro*, they are commonly found to show little or no antimicrobial activity. Most commonly this arises through inability of the putative inhibitors to enter the microorganism to a degree that achieves a concentration sufficient to inhibit the intracellular target enzyme. In such situations, the designer of antimicrobials may try to overcome the impermeability problems by subverting the nutrient transport systems of the organism for the uptake of the putative inhibitors. It is for this purpose that microbial peptide permeases offer the most generally applicable opportunity for uptake of varied antimicrobial compounds.

2. Microbial Peptide Permeases

Many studies, over a number of years, indicate that systems for accumulating small (protein-derived) peptides occur universally amongst heterotrophic microorganisms, ranging from diverse prokaryotic species through to unicellular yeasts and more complex fungi [1–6]. Indeed, such is the ubiquity and nutritional value of peptides that comparable systems are found in all forms of life, ranging from various cells and tissues in animals and mammals through to higher plants [7].

In prokaryotic microorganisms, different species have been characterised to various extents, but it is generally found that several, complementary peptide transport systems coexist [5]. In the best studied species, such as *Escherichia coli* and *Salmonella typhimurium*, three systems occur [8, 9]. The dipeptide permease, Dpp, handles dipeptides selectively, with a low level ability to transport tripeptides [9]. The oligopeptide permease, Opp, transports tri-, tetra- and pentapeptides, with a somewhat lower capacity for dipeptides and for higher oligopeptides [5, 10]. The tripeptide permease, Tpp, handles tripeptides optimally, with a low level ability towards dipeptides [11]. Their substrate preferences have permitted a variety of mutants in each permease to be isolated by selecting for resistance to particular, toxic peptide analogues, e.g. Dpp using bacilysin and lysyl aminoxy alanine (LysOAla) [8, 12]; Opp using triornithine [1, 13]; and Tpp using alafosfalin [11, 14]. Such mutants have aided the genetic characterisation of these systems, and the biochemical description of their protein components and of their substrate specificities. Similarly, in eukaryotic fungi such as *Candida albicans*, analogous Dpp and Opp systems have been described [15].

3. Smugglins and Illicit Transport

As information on the substrate specificities of the bacterial transport systems accumulated it became clear that, depending on the specific system, rather strict requirements existed towards the amino terminus and for the normal α-linkage and L-stereochemistry of the peptide bond, but that otherwise much structural variation could be tolerated [5]. Such conclusions are compatible with the need uniquely to recognise a peptide structure, whilst also being able to accommodate the immense variety of peptide structures inherent in the permutations of protein amino acid sequence. This realisation led to the proposal that modified protein amino acids and non-natural analogues might also be transported in peptide form [2]. Furthermore, it provided the possibility of transporting normally impermeant moieties, if these were attached to or incorporated into a peptide structure in a manner compatible with the structural requirements of a peptide permease. Various terms have been coined to describe this peptide-mediated accumulation of impermeant residues, including: illicit transport [16], portage transport [17, 18] and warhead delivery [19]. The peptide-carrier complexes themselves have been given the name smugglins [3, 20]. Experimental confirmation of the principle of illicit transport was demonstrated early on [16, 17].

Clearly, if the impermeant moiety were potentially toxic, it provides an opportunity for the development of a class of peptide carrier prodrugs. A crucial additional step is that once the smugglin has been accumulated intracellularly, then its inhibitory warhead must be released through enzymic action. The toxic moiety, being unable to enter the cell except as a smugglin, cannot undergo exodus once cleaved off and accumulates to a high concentration intracellularly. This overall process is shown schematically in Fig. 1.

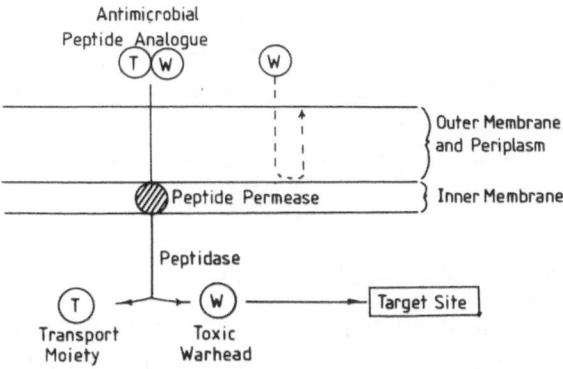

Fig. 1. The concept of warhead delivery—the use of peptides to transport impermeant molecules into the cell via the peptide permeases

4. Natural and Synthetic Smugglins

Although unbeknown to those researchers first proposing the smugglin concept and preparing the first synthetic examples, Nature had for long endorsed the concept and produced many examples of its own. Amongst the very first characterised was bacilysin (tetaine), which shows both antibacterial and antifungal activity deriving from action of its *C*-terminal glutamine analogue anticapsin [21]. Among other natural, antimicrobial smugglins are lindenbein [22], phosphinothricyl AlaAla and analogues [23, 24], phaseolotoxin [25], polyoxins and nikkamycins [26] and others [6, 18, 19, 23, 27, 28]. Further examples are reported at regular intervals and it is certain that the synthetic chemists will have much to learn from detailed consideration of the type and range of chemical structures that have competed successfully for inclusion in Nature's portfolio.

In the last decade, a wide range of synthetic smugglins has been produced and tested for antimicrobial activity. Amongst the most extensively studied are those incorporating the toxic alanine mimetic, L-1-aminoethylphosphonic acid, exemplified by alafosfalin and its analogues [29, 30]. Other examples are: *m*-fluorophenylalanyl peptides [31]; pyrimidine-peptide complexes [32]; peptides containing novel inhibitors of glucosamine-6-phosphate synthetase [33] and of CMP-KDO synthetase [34]; inhibitors linked to the α-carbon of glycine [35] or to side-chain thiols [36]; and rather than forming part of the side chains or *C*-terminal appendages, residues have been incorporated into the backbone e.g. aminoxy-, aza- and retro-derivatives [6, 12]. The range of compounds encompassed in these studies has done much to reveal the potential and the limitations of the smugglin approach. However, consideration of the overall design of these smugglins has been minimal, for the studies have been largely *ad hoc*, aimed at quickly producing a carrier to transport a particular warhead. They illustrate rather well the fundamental distinction between the "industrial" or "applied" approach and the "academic" or "pure" approach, in which the former seeks "an answer now" and the latter "the answer sometime". Thus, if, as an academic, one seeks to establish principles for optimal design and use of smugglins, what factors need to be considered and what experimental approaches adopted? These questions are addressed in the remainder of this article.

5. Peptide Permease Characterisation and Smugglin Design

The following features are amongst the most important that need to be considered when designing peptide carriers for applications to smugglins. For present purposes, these topics are restricted to consideration of the microbial target, and exclude related aspects, e.g. oral bioavailability and pharmacokinetics of the smugglin in the human. Some aspects of these topics have been considered previously [6].

    i.   Number of peptide permeases.
    ii.   Permease expression *in vivo*.
    iii. Substrate specificities of the permeases.
    iv. Microbial resistance mechanisms.

For several important classes of microbial pathogens, the number of genetically distinct permeases is established. Given this background, and the availability of a range of toxic peptides, it should prove relatively simple with any other species to characterise its systems by selection of permease mutants. This level of information may be adequate for many instances of smugglin use (see later). Further studies, e.g. immunocross-reactivity, Western blotting using antibodies raised against peptide permease proteins from well characterised species, would provide valuable additional information in establishing relatedness in less studied organisms. Whilst ultimately, cloning and sequence information would provide detailed data on structural and functional similarities.

The possibility that the expression of peptide permease activity may be markedly different between that measured *in vitro* in typical laboratory culture, and that existing *in vivo* in the chemotherapeutic context, must always be considered. *A priori*, the desirability that the permeases be strongly expressed *in vivo* seems likely to be generally realised, because of the excellent nutritional value of peptides and their ubiquity. There is much evidence that peptide permeases may be expressed constitutively at high levels with some nutrient-dependent fine tuning of this situation. For example, in *E.coli* the Opp and Dpp binding proteins (OppA, DppA) are normally the most abundant of all periplasmic proteins. Opp expression is generally reported as being constitutive, although we have observed that *E.coli* OppA production can be markedly decreased under conditions of phosphate starvation [C.A. O'Neill and J.W. Payne, unpublished results]. Evidence exists that availability of amino acids affects peptide permease activity, particularly with Dpp [5, 38]. It has been reported that Tpp expression in *S.typhimurium* is enhanced under anaerobic conditions [39], a condition that also appears to influence activity of other peptide permeases in *Salmonella* and *E.coli* (J.W. Payne, unpublished observations). In fungi, such as *Saccharomyces cerevisiae* and *C.albicans* nitrogen limitation increases peptide permease activity about 3-fold. In the latter organism, it has been shown that peptide substrates actually induce enhanced expression of Dpp and Opp [40], a potentially suicidal tendency for smugglin application, and differential permease expression has been reported for the yeast and mycelial forms of this organism [41].

Thus, although the quantitative expression of particular permeases may be moderated by nutrient status and physiological condition, all the evidence indicates that microbial peptide transport capacity *in vivo* is always likely to be sufficient for extensive smugglin accumulation and potential antimicrobial activity.

The third feature, namely determination of the substrate specificities of the peptide permeases is clearly crucial to the design of optimal peptide carriers. Characterisation of the affinities and capacities of the permeases from measurement of transport kinetics has been successfully performed with a number of species [5, 42–45]. However, such studies are inherently susceptible to error when using conventional assays based upon accumulation of radioactively-labelled substrate, for transported substrates can be very rapidly cleaved and metabolised, with accompanying efflux [43, 44]. Alternatively, assays based on fluorescent-labelling of peptides have been applied successfully [43-45]. From such studies, broad principles regarding the structural features important for effective transport have emerged; e.g. requirements towards the N-, and C-termini, side chains, α-peptide bond and stereochemistry, together with quantitative data on uptake rates and binding affinities for a limited range of di- and oligopeptides.

Our recent studies, using ligand binding assays with purified binding proteins, to obtain quantitative data on substrate-permease recognition are described below.

Finally, the potential ways in which resistance might arise to smugglin therapy needs consideration. Three main possibilities exist. The first, a mutation in the target enzyme, is a feature common to all chemotherapeutic applications. Its effect in any instance is impossible to predict, being dependent upon the sequence and function of the particular protein. Secondly, in the case of smugglins, like other forms of prodrug, resistance may arise from mutation in the enzyme(s) needed to convert the substance into its active form. With peptide smugglins, such enzymes are likely to be peptidases. In general, microbial cells possess many intracellular peptidohydrolases of high activity and overlapping specificities. Therefore, this form of resistance would not appear to be a particular problem, for mutation in several peptidases may be needed. Nevertheless, information on the peptidohydrolase specificity of target microorganisms would be helpful to smugglin design. This form of resistance should also be countered by using smugglins in combinative therapy (see below). Thirdly, resistance may arise from mutational inactivation of permease function. Clearly, this potential problem is alleviated in principle by the fact that two or more peptide permeases usually occur together, and these can show some measure of substrate overlap. However, the potential of this particular problem has been highlighted by the unusually high frequency at which opp mutants are found to occur in strains of *E.coli* and *S.typhimurium*. These opp operons occur in a region of the chromosome that is a "hot spot" for deletions. It is a matter of some debate whether this opp locus occurs here "by chance", or whether it is purposefully sequestered so as to provide a population balanced between, on the one hand, the advantages of transporting peptide nutrients and peptidoglycan degradation products [46], and on the other, having a built-in resistance mechanism to natural smugglins. To date, there have been no reports to indicate that this high mutational frequency is of widespread occurrence for other peptide permeases in various species. Ways in which the use of smugglins in mixtures might provide a means to overcome this type of resistance are discussed later.

6. Assays with Peptide Binding Proteins to Determine Substrate Specificities of Peptide Permeases

The Opp and Dpp of *E. coli* were chosen as models. These permeases are examples of the shock-sensitive, binding protein (BP)-dependent systems that are energised by phosphate bond energy [47, 48]. Each comprises three or four membrane-bound proteins together with a specific, periplasmic, recognition or binding protein (OppA and DppA).

In the overall transport sequence, it is considered that a peptide enters the periplasm via one or more of the outer membrane porins [49]. It interacts with an appropriate peptide BP, thereby inducing a conformational change that allows the substrate-BP complex to bind to the membrane-bound permease protein(s) [47, 48]. By an essentially unknown mechanism, the peptide is then translocated into the cytoplasm and the BP released into the periplasm, to repeat the cycle. In this model, it is not clear whether the substrate specificity of the system resides solely in the BP, or whether the membrane-bound protein(s) of the translocation pore might also impose structural constraints. In seeking to develop a form of ligand-receptor assay with peptides and the BP, it seemed likely that a comparison of binding data obtained in this way, with the results on substrate specificities/affinities obtained previously for the overall transport process, should indicate whether or not the BP alone conferred specificity on the permease. If the BP does determine permease specificity, then a quantitative assay for peptide-BP interaction would directly provide information of use in the design of general peptide carriers.

7. Substrate-induced Conformational Changes in OppA and DppA Studied by Isoelectric Focussing

We developed procedures for purifying large amounts of OppA and DppA to homogeneity from cold osmotic shockates of *E.coli,* using gel filtration and ion exchange (Mono Q and Mono S) chromatography on a Pharmacia FPLC system (unpublished results). With both proteins in the final stages of purification on ion-exchange, the major peptide BP peak invariably had one or more clearly defined shoulder peaks, which we initially assumed to be contaminating proteins. However, on SDS-PAGE and *N*-terminal protein sequencing all the overlapping peaks for each BP proved to be pure OppA or DppA. Furthermore, if the major and minor peaks were combined in each case and subjected to reverse-phase chromatography (RPC) on a Pro RPC column, a single, sharp symmetric peak was obtained for each protein. When, for example, the OppA fractions from the Mono Q column were examined on an isoelectric focussing (IEF) gel on a Pharmacia Phastsystem, typically three bands were seen, with pI of 6.20, 6.26 and 6.55. The pI 6.20 band was invariably the main species, but the distribution varied somewhat from batch to batch. OppA from the RPC column ran as a single species with pI 6.20. These forms prove to represent the free form of OppA (pI, 6.20) and two liganded forms (pI, 6.26, 6.55). Analogous observations were made for DppA samples, but with different characteristic pI values. Thus, these proteins, as isolated, contain variable amounts of free and (unknown) ligand-bound forms; these ligands can be displaced through solubilisation and chromatography for RPC.

This explanation for the occurrence of the different forms was tested by incubating samples of purified BP with various peptides and then examining the mixtures on IEF. With OppA, for example, incubation with Ala_3 shifted the sample from the form with pI 6.20 to that of pI 6.26. The conversion was stoichiometric, with a progressive shift occurring, which reached complete conversion at a 1 : 1 ratio of BP: Ala_3 (Fig. 2). The effect is both stereospecific, with e.g. DDD–Ala_3 at 200 : 1 causing no conversion, and structurally specific, with various other tripeptides showing similar effects, but with higher molar ratios of peptide: BP often being needed, reflecting a range of different substrate affinities. The relative intensities of the protein bands can be measured using laser densitometry so as to produce quantitative data for peptide-induced shifts; from such data, relative substrate binding affinities can be calculated. Certain tripeptides, e.g. $LysAla_2$, shifted the unliganded pI 6.20 form completely into the pI 6.55 form. Dipeptides, in general, were markedly less effective than tripeptides in causing band changes; the changes

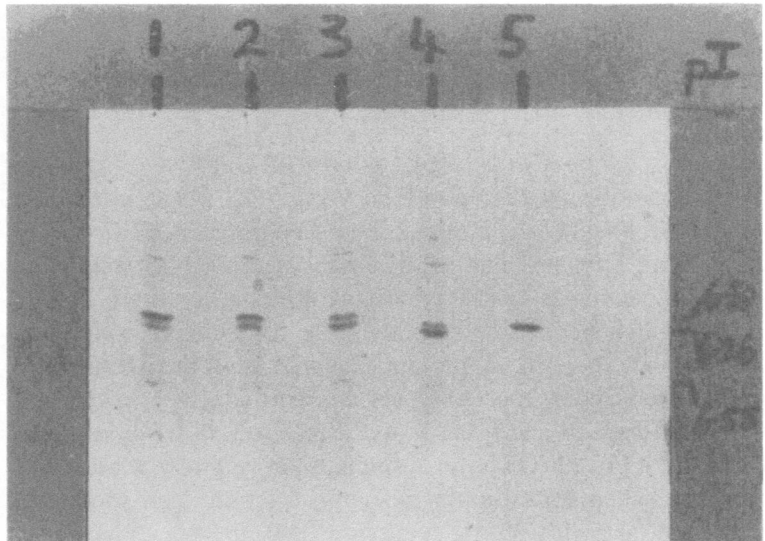

Fig. 2. IEF analysis of the peptide binding ability of the purified Opp Bp. (15) 0 : 1, 0.1 : 1, 0.2 : 1, 0.4 : 1 and 0.8 : 1 molar rations of Ala_3 : Opp BP, respectively

produced by higher oligopeptides reflected a wide range of affinities towards OppA.

Exactly analogous observations were made when DppA was incubated with peptides, but here, as expected, dipeptides showed greater affinities than tripeptides. However, in this case, binding of a high affinity substrate such as Ala_2 induced a shift from the unliganded form of higher pI, 6.11, to a liganded band of lower pI, 6.01 (Fig. 3). A further liganded variant of pI 5.90 was seen with certain dipeptides, e.g. AlaPhe, analogous to the situation observed with OppA.

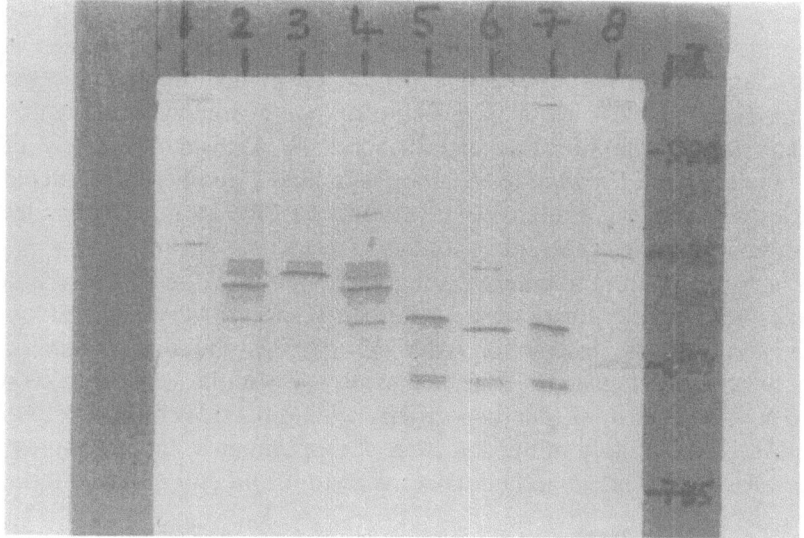

Fig. 3. IEF analysis of the peptide binding specificity of the purified Opp and Dpp BPs. (1 and 8) pI markers; (2) purified Dpp Bp; (3, 4) 1 : 1 molar ratio of Ala_2 and $DDAla_2$: Dpp BP, respectively; (5) purified Opp BP; (6, 7) 1 : 1 molar ratio of Ala_3 and $DDDAla_3$: Opp Bp, respectively

The various pI forms of these BPs, observed on the IEF gels, provide compelling evidence for the substrate-induced conformational rearrangements that is an essential feature of their proposed mode of action. No ready explanation is available for the surprising observation that the binding of rather similar peptides can give rise to liganded forms of markedly different pIs, implying significantly different conformational rearrangements. Whether the existence of the two liganded forms reflects a capacity for an unexpected diversity of biological function is presently being explored.

On the other hand, the general conclusion from these studies is that structural specificities and relative binding affinities closely follow those reported for transport. This important feature, therefore, supports the premise that studies with purified binding proteins can provide information paralleling the overall properties of the permease.

8. Assays for Radioactive Ligand Binding to Binding Proteins

Quantitative analysis of the peptide-induced, conformational changes discussed above can provide much information useful to the design of the peptide carrier function of smugglins, although monitoring changes with peptides of low affinity is difficult. To improve and to extend the type of structural data needed, we have developed (unpublished results) a modified version of an earlier assay [50]. In this, purified BP (OppA or DppA) is incubated with a radioactively-labelled peptide substrate, with or without a peptide competitor. The substrate-BP complex is precipitated by ammonium sulphate and collected on a membrane filter; the radioactivity retained on the filter is a measure of bound ligand, from which quantitative data on peptide substrate affinities can be derived. A variety of radioactively-labelled peptides were used, amongst which a number of iodinated tyrosine peptides, which were of relatively low affinity, allowed study of peptides with a wide range of affinities [51, and unpublished results].

Here again, a crucial feature to emerge is that the quantitative results obtained parallel the data obtained previously for transport studies [5, 43–45], i.e. it endorses the conclusion that the interaction with BP dictates the overall kinetic parameters of the transport system. Thus, the same specificities were observed with respect to N- and C-termini, stereospecificity, etc.; dipeptides bind poorly to OppA; triornithine, used widely in the selection of Opp mutants, binds competitively with Ala$_3$. However, novel and unexpected features did emerge from these

Table 1. Relative peptide affinities for Opp binding protein. Inhibition of $[^{125}I]$ TyrGlyGly by competitor at 1 : 1

LLL-Ala$_3$	96	AlaAla	1
LLD-Ala$_3$	32	LysAla	11
LDL-Ala$_3$	0	Ala$_6$	94
LysLysLys	89	Gly$_6$	38
LeuLeuLeu	81	Met Enkephalin (5aa)	13
OrnOrnOrn	64	Bradykinin (9aa)	18
AlaGlyGly	54	Angiotensin (10aa)	17
GlyGlyPro	5	(I)Angiotensin (10aa)	19
Ala$_3$OMe	87	Substance P (11aa)	25
Ala$_3$P	67	Somatostatin (14aa)	34
N-AcAla$_3$	0	α–Endorphin (16aa)	17
N-MeGly$_3$	21		

Data were obtained using the ammonium sulphate assisted filter binding assay.
Ala$_3$P is alanylalanyl-L-1-aminoethylphosphonic acid.

studies (Table 1). For example, with OppA, hexapeptides compete very well for binding with tripeptides, and yet, in general, they are poorly transported [5]. This result endorses earlier conclusions that cytoplasmic transport of exogenous peptides is controlled by the sieving effect of the outer membrane porins [49]. More surprisingly, even larger peptides showed good binding to OppA (Table 1). This conclusion was based not only on their competitive ability against radioactive tripeptides, but also from direct binding studies using [^{125}I] Angiotensin, which could be competed against by LLL- but not DDD-tripeptides. If these binding characteristics also reflect the transport ability of Opp, then this permease may serve other functions in addition to absorption of small peptides. Various possibilities come to mind. For example, the uptake of fragmented signal peptides from secreted proteins, absorption of the early products of hydrolysis of periplasmic proteins, etc.

In conclusion, therefore, results from these assays have confirmed the validity of using BPs to characterise permease specificities. The quantitative data obtained can help to direct the design of smugglin carrier moieties. Combining such structural information with computer molecular modelling, will lead us towards our eventual goal of producing non-peptide mimetics [52] of microbial carrier moieties that will retain the desirable property of oral bioavailability, but will not suffer from the limitations of susceptibility to host peptidohydrolases.

9. Use of Smugglins in Antimicrobial Chemotherapy

The use of smugglins potentially makes accessible a vast range of intracellular targets, and to capitalise on this, novel compounds with good inhibitory indices will be sought. Whereas it is desirable that compounds with very good MIC values are selected, it is our contention that much more attention should be paid to consideration of resistance mechanisms than is currently the case when evaluating new compounds. When considering theoretical resistance to smugglins (see above), permease mutation is most important. In this regard, it is fortunate that several peptide permeases occur together. We are approaching this matter in several ways, in some respects adopting the lessons of Nature (see below). Firstly, by aiming to design smugglin carriers that effectively use several peptide permeases, making use of the structural information obtained from the assays described above *inter alia*. Secondly, a smugglin mixture could be used, incorporating a common warhead with two or more carrier moieties that provide complementary transport via several permeases. Finally, we are exploring the potential of also using smugglins in mixtures, but in this case together with other types of antimicrobials, which have different modes of action and act against other targets, e.g. in the cell envelope or membrane.

To evaluate this last possibility, we have tested a variety of smugglins, which have been characterised as using predominantly one permease, in combination with various antibacterials [53]. Using a microtitre plate assay, a fractional inhibitory concentration (FIC) was determined in each case that represented the ratio of the concentration of inhibitor in the mixture, to the concentration of inhibitor alone that gave the same extent of inhibition in each case. When these were plotted (Fig. 4) the sum of the FICs at the minimum yields a fractional inhibitory concentration index (FICI); a figure that indicates whether synergy (<0.2), additivity ($\approx$1) or antagonism (>1) occurs between the components. In this limited study

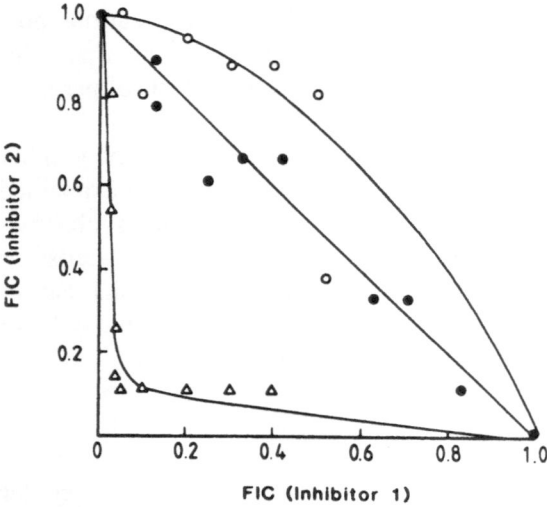

Fig. 4. Determination of fractional inhibitory concentration index (FICI) from FIC isobolograms for smugglin combinations. BocProAla-OAla with Val-D-OAla ($\bullet$); Lys-OAla with Val-D-OAla ($\circ$); BocProAla-OAla with Ala$_2$P ($\triangle$). Results indicate antagonism, additivity and synergism, respectively, with corresponding FICI values of 1.3, $\approx$1 and 0.23. Boc is t-butyloxycarbonyl, other abbreviations are as in Table 2

[53], several examples of good synergy were noted (Table 2). More importantly, no resistant (double mutants) were observed. Thus, these studies provide support for the principle of smugglin combinative therapy, rather than indicating a particular combination of therapeutic potential. However, the interesting observation did emerge that smugglins and β-lactams may be a particularly efficacious combination.

10. Applications of the Lessons of Nature to Smugglin Design

The existence of natural smugglins endorses the value of the principle of illicit transport, and should provide encouragement for the designers of antibacterial agents. It is noteworthy when scrutinising the structures that Nature has evolved,

Table 2. Effectiveness of antibacterial combinations

Inhibitors		FICI	Interaction
Orn₃ (Opp)	Ala₂P(Tpp)	0.6	Weak Synergy
Orn₃ (Opp)	Ala₃P(Opp)	>1	Antagonism
Orn₃ (Opp)	Val-OAla(Dpp)	0.7	Weak Synergy
Ala₂P (Tpp)	ProAla-OAla(Opp)	0.2	Strong Synergy
Ala₂P (Tpp)	Ala₃P(Opp)	≈1	Additivity
Val-D-OAla(Dpp)	LysLeuNHdKDO (Opp)	0.8	Weak Synergy
Orn₃ (Opp)	Cephalexin	0.2	Strong Synergy
Ala₂P (Tpp)	Cephalexin	0.2	Strong Synergy
Clavulanate	Amoxycillin	0.7	Weak Synergy

Fractional inhibitory concentration indices (FICI) were determined as shown in Fig. 4. Abbreviations for inhibitors are: Ala₂P and Ala₃P are alanyl- and alanylalanyl-L-1-aminoethylphosphonic acid, respectively [29]; -OAla is aminoxyalanine [12]; LysLeuNHdKDO is N-(LysLeu)-8-amino-2,6-anhydro-3,8-dideoxy- D-glycero-D-talo-octonic acid [34].

that these smugglins often occur as mixtures of various forms, e.g. with different amino acid carrier residues, varied chain length, etc. It is our view that this indicates not the existence of biosynthetic enzymes of poor specificity, but rather the purposeful recruitment of "relaxed" enzymes to provide smugglin combinations that offer beneficial variation in routes of uptake, as well as varied mechanisms of peptidase activation, and differing susceptibilities to enzymic inactivation. In short, they illustrate the advantages of using smugglin combinations against varied microorganisms and offer criteria for consideration by the designer of an antimicrobial agent. Furthermore, such smugglin mixtures are often produced together with other natural antimetabolites, illustrating the potential benefits of combinative therapy.

11. A Superfamily of Membrane Proteins

Finally, sequence studies upon the prokaryotic Opp membrane proteins have indicated that they share extensive homology with a variety of membrane proteins in higher organisms [47, 54]. Most interestingly, perhaps, is the existence of the cystic fibrosis gene product and the multiple drug resistance protein in this "superfamily". Thus, further detailed studies upon the model peptide permeases of bacteria may not only prove valuable for smugglin design, but may also provide information relevant to several other critical areas of current biological enquiry.

References

1. Payne JW and Gilvarg C, Peptide transport. *Adv. Enzymol. Relat. Areas Molec. Biol.* **35**: 187–244, 1971.
2. Payne JW, Mechanisms of bacterial peptide transport. *CIBA Found. Symp.* **4**: 17–32, 1972.

3. Payne JW, Peptides and microorganisms. *Adv. Microbiol. Physiol.* **13**: 55–113, 1976.
4. Becker JM and Naider F, Transport and utilization of peptides by yeast. In: *Microorganisms and Nitrogen Sources* (ed Payne JW), pp.257–279. John Wiley and Sons, Chichester, London, 1980.
5. Payne JW, Transport and utilization of peptides by bacteria. In: *Microorganisms and Nitrogen Sources* (ed Payne JW), pp. 211–256. John Wiley and Sons, Chichester, London, 1980.
6. Payne JW, Drug delivery systems: optimising the structure of peptide carriers for synthetic antimicrobial drugs. *Drugs Exptl. Clin. Res.* **12**: 585–594, 1986.
7. Matthews DM and Payne JW, Transmembrane transport of small peptides. *Curr. Topics Membr. Transp.* **14**, 331–425, 1980.
8. Higgins CF and Gibson MM, Peptide transport in bacteria. *Meth. Enzymol.* **125**: 365-377, 1986.
9. Payne JW, Peptide transport in bacteria: methods, mutants and energy coupling. *Biochem. Soc. Trans.* **11**: 794–798, 1983.
10. Payne JW, Oligopeptide transport in *Escherichia coli*. *J. Biol. Chem.* **243**: 3395–3403, 1968.
11. Alves RA and Payne JW, The number and nature of the peptide-transport systems of *Escherichia coli*: characterization of specific transport mutants. *Biochem. Soc. Trans.* **8**: 704–705, 1980.
12. Payne JW, Morley JS, Armitage P and Payne GM, Transport and hydrolysis of antibacterial peptide analogues in *Escherichia coli*: backbone-modified aminoxy peptides. *J. Gen. Microbiol.* **130**: 2253–2265, 1984.
13. Barak Z and Gilvarg C, Triornithine-resistant strains of *E. coli*: isolation, definition and genetic studies. *J. Biol. Chem.* **249**: 143–148, 1974.
14. Gibson MM, Price M and Higgins CF, Genetic characterization and molecular cloning of the tripeptide permease (tpp) genes of *Salmonella typhimurium*. *J. Bacteriol.* **160**: 122–130, 1984.
15. Payne JW and Shallow DA, Studies on drug targetting in the pathogenic fungus *Candida albicans*: peptide transport mutants resistant to polyoxins, nikkomycins and bacilysin. *FEMS Microbiol. Lett.* **28**: 55–60, 1985.
16. Ames BN, Ames GF-L, Young JD, Isuchiya D and Lecocq J, Illicit transport: the oligopeptide permease. *Proc. Natl. Acad. Sci. USA* **70**: 456–458, 1973.
17. Fickel TE and Glivarg C, Transport of impermeant substances in *E. coli* by way of oligopeptide permease. *Nature New Biol.* **241**, 161–163, 1973.
18. Gilvarg C, Portage transport. In: *The Future of Antibiotherapy and Antibiotic Research* (eds Ninok L et al.), pp. 352–362. Academic Press, London, 1981.
19. Ringrose PS, Peptides as antimicrobial agents. In: *Microorganisms and Nitrogen Sources* (ed Payne JW), pp. 641–692. John Wiley and Sons, Chichester, London, 1980.
20. Matthews DM and Payne JW, Peptides in the nutrition of microorganisms and peptides in relation to animal nutrition. In: *Peptide Transport in Protein Nutrition* (eds Matthews DM and Payne JW), pp. 1–60. North Holland and American Elsevier, Amsterdam, 1975.
21. Kenig M and Abraham EP, Antimicrobial activities and antagonists of bacilysin and anticapsin. *J. Gen. Microbiol.* **94**: 37–45, 1976.
22. Molloy BB, Lively DH, Gale RM, Gorman M, Boeck LD, Higgens CE, Kastner RE, Huckstep LL and Neuss N, New dipeptide antibiotic from *Streptomyces collinus*, lindenbein. *J. Antibiot.* **25**: 137–140, 1972.
23. Diddens H, Dorgerloh M and Zahner H, Metabolic products of microorganisms 176. On the transport of small peptide antibiotics in bacteria. *J. Antibiot.* **32**: 87–90, 1979.
24. Murakami T, Anzai H, Imai S, Safoh A, Nagaoka K and Thompson CJ, The bialaphos biosynthetic genes of Streptomyces hygroscopicus: molecular cloning and characterization of the gene cluster. *Mol. Gen Genet.* **205**: 42–50, 1986.
25. Staskawicz BJ and Panopoulos NJ, Phaseolotoxin transport in *Escherichia coli* and *Salmonella typhimurium* via the oligopeptide permease. *J. Bacteriol.* **142**: 474–479, 1980.

26. Muller H, Furter R, Zahner H and Rast DM, Metabolic products of microorganisms 203: Inhibition of chitosomal chitin synthetase and growth of *Mucor rouxii* by nikkomycin Z, nikkomycin X and polyoxin A. A comparison. *Arch. Microbiol.* **130**: 195–197, 1981.

27. Scannell JP and Pruess DL, Naturally occurring amino acid and oligopeptide antimetabolites. In: *Chemistry and Biochemistry of Amino Acids, Peptides and Proteins* (ed Weinstein B), pp. 189–243. Marcel Dekker, New York, 1974.

28. Ringrose PS, Warhead delivery and suicide substrates as concepts in antimicrobial drug design. In: *Scientific Basis of Antimicrobial Chemotherapy* (eds Greenwood D and O'Grady F), pp. 219–266. Cambridge Univ. Press, 1985.

29. Atherton FR, Hall MJ, Hassall CH, Lambert RW and Ringrose PS, Phosphonopeptides as antibacterial agents: rationale, chemistry and structure-activity relationships. *Antimicrob. Agents Chemother.* **15**: 677–683, 1979.

30. Atherton FR, Hall MJ, Hassall CH, Lambert RW, Lloyd WJ, Lord AV, Ringrose PS, and Westmacott D, Phosphonopeptides as substrates for peptide transport systems and peptidases of *Escherichia coli*. *Antimicrob. Agents Chemother.* **24**: 522–528, 1983.

31. Kingsbury WD, Boehm JC, Mehta RJ and Grappel SF, Transport of antimicrobial agents using peptide carrier systems: anticandidal activity of *m*-fluorophenylalanine-peptide conjugates. *J. Med. Chem.* **26**: 1725–1729, 1983.

32. Steinfeld AS, Naider F and Becker JM, Anticandidal activity of 5-fluorocytosine-peptide conjugates. *J. Med. Chem.* **22**: 1104–1109, 1979.

33. Andruskiewicz R. Chmara H, Milewski S and Borowski E, Synthesis and biological properties of N^3-(4-methoxyfumaroyl)-L-2,3-diaminopropanoic acid dipeptides, a novel group of antimicrobial agents. *J. Med. Chem.* **30**: 1715–1719, 1987.

34. Hammond SM, Claesson A, Jansson AM, Larsson L-G, Pring BG, Town CM and Ekström B, A new class of synthetic antibacterials acting on lipopolysaccharide biosynthesis. *Nature* **327**: 730–732, 1987.

35. Kingsbury WD, Boehm JC, Perry D and Gilvarg C, Portage of various compounds into bacteria by attachment to glycine residues in peptides. *Proc. Natl. Acad. Sci. USA* **81**: 4573–4576, 1984.

36. Boehm JC, Kingsbury WD, Perry D and Gilvarg C, The use of cysteinyl peptides to effect portage transport of sulfhydryl-containing compounds in *Escherichia coli*. *J. Biol. Chem.* **258**: 14850–14855, 1983.

37. Morley JS, Payne JW and Hennessey TD, Antibacterial activity and uptake into *Escherichia coli* of backbone-modified analogues of small peptides. *J. Gen. Microbiol.* **129**: 3701–3708, 1983.

38. Olson ER, Dunyak DS, Jurss LM and Poorman RA, Identification and characterization of dppA, an *Escherichia coli* gene encoding a periplasmic dipeptide transport protein. *J. Bacteriol.* **173**: 234–244, 1991.

39. Jamieson DJ and Higgins CF, Anaerobic and leucine-dependent expression of a peptide transport gene in *Salmonella typhimurium*. *J. Bacteriol.* **160**: 131–136, 1984.

40. Payne JW, Barrett-Bee KJ and Shallow DA, Peptide substrates rapidly modulate expression of dipeptide and oligopeptide permeases in *Candida albicans*. *FEMS Microbiol. Lett.* **79**, 15–20, 1991.

41. Milewski S, Andruskiewicz R and Borowski E, Substrate specificity of peptide permeases in *Candida albicans*. *FEMS Microbiol. Lett.* **50**: 73–78, 1988.

42. Payne JW and Bell G, The transport and utilization of peptides by *Escherichia coli*: interrelations between peptide uptake and amino acid exodus and biosynthesis. *FEMS Microbiol. Lett.* **2**: 259–262, 1977.

43. Payne JW and Bell G, Direct determination of the properties of peptide transport systems in *Escherichia coli*, using a fluorescent-labelling procedure. *J. Bacteriol.* **137**: 447–455, 1979.

44. Payne JW and Nisbet TM, Limitations to the use of radioactively labelled substrates for studying peptide transport in microorganisms. *FEBS Letts.* **119**: 73–76, 1980.

45. Payne JW and Nisbet TM, Continuous monitoring of substrate uptake by microorganisms using fluorescamine: application to peptide transport by *Saccharomyces cerevisiae* and *Streptococcus faecalis*. *J. Appl. Biochem.* **3**: 447–458, 1981.
46. Goodell EW and Higgins CF, Uptake of cell wall peptides by *Salmonella typhimurium* and *Escherichia coli*. *J. Bacteriol.* **169**: 3861–3865, 1987.
47. Higgins CF, Hyde SC, Mimmack MM, Gileadi U, Gill DR and Gallagher MP, Binding protein-dependent transport systems. *J. Bioenerg. Biomembr.* **22**: 571–592, 1990.
48. Ames GFL, Bacterial periplasmic transport systems: structure, mechanism and evolution. *Annu. Rev. Biochem.* **55**: 397–425, 1986.
49. Alves RA, Gleaves JT and Payne JW, The role of outer membrane proteins in peptide uptake by *Escherichia coli*. *FEMS Microbiol. Lett.* **27**: 333–338, 1985.
50. Richarme G and Kepes A, Study of binding protein ligand interactions by ammonium sulphate assisted adsorption on cellulose ester filters. *Biochim. Biophys. Acta* **742**: 16–22, 1983.
51. Tyreman DR, Peptide transport systems and antibiotic design. Ph.D. Thesis, University of Wales, Bangor, 1990.
52. Freidinger RM, Non-peptide ligands for peptide receptors. *Trends Pharmacol. Sci.* **10**: 270-274, 1989.
53. Smith MW and Payne JW, Simultaneous exploitation of different peptide permeases by combinations of synthetic peptide smugglins can lead to enhanced antibacterial activity. *FEMS Microbiol. Lett.* **70**: 311–316, 1990.
54. Ames GFL, Mimura CS and Shyamala V, Bacterial periplasmic permeases belong to a family of transport proteins operating from *Escherichia coli* to human: Traffic ATPases. *FEMS Microbiol. Rev.* **75**: 429–446, 1990.

Inhibitors of 2,3-Oxidosqualene-Lanosterol Cyclase as Antifungal Agents

S. Jolidon, A. Polak-Wyss, P. G. Hartman and P. Guerry

Pharma Division, Preclinical Research, F. Hoffmann-La Roche Ltd., CH-4002 Basel, Switzerland

Abstract. Most known antifungals act on enzymes involved in the biosynthesis of ergosterol, the main sterol in fungi. 2,3-Oxidosqualene-lanosterol cyclase is one of these enzymes; several inhibitors of the enzyme are known, which however have poor antifungal activity. A simple approach based on mechanistic considerations led to the synthesis of potent bifunctional inhibitors of 2,3-oxidosqualene-lanosterol cyclase. These amino-ketones represent the first cyclase inhibitors which also show an effective *in vitro* activity against a wide range of medically important fungi. A cell-free assay is presented, which allows accurate determination of IC_{50}-values.

1. Introduction

Ergosterol is an important constituent of fungal plasma cell membranes; most known antifungals act on enzymes involved in its biosynthesis (Fig. 1):
- allylamines and related compounds inhibit the expoxidation of squalene [1];
- azoles and triazoles inhibit demethylation of the sterol at position 14 [2];

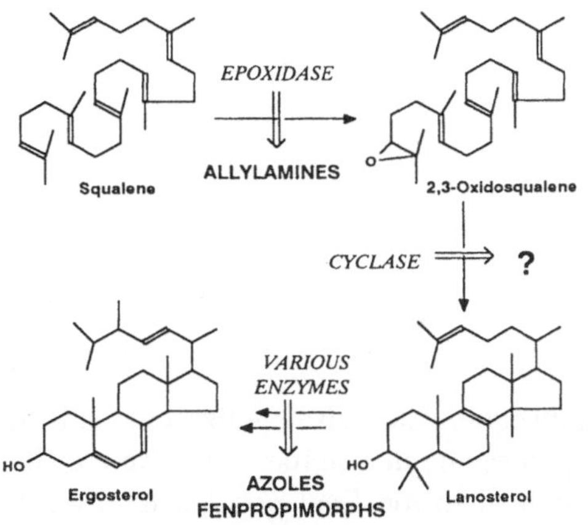

Fig. 1. Sterol biosynthesis pathway in fungi; key enzymes are shown together with some inhibitors showing antifungal activity

– fenpropidines and fenpropimorphs inhibit isomerisation of the delta8–delta7 double bond as well as reduction of the delta14-double bond in sterols [3].

2,3-Oxidosqualene-lanosterol cyclase [EC 5.4.99.7] is a key enzyme on the pathway to the biosynthesis of ergosterol, as it is responsible for the formation of lanosterol, the first sterol to appear in the biosynthetic cascade. It has been shown that this enzyme is essential for yeasts; mutants of *Saccharomyces cerevisiae* which are deficient in this enzyme can only survive in the presence of exogenous sterols [4]. This enzyme thus appears to be a reasonable target for the design of an antifungal agent with a new mode of action [5]. As this enzyme is microsomal and membrane bound, its isolation and purification is rather tedious. Partial purification of the hog liver enzyme has been carried out [6] and, more recently, the cyclase from rat liver has been purified to homogeneity [7]. The gene coding for the cyclase of the fungus *Candida albicans* has recently been cloned and expressed in *Saccharomyces cerevisiae* [8]. But as yet no three-dimensional structure of a 2,3-oxidosqualene-lanosterol cyclase, which would greatly facilitate the rational design of inhibitors, is available.

We therefore decided to design inhibitors of this enzyme based on simple mechanistic considerations. The postulated mechanism of action of this enzyme is shown in Fig. 2; the prefolded 2,3-oxidosqualene in its chair/boat/chair-conformations is embedded in a highly lipophilic pocket and is thought to be protonated at

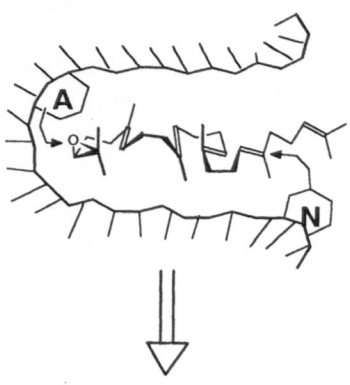

Lanosterol

Fig. 2. Postulated mechanism of action of 2,3-oxidosqualene-lanosterol cyclase assuming an acidic and a nucleophilic site in the enzyme cavity

the epoxide ring by an acidic residue **A**. Subsequent stepwise cyclization leads to a protosterol cation which is probably trapped as an intermediate by an active site nucleophile **N**. Subsequent rearrangements lead to the final product lanosterol. The interaction with the nucleophile however appears to be not essential for substrate recognition, as many analogs of 2,3-oxidosqualene, which lack the possibility of being trapped by the postulated group **N** (missing double bonds), still are cyclized by the target enzyme [9].

Many amines with lipophilic side chains are known to act as inhibitors of the cyclase. Azasqualene (Fig. 3), for instance, shows an IC_{50}-value of 4 μM against

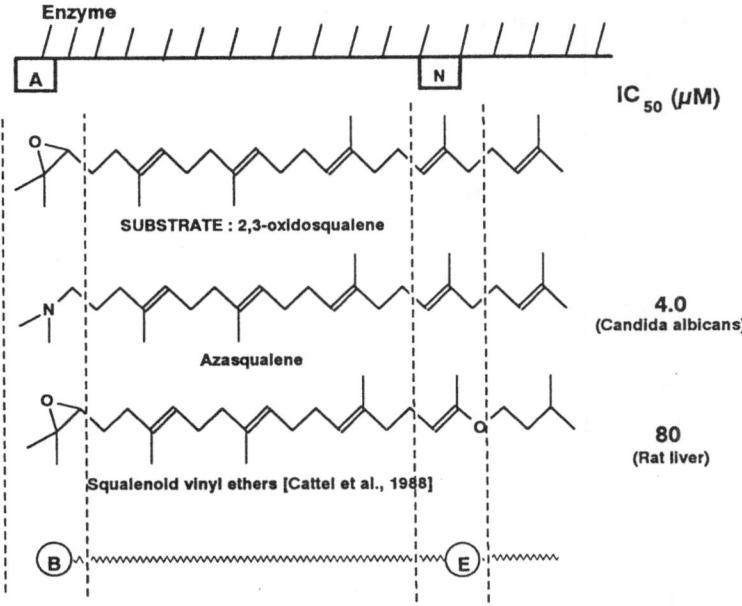

Fig. 3. Known inhibitors and their postulated mechanism of action. The substrate 2,3-oxidosqualene is shown together with the inhibitor azasqualene and a squalenoid vinyl ether inhibitor. Azasqualene is thought to interact with the acidic region **A** of the active site, whereas the vinyl ether is thought to interact mainly with the nucleophilic part **N** of the active site

the cyclase of *Candida albicans;* this compound is thought to block the enzyme as a result of the basic amino group strongly binding to the acidic group **A** of the active site, the lipophilic side chain mimicking the rest of the 2,3-oxidosqualene molecule. It is interesting to note that a less sophisticated molecule like *N,N*-dimethyldodecylamine still shows potent enzyme inhibition (IC_{50} = 11.7 μM against the cyclase of *Candida albicans*). This demonstrates that the enzyme can accomodate many lipophilic side chains. Unfortunately, azasqualene shows only marginal antifungal activity (Fig. 4), a fact which is so far not fully understood. It should be mentioned that Cattel et al. [10] have already synthesized squalenoid vinyl ethers able to interact with the nucleophilic region **N** of the cyclase; such compounds are thought to lead to an enzyme bound ketal-type intermediate.

	IC$_{50}$ (μM)	MIC (μg/ml)			
	Candida albicans	Candida albicans	Histoplasma capsulatum	Aspergillus fumigatus	Trichophyton mentagrophytes
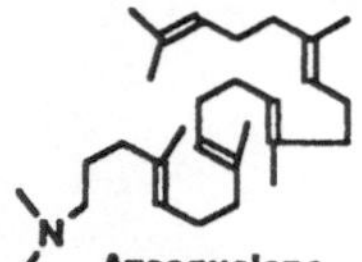 Azasqualene	4.0	>100	100	>100	>100
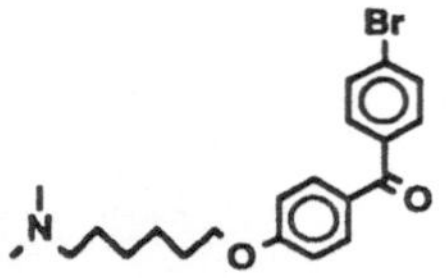 Ro 43-8212	1.1	100	10	100	1
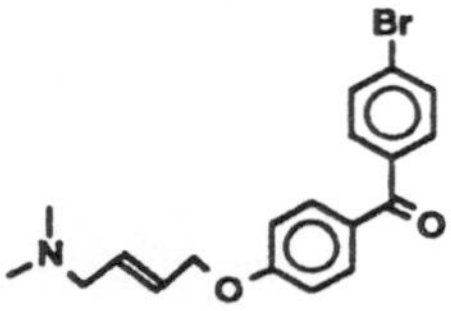 Ro 44-4281	0.32	100	10	10	1
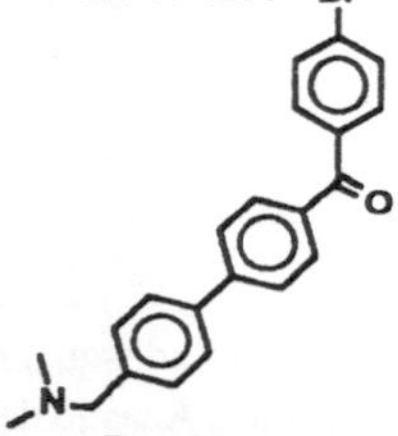 Ro 44-2103	0.11	10	10	1	1

Fig. 4. Enzyme inhibiting activity and antifungal spectrum of selected bifunctional cyclase inhibitors, compared to properties of a monofunctional inhibitor, azasqualene

However, these inhibitors lack a strongly basic group for interaction with the enzyme's acidic site **A** and the inhibition constant was moderate (IC$_{50}$ = 80 μM against rat liver cyclase).

We have tried to improve the properties of cyclase inhibitors by the design and synthesis of compounds able to interact both with the acidic **A** and the nucleophilic region **N** of the active site.

Molecular modelling studies showed that the distance between the oxygen atom of the protosterol intermediate and the position of the newly formed cation (C-20 of the sterol skeleton) is approximately 10.7 Å. We decided to keep an amine as the basic group and to replace the vinyl ether electrophile of Cattel's group by some physiologically and chemically more stable electrophile; a ketone was thought to be the appropriate functional group for this purpose. We therefore turned our attention to aminoketones of the type I (Fig. 5) and modified the molecule in order to reach a distance of nearly 10.7 Å between the nitrogen atom of the amine (interaction with the acidic site **A**) and the ketone (electrophile for interaction with the nucleophilic site **N**). Ketones of the benzophenone-type emerged as powerful functional groups for our purpose.

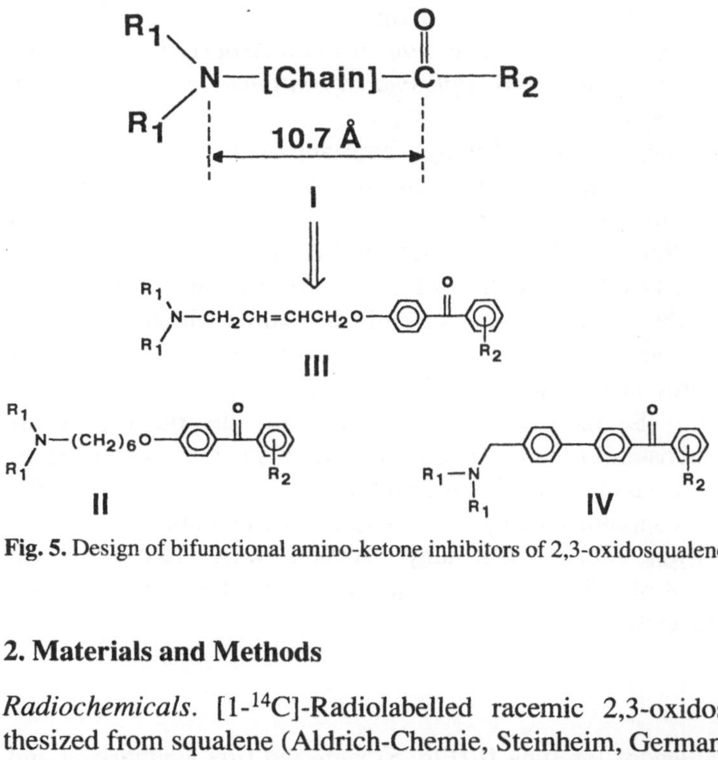

Fig. 5. Design of bifunctional amino-ketone inhibitors of 2,3-oxidosqualene-lanosterol cyclase

2. Materials and Methods

Radiochemicals. [1-^{14}C]-Radiolabelled racemic 2,3-oxidosqualene was synthesized from squalene (Aldrich-Chemie, Steinheim, Germany) in analogy to the method described by Nadeau and Hanzlik [11]. The specific activity was 456 MBq/mmol.

Compounds. Azasqualene was prepared from squalene [12]. The 4-[[6-(dimethylamino)hexyl]-oxy]-benzophenone of the type II can be prepared by a phase-transfer catalyzed alkylation of 4-hydroxybenzophenone with an excess of 1,6-dibromohexane (both Fluka AG, Buchs, Switzerland). Treatment of the intermediate 6-bromo-ether with an excess of dimethylamine in ethanol led to the desired compound II. The 4-[[4-(dimethylamino)-2-butenyl]-oxy]-benzophenone of the type III can be obtained in analogy to the aforementioned procedure by replacing 1,6-dibromohexane with 1,4-dibromo-2-butene (Fluka AG, Buchs,

Switzerland). The 4-[(dimethylamino)methyl]-4-biphenylyl phenyl ketone of the type IV can be prepared by a Friedel-Crafts acylation of 4-methylbiphenyl (Aldrich-Chemie, Steinheim, Germany) with benzoyl chloride using aluminium chloride in carbon disulfide or nitrobenzene. The compound is brominated with N-bromosuccinimide in carbon tetrachloride and the intermediate benzyl bromide is reacted with excess dimethylamine in ethanol to give the desired compound IV. Derivatives of the compounds II to IV were prepared in analogous manner [13]. All compounds were fully characterized and show correct spectroscopical data.

Antifungal activity. The antifungal activity of the cyclase inhibitors was obtained by determination of the minimum inhibitory concentrations (MIC-values) on Rowley-agar against strains of *Candida albicans* and *Aspergillus fumigatus* (after 2 days) and against strains of *Histoplasma capsulatum* and *Trichophyton mentagrophytes* (after 7 days).

Cell free assay. To measure the inhibition constant of the test compounds against the 2,3-oxidosqualene-lanosterol cyclase of *Candida albicans,* 1.0 g of fungal cells are treated with 1 mg Zymolase 100T (Seikagaku Kogyo, Japan) and 12.5 µl of β-mercaptoethanol in 5 ml digestion buffer for 30 min at 30°. The protoplasts obtained are lysed by addition of 2 ml 100 mM phosphate buffer. Subsequent centrifugation at 15000 g yields a cell free extract which retains full cyclase activity as shown by a 42% incoporation of racemic [1-^{14}C]-2,3-oxidosqualene[1] in the presence of the non ionic detergent Decyl Poe (*n*-decylpentaoxyethylene; Bachem, Switzerland). Interestingly, this detergent prevents further conversion of lanosterol to fungal sterols, thus allowing accurate measurements of IC_{50}-values if the assay is done in the presence of varying amounts of the test compound. The non-saponifiable lipids were extracted and applied to TLC-plates (silica gel, F_{254}, Merck, Germany), which were run twice in dichloromethane. The radiolabelled spots were quantified with an automatic TLC-scanner (Rita-3200, Raytest).

3. Results

Benzophenone derivatives of the type II (Fig. 5) were the first synthesized and despite being rather flexible show impressive inhibition constants for the target enzyme and furthermore strongly inhibit fungal growth (Fig. 4).[2] This result indicated that our approach was sound. In order to decrease the rotational freedom of the inhibitor molecules, we synthesized compounds containing double bonds (type III in Fig. 5) or even more rigid inhibitors containing the biphenyl moiety (type IV in Fig. 5); the distance between the N-atom and the carbonyl C-atom of compound IV was shown to be very close to the critical distance of 10.7 Å (see

---

[1] Only the (3S)-enantiomer of the racemic 2,3-oxidosqualene is cyclized by the enzyme [14].

[2] Compounds of the type II, where the oxygen atom of the ether has been replaced by another –CH$_2$– group, have similar properties.

Fig. 6). As expected, an increase in the rigidity of the inhibitors is accompanied by a substantial decrease in the IC_{50}-value; rigid biphenyl compounds of the type IV usually show a ten-fold better IC_{50}-value as compared to the more flexible compounds of the type II. The potent enzyme inhibitor Ro-44-2103 also shows an interesting spectrum of activity against several pathogenic and opportunistic fungi (Fig. 4).

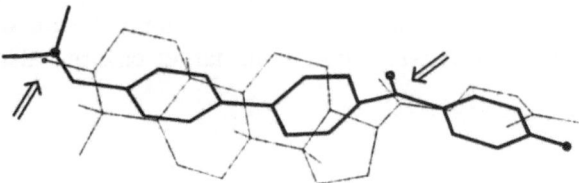

Fig. 6. Molecular modelling graphic showing the matching of the rigid inhibitor Ro-44-2103 (heavy line) to the protosterol cation (light line). The distance between the nitrogen atom and the carbonyl-C atom of the inhibitor is very close to the distance expected in the protosterol cation between the oxygen atom and the position (C-20 on the steroid skeleton) where the carbocation is formed. (Picture from the MOLOC-program, Hoffmann-La Roche Inc.)

Structure/activity studies are in general agreement with the postulated mechanism of action of these compounds; i.e. large substituents R_1 in compounds II to IV are deterimental to the activity against the target enzyme (Fig. 7). In fact,

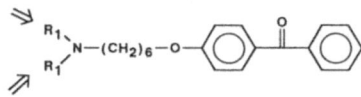

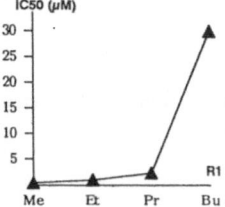 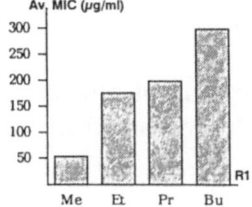

Fig. 7. Structure/activity relationship showing the influence of the size of the amino-substituent R_1 on enzyme inhibition (IC_{50}-value) and on the antifungal activity (average MIC-values against 4 fungal strains: *Candida albicans, Aspergillus fumigatus, Histoplasma capsulatum* and *Trichophyton mentagrophytes*). A good correlation is found between data sets

according to Fig. 2, large substituents at the amino-group (interaction with the acidic site **A** of the enzyme) will probably interfere with the enzyme cavity. Interestingly, the *in vitro* antifungal activity of these compounds reflects exactly the activity against the 2,3-oxidosqualene-lanosterol cyclase (Fig. 7).

Figure 8 shows the influence of the chain length in compounds of the type II. Best enzyme inhibiting activity was found for compounds with a chain length between 5 and 7 carbon atoms ($n = 5$ to 7). The *in vitro* antifungal activity, however, does not fully correlate with the IC_{50}-values; the compound with $n = 5$, for example, shows a lower antifungal activity than expected from its IC_{50}-value. The reason for this discrepancy is so far not clearly understood. It should be noted, though, that only compounds which effectively inhibit the target enzyme also show antifungal activity.

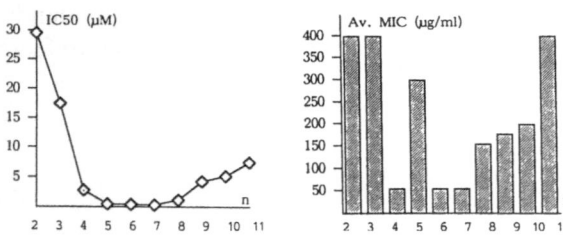

Fig. 8. Structure/activity relationship showing the influence of the chain length on enzyme inhibition (IC_{50}-values) and the antifungal activity (average MIC-values against four fungal strains: *Candida albicans, Aspergillus fumigatus, Histoplasma capsulatum* and *Trichophyton mentagrophytes*). The correlation between data sets is not perfect, but only inhibitors with low IC_{50}-values show antifungal activity

Figure 9 shows the broad spectrum of activity of selected cyclase inhibitors; these MIC-values compare favourably with many known antifungals.

4. Discussion

This study of structurally new inhibitors of 2,3-oxidosqualene-lanosterol cyclase shows that it is possible to design and synthesize potent inhibitors based on simple mechanistic considerations. The structure/activity pattern observed within these

Structure	$IC_{50}(\mu M)$ C.albicans	MIC C.albicans	MIC C.species	MIC Cryptococc.	MIC Dermatoph.	MIC Dimorph.
RO-43-8212 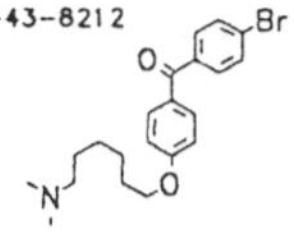	1.09	5.00	1.25	6.48	1.25	0.60
RO-44-2103 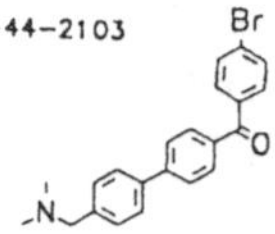	0.11	1.25	2.00	3.25	0.60	1.25
RO-43-6913 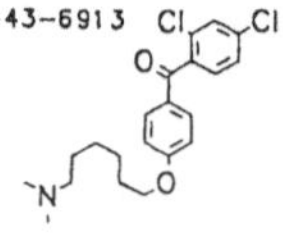	0.66	0.60	0.11	0.35	0.30	0.30

Fig. 9. The IC_{50}-values of selected cyclase inhibitors shown together with their antifungal activity. The MIC-values shown are average values (in µg/ml) measured against several strains of the fungal species (16 strains of *Candida albicans*; 10 strains of *Candida* species; 7 strains of *Cryptococcus* species; 6 strains of dermatophytes and 6 strains of dimorphic fungi)

series of amino-ketones, even if not completely understood, roughly parallels the activity against the target enzyme.

It can thus be reasonably assumed that the observed antifungal activity is the result of the inhibition of 2,3-oxidosqualene-lanosterol cyclase, but it remains unclear whether the antifungal activity of these compounds is due to the accumulation of 2,3-oxidosqualene or to a depletion in ergosterol in the fungal cells. These amino-ketones thus represent the first cyclase inhibitors which show an effective *in vitro* activity against a wide range of medically important pathogenic and opportunistic fungi. Even though many aspects of the inhibition of this enzyme remain to be clarified, this approach could lead to interesting antifungals with a new mode of action.

References

1. Ryder NS, Mode of action of allylamines. In: *Sterol Biosynthesis Inhibitors,* eds Berg D and Plempel M, VCH Weinheim, 151–167, 1988.
2. Vanden Bossche H, Mode of action of pyridine, pyrimidine and azole antifungals. In: *Sterol Biosynthesis Inhibitors,* eds Berg D and Plemple M, VCH Weinheim, 79–119, 1988.
3. Mercer AI, The mode of action of morpholines. In: *Sterol Biosynthesis Inhibitors,* eds Berg D and Plempel M, VCH Weinheim, 120–150, 1988.

4. Karst F and Lacroute F, Ergosterol biosynthesis in *Saccharomyces cerevisiae*. *Molec. Gen. Genet.* **154**: 269–277, 1977.

5. Cattel L, Ceruti M, Viola F, Delprino L, Balliano G, Duriatti A and Bouvier-Navé P, The squalene-2,3-epoxide cyclase as a model for the development of new drugs. *Lipids* **21**: 31–38, 1986.

6. Duriatti A and Schuber F, Partial purification of 2,3-oxidosqualene-lanosterol cyclase from hog liver. Evidence for a functional thiol residue. *Biochem. Biophys. Res. Commun.* **151**: 1378–1385, 1988.

7. Kusano M, Abe I, Sankawa U and Ebizuka Y, Purification and some properties of squalene-2,3-epoxide-lanosterol cyclase from rat liver. *Chem. Pharm. Bull.* **39**: 239–241, 1991.

8. Kelly R, Miller SM, Lai MH and Kirsch DR, Cloning and characterization of the 2,3-oxidosqualene cyclase-coding gene of *Candida albicans*. *Gene* **87**: 177–183, 1990.

9. van Tamelen EE and Freed JH, Biochemical conversion of partially cyclized squalene-2,3-oxide types to the lanosterol system. Views on the normal enzymic cyclization process. *J. Amer. Chem. Soc.* **92**: 7206–7207, 1970.

10. Ceruti M, Viola F, Dosio F, Cattel L, Bouvier-Navé P and Ugliengo P, Stereospecific synthesis of squalenoid epoxide vinyl ethers as inhibitors of 2,3-oxidosqualene cyclase. *J. Chem. Soc. Perkin Trans.* **I**: 461–469, 1988.

11. Nadeau RG and Hanzlik RP, Synthesis of labeled squalene and squalene 2,3-oxide. *Methods Enzym.* **15**: 346–351, 1969.

12. Ceruti M, Balliano G, Viola F, Cattel L, Gerst N and Schuber F, Synthesis and biological activity of azasqualenes, bis-azasqualenes and derivatives. *Eur. J. Med. Chem.* **22**: 199–208, 1987.

13. F. Hoffmann-La Roche Ltd., Europ. Pat. Appl., Publ. No A-0401798 and A-0410359; Europ. Pat. Appl. No 911110016.2.

14. van Tamelen EE, Bioorganic characterization and mechanism of the 2,3-oxidosqualene-lanosterol conversion. *J. Amer. Chem. Soc.* **104**: 6480–6481, 1982.

The Use of the 3-Dimensional Structures of Rhinoviruses in the Design of Antiviral Agents

G. D. Diana, A. M. Treasurywala, T. R. Bailey and R. C. Oglesby

Department of Medicinal Chemistry, Sterling Research Group, Rensselaer, NY 12144, USA

Abstract. The X-ray crystal structure of several compounds related to disoxaril, bound to human rhinovirus-14, has provided the opportunity to develop a model which describes the nature of the binding of these compounds and has assisted in the design of more potent compounds. The structure of HRV-14 has been compared with other serotypes, which have been sequenced suggesting that certain changes of specific residues affect the size of the compound binding site. These changes, examined in an effort to determine their effect on the interaction of the antiviral compound within the binding site, suggest the importance of occupying maximum space and consequently enhancing hydrophobic interactions, despite the opportunity for hydrogen bonding existing due to the presence of tyrosine and asparagine residues within the pocket. The stacking phenomena of the aromatic ring of several analogues with Tyr[128] and Tyr[152] in rhinovirus-14 was examined with respect to both electrostatic and π stacking. Electronics were found to play no part in the interactions. These results were confirmed using the program CoMFA which suggested that steric effects were the most important factors relating to activity.

1. Introduction

Over the past several years, a series of compounds encompassed in Structure I have been shown to demonstrate potent activity against a wide spectrum of rhino- and enteroviruses [1–5]. These compounds inhibit viral replication by preventing uncoating in the case of rhinovirus-2 and polio 2 [6, 7], and by blocking adsorption

of the virus to the cell in the case of rhinovirus-14 [8]. Although the mode of action has been clearly established, the mechanism by which the drugs bind to the viral capsid was unknown.

Recently, the 3-dimensional structure of HRV-14 was elucidated [9] and, subsequently, several oxazolinylphenyl isoxazoles I, bound to the capsid protein of HRV-14 were examined by X-ray crystallography [10, 11]. The results of these studies have allowed us to examine interactions of these compounds with the virus on a molecular level and also to develop some hypotheses concerning the mode of binding of these compounds to the viral capsid.

The compound-binding site resides on the surface of the viral capsid protein VP 1 below a depression, referred to as a canyon, and which is considered to be the putative cell receptor binding site. The site consists of a hydrophobic pocket ≈25Å long and 8 to 10Å wide, which undergoes a conformational change of

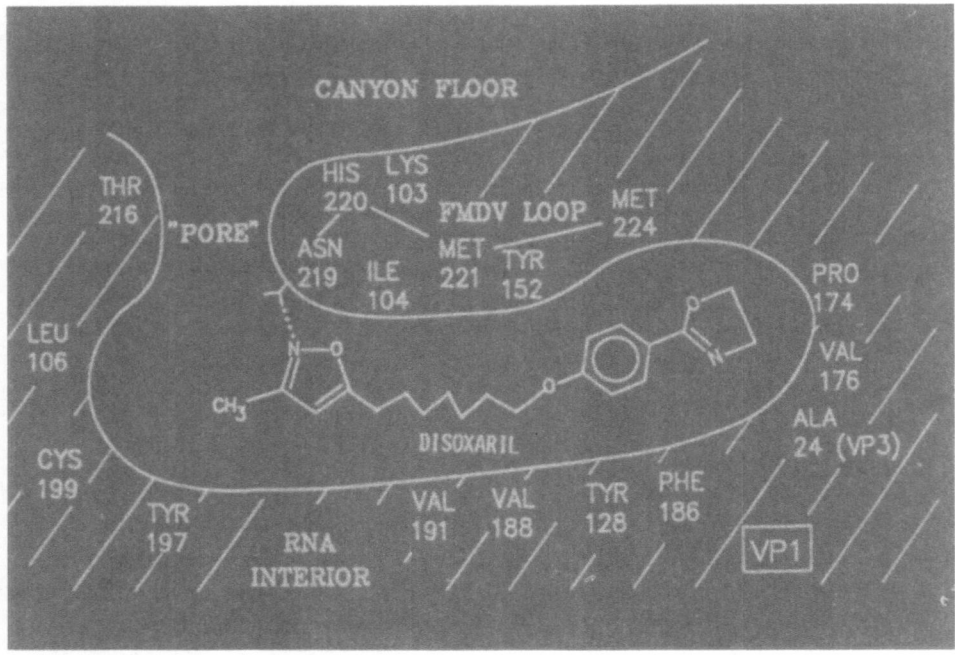

Fig. 1. A model of the hydrophobic pocket binding-site of HRV-14 with disoxaril included

between 3–5Å following insertion of the compound (Fig. 1). It has been shown that this conformational change extends to the canyon or cell receptor binding site in HRV-14, which may account for the inhibition of adsorption of the virus to the cell by these compounds [8]. With the X-ray coordinates in hand, a systematic evaluation of the molecular interaction of portions of the molecule with the receptor was performed and the effect of these interactions on activity examined.

2. Enantiomeric Effects

One of the initial compounds examined by X-ray crystallography was Compound II which inserts into the pocket of VP 1 with the isoxazole end residing in the

II

interior of a β-barrel. The methyloxazolinyl moiety is located below the canyon floor with the nitrogen of the oxazoline ring within hydrogen bonding distance of Asn 219. Compound II has an asymmetric center at the 4-position of the oxazoline ring; it was subsequently determined that the S isomer was 10 times more active than the R. This result was consistent with the fact that preferential binding of the S isomer occurred at the binding site in VP 1 [11].

Table 1. Comparative evaluation of enantiomers against HRV-14

	MIC (μM)[a]	
X	S*	R*
CH_3	0.056	0.56
C_2H_5	0.03	0.16
n-C_3H_7	0.03	0.18
i-C_3H_7	0.08	1.57
n-C_4H_9	0.15	1.31

[a] Minimum inhibitory concentration obtained from a plaque reduction assay. * Absolute configuration.

Additional homologs (Table 1) were prepared and tested against HRV-14 in order to determine the extent of this enantiomeric effect, and to examine the interaction of the extended chain with residues within the pocket. In every case, the S isomer was more active than the R. Optimum activity was achieved with the ethyl and propyl homologs.

3. Molecular Graphics

The interactions of the R and S conformers in the binding site were analyzed by performing an energy profiling study using the X-ray structure of II in the HRV-14 binding site as a starting point, and examining the interaction of the compounds in Table 1 with residues within 8Å. The purpose of this study was to determine the location of energy minima as the oxazolinyl ring was rotated through 360°. All calculations were performed on a VAX 11/785. Hydrogen atoms were removed and charges were set on the atoms in the pocket and on the compounds.

The coordinates of the S conformer of II were displayed on an Evans and Sutherland graphics device using the program Chem-X. The difference in activity of the S and R conformers could be rationalized by the observed hydrophobic interaction of the S-methyl group with a pocket formed by Leu 106 and Ser 107,

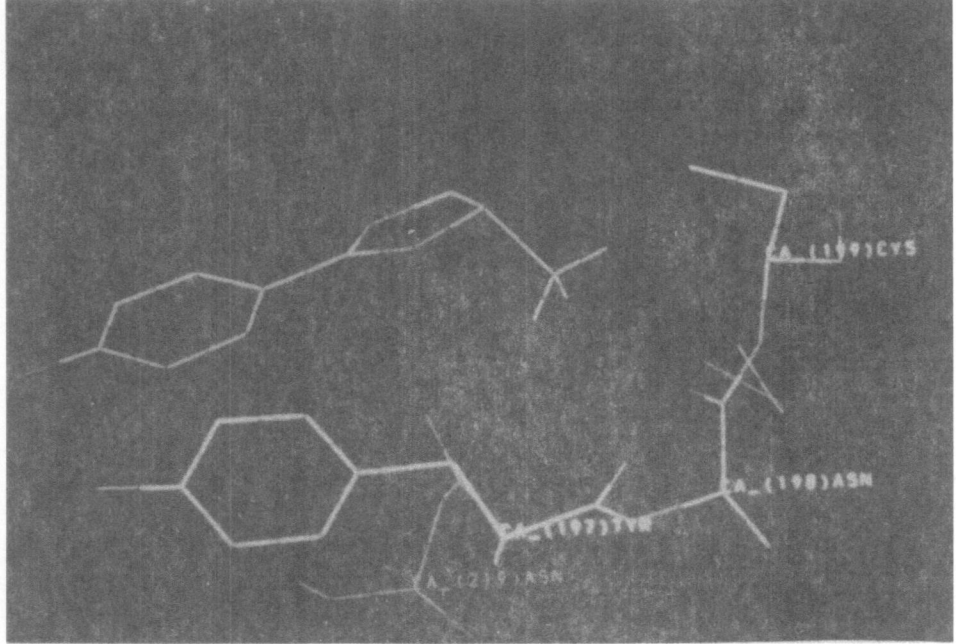

Fig. 2. The interaction of the (S)-methyl group of II with a hydrophobic pocket formed by Leu 106 and Ser 107

(Fig. 2). The R isomer, on the other hand, is pointing away from this pocket. Assuming the same torsion angle between the phenyl and oxazoline rings for the R isomer as was determined for the S isomer by X-ray crystallography, the former would interact unfavorably with the carbonyl group of Asn 198 (Fig. 3). The results of this study suggested that a rigidly confined conformation with a twist angle of 10–30° is conducive to high levels of activity. This was consistent with the results of the X-ray data for Structure II which revealed a torsion angle of 10–15° between these rings.

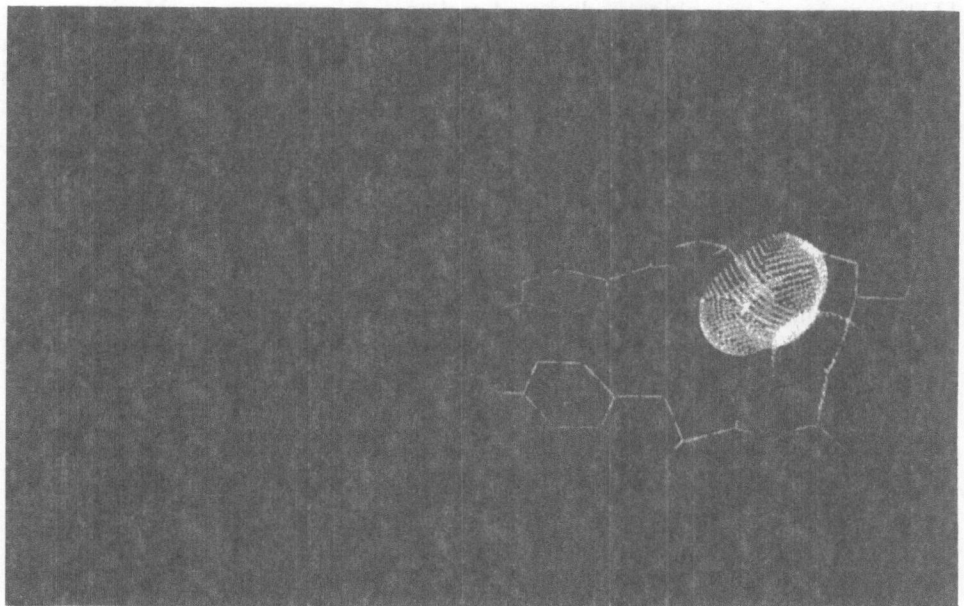

Fig. 3. The interaction of the (R)-methyl group of II with the carbonyl of Asn 198

In performing the energy profiling studies, hydrophobic interactions were taken into account and considered as the major contributioning factors to the binding energy.

4. Structure-Activity Studies

X-ray crystallography studies have been performed on several compounds in the isoxazole series with 5 and 7 carbon chains connecting both ends of the molecule

Table 2. *In vitro* evaluation against HRV-14

	n =	4	5	6	7	8
X =	H	NA	0.73	2.9	0.41	3.92
	Cl	9.2	2.41	3.86	1.06	14.32

MIC (μM)

[11]. A series of homologs of disoxaril and analogs containing a chlorine on the phenyl ring (Table 2) were evaluated against HRV-14 in an effort to establish a structure-activity relationship between chain length and activity. In both series, a chain length of 7 carbon atoms was required for optimum activity. The significance of these findings will become more obvious further on in the discussion.

5. A Model for HRV-14 Activity

The elucidation of the 3-dimensional structure of HRV-14, the identification of the compound-binding site on the capsid protein, and the X-ray structures of several compounds bound to HRV-14 have allowed for the development of a model representing the requirements for activity for this series of compounds. The model

Table 3. Compounds active against HRV-14[*]

Compd. #		MIC (µM)
1[a]		0.05
2[a]		0.16
3		0.06
4[b]		2.41
5[a]		0.51
6		0.03
7		0.14

[*] The X-ray conformations used in this study were provided by Michael Rossmann et al. (See Refs. 10 and 11). [a] The synthesis and evaluation of these compounds is described in Ref. 1. [b] See Ref. 4.

was developed by examining 7 active compounds (Table 3) and 7 inactive compounds (Table 4). In the former series, the conformation of compounds 1, 3, 4, and 5 bound to HRV-14 was determined by X-ray crystallography. Compounds 2 and 6 were constructed using the program SYBYL from the X-ray conformation of 1, and 7 was constructed from compound 4. The assumption was made, based on the X-ray crystallographic results on a variety of compounds in this series [11], that compounds 1, 2 and 6 would assume the same orientation while 5 and 7 would have the opposite orientation when bound to HRV-14. Since the determination of

Table. 4. Compounds inactive against HRV-14*

Compd. #

8	
9	
10	
11	
12	
13	
14	

*The conformations of these compounds which were used in this study were modeled after compound 4 from Table 3.

the coordinates for the inactive compounds by X-ray methods was not possible, compound 4 (Table 3) was used as a template for all of the inactive compounds in Table 4. The composite of active and inactive structures generated was then compared using volume maps comprised of van der Waals surfaces. The volume maps were examined and differences between the active and inactive structures noted. The results of this comparison revealed two major differences:

1. Inactive compounds display excess bulk around the phenyl ring. Although some bulk in this area appears desirable, exceeding certain limitations results in inactive compounds, possibly due to spatial constraints within the compound binding site.

2. The active compounds occupy space in the binding site beyond that of the inactive compounds. The most active compounds (1 and 3) extend well into the pore area of the binding site, suggesting a space filling requirement in this area. Compounds 1 and 3 are in the opposite orientation to each other such that the methyl group on the oxazoline ring of compound 1 extends into the pore, while the hydroxylated side chain of 3 is in the same area (Fig. 1).

The conclusions drawn from this model substantiate the results of the SAR studies on the chain length previously described [13], where a chain length of 7 carbon atoms appears to satisfy the space-filling requirements of the model. In addition, it has been shown by regression analysis that excessive bulk on the phenyl ring is detrimental to activity [5], which also confirms the results of this study.

The importance of bulk with respect to activity was supported by the use of a QSAR method called CoMFA (Comparative Molecular Field Analysis, Tripos Associates). The concept embodied in this program is that differences in an objective property are often related to differences in the shapes of the non-covalent field surrounding the molecules in question. Steric and electrostatic fields are taken into consideration.

Structures obtained from X-ray analysis of eight compounds bound to HRV-14 and whose activity against HRV-14 was determined were used for this analysis. The structures were extracted into a SYBYL database and charges were calculated for them using the AM1 Hamiltonian in a single point calculation. The CoMFA analysis consisted of placing these molecules collectively into an imaginary box and then dividing the box into grid points which were evenly spaced. The molecules were then placed individually into the box and values of a given property (or set of properties) contributed by the molecule were evaluated at each grid point. From this analysis a table was generated (Table 5), consisting of values obtained at each grid point (designated by $*$), in addition to measured parameters (MIC, $\log p$) and calculated values (CMR and dipole moments). The data in the table were suited for a classical QSAR analysis and the program sought a statistically significant correlation between the value of some property somewhere in space around a set of molecules and a dependent variable, in this case, biological activity. A Partial Least Squares (PLS) analysis was performed on all entries in the

Table 5. Parameters used in the CoMFA analysis

No.	MIC	log (1/MIC)	log p	CMR[a]	XDIP[b]	YDIP[b]	ZDIP[b]	DIP[b]	CoMFA[c]
1	2.41	−0.30	4.32	9.340	0.58	−2.78	−5.13	5.86	*
2	0.05	1.25	5.49	9.313	0.38	3.17	2.43	4.01	*
3	0.16	0.79	6.20	9.770	0.28	3.22	2.48	4.08	*
4	0.06	1.20	3.07	12.146	4.80	2.04	−2.70	5.88	*
5	0.51	0.14	3.90	8.849	0.10	−1.36	−3.82	4.05	*
6	0.03	1.47	6.02	10.268	1.15	0.98	1.61	2.20	*
7	0.14	0.85	6.11	10.175	0.18	−0.38	−3.26	3.29	*

[a] Molar refractivity; [b] Dipole moments generated through AM1 calculations; [c]CoMFA generated data using a hydrogen ion as a probe. Partial least squares was used for analysis of data, employing five cross validation groups, three components and 100 iterations.

table with cross validation. The significant conclusions drawn from this analysis were the following:

1. There was no statistically significant correlation between the attraction for a unit positive or negative charge (electrostatic parameter) around the molecules and biological activity.

2. There was a statistically significant correlation between the presence of bulk in some regions (particularly in the pore region) and the absence of bulk in other regions, and the antiviral activity of the compounds. These results are illustrated graphically by a plot of the so-called standard deviation residuals in a contour map around the molecules. These results are consistent with the data obtained from direct inspection of the virus pocket and from other computational experiments.

In order to further substantiate the conclusions drawn from the volume map study as well as from the results of the CoMFA analysis concerning the importance of the location of bulk on activity, three compounds shown in Table 6 were synthesized and tested against HRV-14. The prediction would be that the addition of bulk at the terminus of the isoxazole ring should result in an increase in antiviral activity. The results in Table 6 confirm these expectations where there is an

Table 6. Effect of chain length on activity against HRV-14

R	MIC (μM)
CH_3	1.27
C_2H_5	0.43
C_3H_7	0.16

increase in activity as the length of the hydrocarbon chain is increased from methyl to propyl. It is interesting to note that the propyl homolog has the same molecular weight as the seven carbon homologs shown in Table 2 but with enhanced activity.

The results of these studies suggest that the use of the three-dimensional structure of rhinoviruses may be a useful tool for the design of more potent antirhinovirus agents. As the structure of other serotypes becomes available, comparable studies will be performed.

References

1. Diana GD, McKinlay MA, Otto MJ, Akullian V and Oglesby RC, [[(4,5-Dihydro-3-oxazolyl)phenoxy]alkyl]isoxazole. Isoxazoles with antipicornavirus activity. *J. Med. Chem.* **28**: 1906–1910, 1985.
2. Otto MJ, Fox MP, Fancher MJ, Kuhrt MF, Diana GD and McKinlay MA, *In vitro* activity of win 51,711, a new broad spectrum antipicornavirus drug. *Antimicrob. Ag. Chemother.* **27**: 883–886, 1985.
3. McKinlay MA, Frank JA, Benziger DP and Steinberg BA, Oral efficacy of win 51711 in mice infected with human poliovirus. *J. Infect. Dis.* **154**: 676–679, 1986.
4. Diana GD, Oglesby RC, Akullian V, Carabateas PM, Cutcliffe D, Mallamo JP, Otto MJ, McKinlay MA, Maliski EG and Michalec SJ, Structure-activity studies of 5-[[4-(4,5-dihydro-2-oxazolyl)phenoxy]alkyl]-3-methylisoxazole: Inhibitors of picornavirus uncoating. *J. Med. Chem.* **30**: 383–386, 1987.
5. Diana GD, Cutcliffe D, Oglesby RC, Otto MJ, Mallamo JP, Akullian V and McKinlay MA, Synthesis and structure-activity studies of some disubstituted phenylisoxazoles against human picornaviruses. *J. Med. Chem.* **32**: 340, 1989.
6. Fox MP, Otto MJ, Shave WJ and McKinlay MA, Prevention of rhinovirus and poliovirus uncoating by win 51,711, a new antiviral drug. *Antimicrob. Ag. Chemother.* **30**: 110–116, 1986.
7. Zeichardt H, Otto MJ, McKinlay MA, Willingmann P and Habermehl KO, Inhibition of poliovirus uncoating by disoxaril. *Virology* **160**: 281–285, 1987.
8. Pevear DC, Fancher MJ, Felock PJ, Rossmann MA and Dutko FJ, Conformational change in the floor of the human rhinovirus canyon blocks adsorption to HeLa cell receptors. *J. Virol.* **23**: 2002–2007, 1989.
9. Rossmann MJ, Arnold E, Erickson JW, Frankenberger EA, Griffith JP, Hecht HJ, Johnson JE, Vriend G, Structure of a human common cold virus and functional relationship to other picornaviruses. *Nature* **317**: 145–153, 1985.
10. Smith TJ, Kremer MJ, Luo M, Vriend G, Arnold E, Kamer G, Rossmann MJ, McKinlay MA, Diana GD, and Otto MJ, The site of attachment in human rhinovirus 14 for antiviral agents that inhibit uncoating. *Science* **233**: 1286–1291, 1986.
11. Badger J, Minor I, Kremer MJ, Oliveira MA, Smith TJ, Griffith JP, Guerin DMA, Krishnaswamy S, Luo M, Rossmann MG, McKinlay MA, Diana GD, Dutko FJ, Fancher M, Reuckert RR and Heinz BA, Structural analysis of a series of antiviral agents complexed with human rhinovirus-14. *Proc. Natl. Acad. Sci.* **85**: 3304–3308, 1988.
12. Diana GD, Otto MJ, Treasurywala AM, McKinlay MA, Oglesby RC, Maliski EG, Rossmann MG and Smith TJ, Enantiomeric effect of homologues of disoxaril on the inhibitory activity of human rhinovirus-14. *J. Med. Chem.* **31**: 540–544, 1988.
13. Diana GD, McKinlay MA, Brisson CJ, Zalay ES, Miralles JV and Salavador UJ, Isoxazoles with antipicornavirus activity. *J. Med. Chem.* **28**: 748–752, 1985.

Design of Virus-Specific Inhibitors of Terminal Glycosylation Enhancing the Antigenicity of Viral Glycoproteins

R. Datema* and S. Olofsson†

*Department of Antiretroviral Therapy, Sandoz Research Institute Vienna, Brunner Strasse 59, A-1235 Vienna, Austria
†Department of Clinical Virology, University of Göteborg, Guldhedsgatan 10B, S-41346 Göteborg, Sweden

Abstract. The studies described here show three findings essential for the design of inhibitors of terminal glycosylation selective for virus-infected cells: (1) terminal glycosyl residues can modulate the antigenicity of viral glycoproteins, for example by masking potential neutralizing epitopes; (2) inhibition of translocation of sugar nucleotides into the Golgi-lumen can lead to interference with terminal glycosylation (branching; galactose addition, sialic acid addition) giving rise to increased antigenicity of viral glycoproteins; (3) terminal glycosylation inhibitors can be generated in virus-infected cells by virus-coded enzymes. Design of inhibitors based on these findings may complement antiviral therapy and increase our understanding of the role of terminal glycosylation of viral glycoproteins in the intact host.

1. Terminal Glycosylation of Viral Glycoproteins

Viral proteins are N-glycosylated initially by the transfer *en bloc* of a glucosylated high-mannose oligosaccharide ($Glc_3Man_9GlcNAc_2$) from the lipid dolichol-diphosphate to a nascent protein. Immediately following transfer, the N-linked oligosaccharides are processed by glycosidases. The extent of processing is dependent on a variety of conditions, which may include the location of a particular oligosaccharide on the protein. The trimming of sugar residues and the subsequent addition of peripheral sugars, to result in so-called hybrid-type and complex-type oligosaccharides, occurs by concerted action of glycosidases and glycosyltransferases, and can result in a plethora of oligosaccharides [1], see Fig.1.

Some viral glycoproteins contain O-linked oligosaccharides [2]. The O-linked oligosaccharides are assembled by stepwise addition of sugar residues onto a protein in transit through the Golgi apparatus. In fact, the synthesis of O-linked oligosaccharides occurs simultaneously with, and is mechanistically identical to the addition of peripheral sugars to N-linked oligosaccharides [3]. These processes are referred to here as "terminal glycosylation", and concern the addition of GlcNAc, Gal, Fuc and NeuAc to the glycoprotein (see Fig.1).

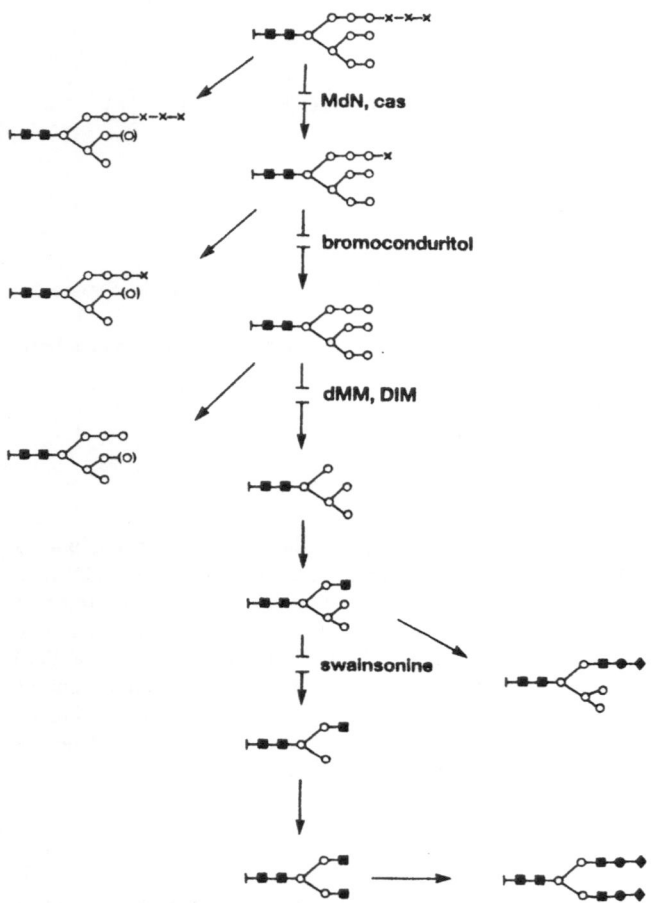

Fig.1. Pathways leading to formation of a biantennary complex-type oligosaccharide, and blocks in the pathway caused by inhibitors. Abbreviations: MdN = *N*-methyl-1-deoxynojirimycin; cas = castanosperimine; dMM = 1,5-dideoxy-1,5-imino-D-mannitol (manno-1-deoxynojirimycin); DIM = 1,4-dideoxy-1,4-imino-D-mannitol; ■ = GlcNAc; ○ = Man; × = GLc; ● = Gal; ♦ = NeuAc

2. Why Inhibitors of Terminal Glycosylation?

Compounds interfering with protein glycosylation have been used videly to study the biological role of oligosaccharides of viral glycoproteins [4]. To prevent glycosylation, cultures of infected cells can be treated with tunicamycin or certain deoxy- or fluorosugars [5, 6]. The compounds prevent the assembly of the dolichol diphosphate-linked precursor of *N*-linked oligosaccharides. To allow glycosylation, but to prevent maturation of the protein-linked oligosaccharides, cultures of infected cells can be treated with castanospermine or N-methyl-1-deoxynojirimycin [2, 7, 8, 9]. These compounds inhibit trimming glycosidases converting the glucosylated high-mannose oligosaccharides to the mature forms of N-linked oligosaccharides.

In different viral systems the biological effects of these inhibitors have been diverse. Interference with glycosylation can affect the intracellular transport, proteolytic stability, proteolytic processing, disulfide bond formation in viral glycoproteins, and in addition, or as a consequence, virion assembly, the site of virus budding, etc., as reviewed for example in Ref. [2]. Nevertheless, the maturation of enveloped viruses in cultures of infected cells was not affected in any viral system by blocking the processing of the N-linked oligosaccharides at the level of mannosidase II, or beyond, that is at the level where the terminal glycosyl residues GlcNac, Gal, Fuc and NeuAc are added to the growing oligosaccharide side-chains. Yet, many oligosaccharides of viral glycoproteins are processed past that level. What, then, is the biological role of terminal glycosylation of viral glycoproteins?

3. Role of Terminal Glycosylation in the Intact Organism

In some complex, non-viral systems interesting biological roles of terminal glycosylation have been observed. For example, terminal glycosylation of T-cell glycoproteins is important for the ability of the cell to react to antigen proliferatively [10]. Indeed, stimulatory effects of the mannosidase II inhibitor swainsonine on the T-cell component of the immune response were observed in treated animals [11]. These effects may, in fact, be responsible for reducing tumor growth and metastasis in swainsonine-treated mice. However, this therapeutic effect of systemically administered swainsonine could also be caused by the drug-induced changes in the terminal glycosylation of cell-surface glycoproteins of the tumor cell: decreased terminal glycosylation (branching, sialylation, or both) is associated with a decrease in metastasis of tumor cells [12].

The above example serves to indicate that assays for the biological function of terminal glycosylation should be extended to conditions *in the intact host*. In other words, terminal glycosylation of viral glycoproteins may play a role at the level of virus-host interaction (see Ref. [2] for a discussion). It is, for example, well appreciated that virions devoid of sialic acid activate the alternate pathway of complement. This mechanism also enhances the infectivity of HIV produced in the presence of an oligosaccharide processing-inhibitor [13]. Further, peripheral sugars modulate the antigenicity of glycoproteins from herpes simplex virus (HSV) type 1 [14, 15] and CAE-virus, a lentivirus [16]. These latter findings may have therapeutic implications for an inhibitor of terminal glycosylation. For instance, exposure of normally hidden peptide epitopes might result in neutralization by group-specific antibodies. Also, carbohydrate antigens may be unmasked by removal of terminal sugars [17].

4. A Target for Design of Virus-Specific Inhibitors of Protein Glycosylation

From the above it is clear that studying the role of terminal glycosylation requires inhibitors selective for virus-infected cells: Only with specific inhibitors, cause-and-effect analysis is feasible in complex organisms. The enzymes involved in terminal glycosylation are obvious targets for inhibitors. However, these glycosyl transferases are not coded for by the virus, so specificity is not easily obtained. Furthermore, there are many different glycosyl transferases, thus hampering targeted inhibitor design. Instead, a transport system translocating sugar nucleotides from the cytoplasm into the lumen of the Golgi apparatus [10] appears a more attractive target [19].

The terminal glycosyl transferases are located inside the trans-Golgi compartment, and so is the acceptor glycoprotein. The substrates of the transferases, the sugar nucleotides are present in the cytosol. Hirschberg and co-workers (reviewed in Ref. [16]) showed that sugar nucleotide translocator proteins in the membranes of the Golgi vesicles transport sugar nucleotides (GDP–Fuc, UDP–Gal, CMP–NeuAc, UDP–GLcNAc) into the Golgi compartment (Fig. 2). These transport

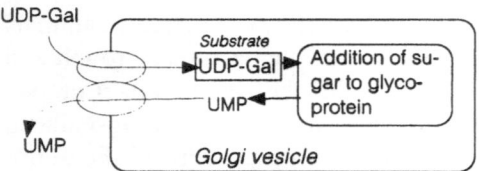

Fig. 2. Schematic diagram of sugar nucleotide translocation in the Golgi system (see Ref. [18])

systems have apparent K_m values (1–10 µM) approximately 100 times lower than the apparent K_m values of glycosyl transferases. The transport of sugar nucleotides into the Golgi lumen is coupled to a counter transport of the luminal nucleoside monophosphate (CMP, UMP, GMP) to the cytoplasm (see Fig. 2). Importantly, Capasso & Hirschberg [20] observed that in cell-free systems certain nucleoside monophosphates can inhibit sugar nucleotide translocation into the Golgi lumen. Such inhibition, when accomplished selectively in infected cells, would lead to selective inhibition of glycosylation of a viral glycoprotein due to lack of substrate. We have demonstrated the feasibility of this approach.

5. Antiviral Agents as Lead-Structures

The antiviral nucleoside analog (E)-5-(2-bromovinyl)-2′-deoxyuridine (BVdU), which is an inhibitor of HSV DNA synthesis [21], also acts as an inhibitor of terminal glycosylation specific for HSV-infected cells [22]. An isolated effect on glycosylation (a block in oligosaccharide chain elongation prior to addition of Gal) can be achieved when BVdU is added to HSV-infected cells after initiation of DNA synthesis. The effect on glycosylation is caused by the 5′-monophosphate of BVdU [22], which is formed only in infected cells by the HSV-specified thymidine kinase. In a cell-free system, this nucleotide inhibits the translocation into the Golgi apparatus of UDP-Gal and CMP-NeuAc, the sugar donors in the galactosyl and sialyl transferase reactions, respectively. It thereby inhibits addition of these sugars to the protein-bound oligosaccharides [22, 23] (see Fig. 3). BVdU-5′-monophosphate did not affect translocation of a purine sugar nucleotide, GDP–

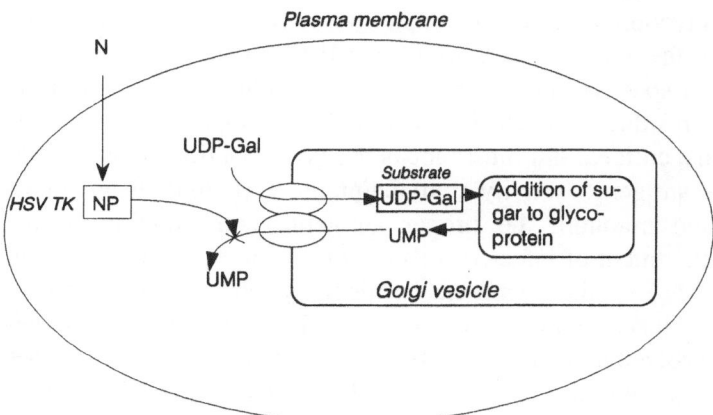

Fig. 3. Inhibition of sugar nucleotide translocation by a nucleoside analog monophosphate (NP)

Fuc. Indeed, carbohydrate analysis of the HSV-encoded glycoprotein gC-1 obtained from infected, treated cells showed a block in the incorporation of Gal and NeuAC into the N- and O-linked oligosaccharides, whereas fucosylation was unimpaired [23]. Thus, in intact cells only the transport of pyrimidine sugar nucleotides and not a purine sugar nucleotide was blocked by a pyrimidine analog, an observation in line with the cell-free studies [20]. These latter studies had also indicated that the translocation systems tolerated 2′-deoxyribose in the nucleoside and substitution in the 5-position of pyrimidines.

Other, trivial, explanations for the observed inhibition could be excluded: inhibition of the glycosyl transferases by the monophosphate, inhibition of glycoprotein synthesis, depletion of sugar nucleotide pools, formation of a BVdU-diphosphate ester of a sugar, inhibition of intracellular transport of the viral glycoprotein [22]. The selectivity for virus-infected cells resides in the phosphorylation of the nucleoside analog by a virus-coded enzyme, HSV thymidine kinase, as the inhibition of glycosylation was not observed in cells infected with a thymidine kinase-negative mutant of HSV-1. In other words, BVdU is a prototype inhibitor of terminal glycosylation specific for HSV-infected cells.

6. Optimization of the Lead

The hydrated, bulky oligosaccharides of glycoproteins can cover considerable portions of the protein surface [24, 25]. This implies that portions of viral glycoproteins are shielded by host cell structures, which are tolerated by immune responses and may mask potential neutralizing epitopes. This can help enveloped viruses escape from immunological clearance. Nevertheless, small changes in the peripheral structures of N-linked oligosaccharides may have consesquences for the antigenic activity of glycoproteins. For example, the removal of terminal sialic acid slightly increased the antigenic activity of site II (designation according to Ref. [26]) of the HSV-1 specified glycoprotein C (gC-1), whereas removal of the penultimate galactose residues resulted in a complete loss of activity of site II epitopes [14, 15]. Thus, an ideal inhibitor blocks the glycosylation steps responsible for masking of antigenic activity, but maintains carbohydrate elements needed to acquire and maintain the proper antigenic conformation of the glycoprotein. BVdU-treatment of infected cells *in vitro* (drug being added late in infection to not affect DNA synthesis) inhibits terminal glycosylation before galactose addition and did not increase the antigenicity of gC-1, in line with the observation that galactose is involved in maintaining the carbohydrate-dependent epitopes of site II [14, 15]. A search amongst analogs of BVdU, revealed that 5-*n*-propyl-2'-deoxyuridine (PdU) increases the antigenicity of gC-1 and induces a change in glycosylation [14, 15]. Both effects were dependent on phosphorylation of PdU by the virally-coded thymidine kinase. Mechanistically, the block in glycosylation exerted by PdU is identical to BVdU: inhibition of sugar nucleotide translocation by the 5'-monophosphate of the nucleoside analog (Olofsson et al., unpublished data). The effect on oligosaccharide structure was different, however. PdU seems to affect primarily brancing of the N-linked oligosaccharides, and blocks galactosylation less extensively than BVdU [27, 28].

Apparently, it is possible through the use of inhibitors to induce exposure of carbohydrate-dependent epitopes without inactivating them. Using monoclonal antibodies, it was shown that the PdU-induced new epitopes were scattered over gC-1 and, importantly, not confined to gC-1 but also found on other HSV-coded glycoproteins, such as gB-1 and gD-1 [28]. The antigenicity-enhancing effect of PdU is caused, at least in part, by demasking of epitopes hidden by terminal glycosyl residues (see below). It is tempting to speculate that the demasking is correlated with the decreased degree of branching of the N-linked oligosaccharides induced by PdU.

7. Nature of PdU-Induced and Galatose-Dependent Epitopes of gC-1

The epitopes of the HSV-1 glycoprotein gC-1 are organized in two antigenic sites [26] (Fig. 4). Antigenic site II contains conformational epitopes, which are highly dependent on $\beta(1-4)$-linked galactose residues of complex-type, N-linked oligosaccharides [14, 15, 30]. Galactosidase treatment eliminates the antigenic activity of site II epitopes, and enzymatic re-addition of the $\beta(1-4)$ linked galactose reconstitutes the antigenic activity [15]. No such epitopes were found in antigenic site I.

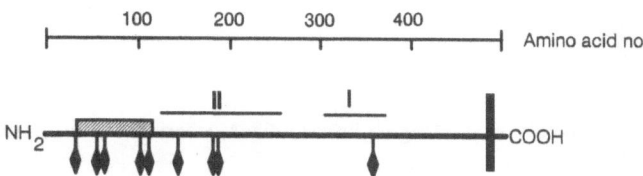

Fig. 4. Location of antigenic sites I and II in the gC-1 molecule. The hatched area indicates location of O-linked glycans in gC-1, the lollipops the location of N-linked glycans. The galactose-dependent epitopes are in site II, and PdU-inducible epitopes occur in site I and site II

None of these carbohydrate-determinants are involved in the PdU-induced increase of antigenic activity. In fact, galactose-dependence (as described above) and PdU-inducibility of epitopes are different antigenic modulations [30]. There are, for example, PdU-inducible epitopes, which are not galactose-dependent. This suggests that the structural changes in carbohydrate composition induced by PdU-treatment of infected cells may not affect the particular galactose residue engaged in stabilizing site II epitopes. As mentioned above PdU induces a decrease of multiantennary N-linked glycans and an increase of biantennary structures, combined with only a partial inhibition of galactosylation. The most simple mechanism, supported by all experimental data, is that certain antennae of the multiantennary glycans mask some of the epitopes in sites I and II. PdU treatment reduces the degree of branching, without eliminating the particular galactose units, being critical for positive modulation of the galactose-dependent epitopes [27], and represents probably a general mechanism for increasing antigenic activity of viral glycoproteins.

8. Future Directions

It is important to determine the exact structures of the oligosaccharides and peptides involved in the PdU-altered antigenicity in several glycoproteins. Whether interference with terminal glycosylation to demask hidden epitopes in virus-infected cells is worth pursuing as an antiviral strategy remain to be seen. We need nucleoside analogs exclusively targeted to the sugar-nucleotide translocation system, and not to nucleic acid synthesis. We can then extend our studies to encompass work in animals. We also need to find out whether other viral enzymes can be exploited to selectively generate inhibitors in cells infected with, for example, retroviruses.

Acknowledgement. This work was supported by grants to S.O. from the Swedish Cancer Society (Grant 2962-B89-OIXA), the Swedish Medical research Council (Grant 9083) and the National Swedish Board for Technical Development (Project 87-0256P).

References

1. Kornfeld R and Kornfeld S, Assembly of asparagine-linked oligosaccharides. *Annu. Rev. Biochem.* **54**: 631–664, 1985.
2. Datema R, Olofsson S and Romero PA, Inhibitors of protein glycosylation and glycoprotein processing in viral systems. *Pharmac. Ther.* **33**: 221–286, 1987.
3. Schachter H, Coordination between enzyme specificity and intracellular compartmentation in the control of protein-bound oligosaccharide biosynthesis. *Biochem. Cell Biol.* **64**: 163–181, 1986.
4. Schwarz RT and Datema R, The lipid pathway of protein glycosylation and its inhibitors: the biological significance of protein-bound carbohydrates. *Adv. Carbohydr. Chem. Biochem.* **40**: 287–379, 1982.
5. Elbein AD, Inhibitors of the biosynthesis and processing of *N*-linked oligosaccharides. *Crit. Rev. Biochem.* **16**: 21–49, 1984.
6. Schwarz RT and Datema R, Inhibitors of protein glycosylation. *Trends Biochem. Sci.* **5**: 65–67, 1980.
7. Schwarz RT and Datema R, Inhibitors of trimming: new tools in glycoprotein research. *Trends Biochem. Sci.* **9**: 32–34, 1984.
8. Romero PA, Saunier B and Herscovics A, Comparison between 1-deoxynojirimycin and *N*-methyl-1-deoxynojirimycin as inhibitors of oligosaccharide processing in intestinal epithelial cells. *Biochem. J.* **226**: 733–740, 1985.
9. Pan YT, Hori H, Saul R, Sanford BA, Molyneux RJ and Elbein AD, Castanospermine inhibits the processing of the oligosaccharide portion of the influenza viral hemagglutinin. *Biochemistry* **22**: 3975–3984, 1983.
10. Kino T, Inamura N, Nakahara K, Kiyoto S, Goto Y, Terano H, Koshaka M, Aoki H and Imanaka H, Studies of an immunomodulator, swainsonine. II. Effect of swainsonine on mouse immunodeficient system and experimental murine tumor. *J. Antibiot.* **38**: 936–939, 1989.
11. Humphries MJ, Matsumoto K, White SL and Olden K, Oligosaccharide modification by swainsonine treatment inhibits pulmonary colonization by B16-F10 murine melanoma cells. *Proc. Natl. Acad. Sci.* **83**: 1752–1756, 1986.

12. Dennis JW, Laferté S, Waghorne C, Bretman ML and Kerbel RS, β(1–6)Branching of Asn-linked oligosaccharides is directly associated with metastasis. *Science* **236**: 582–585, 1987.

13. Montefiori DC, Robinson E and Mitchell WM, Antibody-independent, complement-mediated enhancement of HIV-1 infection by mannosidase I and II inhibitors. *Antiviral Res.* **11**: 137–146, 1989.

14. Sjöblom I, Lundström M, Sjögren-Jansson E, Glorioso JC, Jeansson S and Olofsson S, Demonstration and mapping of highly carbohydrate-dependent epitopes in the herpes simplex virus type 1-specified glycoprotein C. *J. Gen.Virol.* **68**: 545–554, 1987.

15. Olofsson S, Sjöblom I and Jeansson S, Activity of herpes simplex virus type 1-specified glycoprotein C antigenic site II epitopes reversibly modulated by peripheral fucose or galactose units of glycoprotein oligosaccharides. *J. Gen.Virol* **71**: 889–895, 1990.

16. Huso DL, Narayan O and Hart GW, Sialic acids on the surface of caprine arthritis encephalitis virus define the biological properties of the virus. *J. Virol.* **62**: 1974–1980, 1988.

17. Feizi I and Childs RA, Carbohydrate structures of glycoproteins and glycolipids as differentiation antigens, tumor associated antigens and components of receptor systems. *Trends Biochem. Sci.* **10**: 24–27, 1985.

18. Hirschberg CB and Snider MD, Topography of glycosylation in the rough endoplasmic reticulum and Golgi apparatus. *Annu. Rev. Biochem.* **56**: 63–88, 1987.

19. Datema R and Olofsson S, Nucleotide analogs as herpesvirus-specific inhibitors of protein glycosylation. In: *Nucleotide Analogs as Antiviral Agents* (Martin JC ed.); American Chemical Society, pp. 116–123, 1989.

20. Capasso JM and Hirschberg CB, Effect of nucleotides on translocation of sugar nucleotides and adenosine 3'-phosphate 5'-phosphosulfate into Golgi apparatus vesicles. *Biochim. Biophys. Acta* **777**: 133–239, 1984.

21. DeClercq E, Specific targets for antiviral drugs. *Biochem. J.* **205**: 1–13, 1982.

22. Olofsson S, Milla M, Hirschberg C, DeClercq E and Datema R, Inhibition of terminal *N*-and *O*-glycosylation specific for herpesvirus-infected cells. Mechanism of an inhibitor of sugar nucleotide transport across Golgi membranes. *Virology* **166**: 440–450, 1988.

23. Olofsson S, Lundström M and Datema R, The antiherpes drug (*E*)-5-(2-bromovinyl)-2'-deoxyuridine (BVdU) interferes with formation of *N*-linked and *O*-linked oligosaccharides of the herpes simplex virus type 1 glycoprotein C. *Virology* **147**: 201–205, 1985.

24. Montreuil J, Structure and conformation of glycoprotein glycans. In: *Vertebrate Lectins* (Olden K and Parent JB eds), Van Nostrand; pp. 1–26, 1987.

25. Rademacher TW, Parekh RB and Dwek RA, Glycobiology. *Annu. Rev. Biochem.* **57**: 785–838, 1988.

26. Marlin SD, Holland TC, Levine M and Glorioso JC, Epitopes of herpes simplex type 1 glycoprotein gC are clustered in two distinct antigenic sites. *J. Virol.* **53**: 128–136, 1989.

27. Olofsson S and Datema R, New virus-selective inhibitor of terminal glycosylation increasing immunological reactivity of a viral glycoprotein. *Antivir. Chem. Chemother.* **1**: 17–24, 1990.

28. Olofsson S, Sjöblom I, Glorioso JC, Jeansson S and Datema R, Selective induction of discrete epitopes of herpes simplex virus type 1-specified glycoprotein C by interference with terminal steps in glycosylation. *J. Gen. Virol.*: **72**: 1959–1966, 1991.

Comparison of Crystal Structures of Inhibitor Complexes of the Human Immunodeficiency Virus Protease

A. Wlodawer, A. L. Swain and A. Gustchina*

Macromolecular Structure Laboratory, NCI-Frederick Cancer Research and Development Center
ABL-Basic Research Program, Frederick, Maryland 21702, USA

Abstract. The structures of four complexes of the human immunodeficiency virus type 1 protease with inhibitors, solved in our laboratory, as well as five structures solved elsewhere, are compared. Four different crystal forms are represented in this comparison. Although considerable similarity of the mode of binding is observed in all subsites occupied by the inhibitors, some differences are noted, particularly for the terminal groups of the inhibitors.

1. Introduction

The aspartic protease (PR) encoded by the human immunodeficiency virus type 1 (HIV-1) presents a promising target for designing novel types of drugs which could be therapeutically beneficial in cases of acquired immunodeficiency syndrome (AIDS). The techniques of rational drug design are being employed in a number of industrial and academic laboratories in the search of such compunds. However, for a rational approach to drug design to succeed, it is helpful to characterize the structures of the target enzyme and of enzyme-inhibitor complexes. Observations based on structural data, when used together with other sources of information such as the results of investigations of binding constants and kinetics parameters, can elucidate precise interactions between lead compounds and the enzyme, and this may be necessary for explaining the basis of effective inhibition. This, in turn, may lead to improvement in the potency and specificity of new inhibitors.

The crystal structure of the native HIV-1 protease has confirmed that the molecule is a homodimer and that its active site resembles closely the active sites of other aspartic proteases such as pepsin, chymosin, or rhizopuspepsin, among others [1–3]. For that reason, inhibitors of cellular aspartic protease were tested early on for their ability to inhibit HIV-1 PR. Subsequently, a large number of complexes of HIV-1 protease with new substrate-based inhibitors have been

---

* On leave from the V. A. Engelhardt Institute of Molecular Biology, Academy of Sciences of the USRR, Vavilova Str. 32, Moscow, USSR.

prepared and many crystal structures of these complexes have been solved. Five structures of HIV-1 protease-inhibitor complexes have already been published [4–8] and other structures have been discussed at scientific meetings, with atomic coordinates becoming available either through the Protein Data Bank, or by direct distribution from the originating laboratories.

The coordinate sets for four protease-inhibitor complex structures solved in this laboratory, as well as for five structures solved elsewhere, are used as the basis of comparisons described here. These structures were solved in four crystal forms, three of them orthorhombic, and one hexagonal. The sequences of the inhibitors whose complexes were studied in our laboratory, as well as relevant crystallographic data, are given in the top part of Table 1, while the sequences of inhibitors whose structures were received from other laboratories are listed in the bottom part of that table. These inhibitors are peptide analogs with different sequences and lengths. Some of the inhibitors are nonspecific and able to interact with almost all aspartic proteases (pepstatin, acetyl-pepstatin), while other inhibitors were designed on the basis of the sequences of known substrates of HIV-1 PR. They all bind to the protease in the same general conformation, making similar contacts with the enzyme [9]. Different nonscissile groups replace a central peptide bond. They include an unusual amino acid statine, as well as linkages involving a reduced peptide bond, hydroxyethylamine, and hydroxyethylene moieties. Herein we describe comparative observations of nine PR bound inhibitors (Tables 1 and 2).

2. Methods

The methods used in this laboratory to solve the structure of the native HIV-1 PR, as well as of the inhibitor complexes, were given in previous pubications [2, 4, 5, 8]. A detailed description of the methods used to design a quasi-symmetric inhibitor A-74704 and to solve its structure was published by Erickson et al. [7]. The unpublished atomic coordinates of the complex of HIV-1 PR with the inhibitor Ro-31-8558 were supplied by Dr. Bradford Graves (Roche, Nutley, NJ), while three sets of coordinates (Pepst. AG2 and AG4) of the complexes of statine-based inhibitors were provided by Dr. Krzysztof Appelt from Agouron Pharmaceuticals (La Jolla, CA). While the structure of a complex of HIV-1 PR with acetyl-pepstatin was published by Fitzgerald et al. [6], coordinates used here were from an analogous structure solved subsequently and independently in our laboratory, as discussed below.

Recombinant PR (BH10 Isolate) was a gift from Glaxo (England). Acetyl-pepstatin was a gift from Dr. K. Oda and Prof. S. Murao of Osaka Prefecture University, via Dr. John Kay of the University of Wales, College of Cardiff. Crystals were grown in hanging drops from a mixture of 5–10 mg/ml PR with excess acetyl-pepstatin in 0.4 M NaCl, pH 4.7, with ≈25% ammonium sulfate

Table 1. Inhibitors used in this comparison. The sequences of the inhibitors solved in this laboratory are listed in the top half of the table, and the sequences for the other inhibitors in the bottom half. Inhibition constants (K_i), resolution of the data, and crystallographic R value are given. (Nle, norleucine; Ac, acetyl; OMe, methoxy; BOC, t-butoxy; CHA, cyclohexylalanine; EPY, ethylpyridine; CBZ, carbobenzyloxy; COR, diaminodibenzylpropanol; IVA, isovaleryl; STA, statine)

Inhibitor	Sequence									K_i (nM)	Resol. (Å)	R
MVT-101			Ac-Thr	Ile	Nle	CH2N	Nle	Gln	Arg	760	2.3	0.176
JG-365		Val	Ac-Ser	Leu	Asn	Phe	CH(OH)CH2N	Pro	Ile	Val-OMe	0.24	2.4	0.146
U-85548e	Val	Ser	Gln	Asn	Leu	CH(OH)CH2	Val	Ile	Val	<1	2.5	0.138
Ac.Pepst.		Ac-Val	Val	STA	Ala	STA	Val			1100	2.5	0.167
Ro-31-8558			BOC	CHA	CH(OH)CH2	Val	Ile	EPY		<1	2.3	0.173
A-74704			CBZ	Val	COR	Val	CBZ			4.5	2.8	0.182
Pepst.		IVA	Val	STA	Ala	STA				1070	2.8	0.147
AG2		Ser	Phe	STA	Gln	Ile	Val			540	2.5	0.158
AG4		Ser	Gln	Asn	Ile	Val	STA	Gln		230	2.5	0.148

Table 2. Pseudonymous sequences of the inhibitors used in this comparison. Since the functional groups (P/P′) which bind in protease subsites (S/S′) are not always equivalent to the residue name, pseudonyms are used in this table in an attempt to describe the functionality bound in the protease subsite. For example, the statine moieties are referred to as a Leu functionality, a hydroxymethyl group (CH(OH)), and a methylene (CH2) group. Also, these sequences are written in order of P to P′ rather than in N-terminal to C-terminal order, and only the functionalities *occupying* the PR subsites are included. This table should be used only in conjunction with Table 1 and the text. (Nle, norleucine; Ac, acetyl; OMe, methoxy; BOC, t-butoxy; CHA, cyclohexylalanine; EPY, ethylpyridine; MVA, methylvaline; m.c., main chain; NE2, epsilon nitrogen)

Inhibitor	P5	P4	P3	P2	P1	Linkage	P1′	P2′	P3′	P4′
MVT-101			Ac-Thr	Ile	Nle	CH2N	Nle	Gln	Arg	
JG-365	Val	Ac-Ser	Leu	Asn	Phe	CH(OH)CH2N	Pro	Ile	Val-OMe	
U-85548e	Val	Ser	Gln	Asn	Leu	CH(OH)CH2	Val	Ile	Val	
Ac.Pepst.		CH(OH)	Leu	Leu	CH2	CH(OH)	Leu	Val	Val	
Ro-31-8558				BOC	CHA	CH(OH)CH2	Val	Ile	EPY	
A-74704				Phe	Phe	CH(OH)	Phe	Val		
Pepst.			Leu	Ala	CH2	CH(OH)	Leu	Val	IVA	
AG2		CH(OH)	Ile	Gln	CH2	CH(OH)	Ile	Asn	Phe	
AG4			Val	Ile	CH2	CH(OH)	Leu	Asn	Ser	m.c., NE2 of Gln

added to the reservoir. X-ray data to 2.5 Å were collected from a crystal measuring $0.4 \times 0.1 \times 0.2$ mm. The space group is $P2_12_12_1$ with unit cell parameters $a = 58.58$, $b = 86.70$, $c = 46.24$ Å, isomorphous with that reported by Fitzgerald et al. [6]. The MERLOT program package [10] was used to solve the structure by molecular replacement, using the PR coordinates from the PR-JG-365 complex as the probe. The molecular replacement solution was refined with PROLSQ [11] to an R-factor of 0.272. A model of acetyl-pepstatin was built to fit the electron density maps from that refinement and the complex was refined to a final R-factor of 0.167 with good geometry (rms deviation from ideal bond lengths = 0.017 Å, bond angles = 3.8°). It became apparent during the refinement that the acetyl-pepstatin is bound to the PR in two different orientations. The electron density in the active site corresponding to the inhibitor was noisy. To test for the dominant orientation, the acetyl-pepstatin was built to fit the density in the reverse direction. This direction filled the positive difference Fourier peaks with atoms and left the negative peaks empty. The resulting R-factor after refinement for this orientation is 0.173 with slightly worse geometry than for the first orientation (rms deviation from ideal bond lengths = 0.018 Å, bond angles = 3.9°). The electron density maps calculated from this refinement looked poorer. Based on the R-factors, geometry, and quality of the electron density maps, we conclude that the first orientation dominates and we use this structure as a final model in the comparison herein.

The structures of inhibitor complexes of HIV-1 PR were superimposed with the programs UPAMA [12] and ALIGN [13] and were displayed on an Evans and Sutherland computer graphics system. The interactions among protease molecules (crystal contacts) were identified by a function CONTACT which is part of the molecular graphics program FRODO [14], used by us for display purposes.

This superposition procedure included only α-carbon atoms of the protease. For those inhibitors which are bound to the protease in the opposite direction with respect to monomers 1 and 2, the monomer atoms were exchanged, with the result that all inhibitors are aligned in the same N-terminal to C-terminal orientation. Because of this, when we refer to the elements of the inhibitors and the corresponding subsites of the protease, we must refer to both P/P′ and S/S′. Differentiating between P/S and P′/S′ is only possible for each inhibitor within its environment of the protease.

The resolutions of the structures vary from 2.3 to 2.8 Å. Many of these structures were solved by molecular replacement, so it must be born in mind that at these resolutions some features of the protease structure itself may have been carried through by these techniques rather than uniquely observed.

3. Results

3.1. *Differences/Similarities Among the Protease Molecules in the Structures of the Complexes*

The structure of the protease is very similar among the nine protease-inhibitor complexes, despite the differences in the crystal parameters between them. As the structures are visually compared using computer graphics, the notable differences among them are not only in side chain positions at the surface of the protein, as expected, but also some rearrangements in the torsion angles of parts of the main chain. These differences are at the outer edge of the protease, involved with crystal lattice contacts, but distant to the active site so they have no apparent bearing on inhibitor binding.

In most of the structures, the peptide bond at the tip of the flap between residues 50 and 51 is turned 180° in molecule 1 relative to its position in molecule 2 (Table 3). With the exception of AG4, the torsion angles identifying monomer 1 are ψ_{50} in the range of –20° to –45° and ϕ_{51} between –60° and –95°. In monomer 2, ψ_{150} is

Table 3. Torsion angles of the peptide bonds at the tips of the flaps. With the exception of AG4, these values have been used to characterize the monomers, thereby distinguishing monomer 1 from monomer 2

Inhibitor	ψ_{50}	ϕ_{51}	ψ_{150}	ϕ_{151}
MVT-101	–35	–95	128	100
JG-365	–41	–68	132	92
U-85548e	–24	–84	134	115
Ac.Pepst.	–27	–64	130	96
Ro-31-8558	–35	–94	122	115
A-74704	–42	–89	157	79
Pepst.	–32	–71	132	107
AG2	–38	–78	134	118
AG4	79	123	155	101

in the range of 120° to 160° and ϕ_{151} between 75° and 120°. This is a detail which may not be accurately reflected in the lower resolution structures. However, this is the convention used in numbering monomers 1 and 2 in the coordinate sets compared herein. We will continue to follow this convention in our discussion. Eventually, high resolution structures of PR-inhibitor complexes may resolve this ambiguity.

3.2. Differences/Similarities Among the Inhibitors in the Structures of the Complexes

All of the inhibitors compared here are bound in the protease active site in an extended conformation so that when they are superimposed upon one another, the functional elements align quite similarly (Fig. 1). The contacts between the main chain of the inhibitor and the protease are very similar for all the complexes with the nonhydrolyzable scissile bond analog of each inhibitor aligned with the active site aspartate carboxyl groups (Asp 25/125). In all cases except for MVT-101 (which has a reduced peptide bond as a non-scissile moiety), the hydroxyl group at that P_1–P_1' junction is positioned between the PR aspartate carboxyl groups, within hydrogen bonding distance to each carboxylate oxygen. The configuration of the tetrahedral carbon of the inhibitors is S. Other similarities and differences will be addressed specifically according to inhibitor residue/protease binding pocket.

Although the HIV-1 PR is a symmetric enzyme, the same enzyme with inhibitor bound is not. The N→C polarity of the inhibitor peptide induces asymmetry on the PR, though it is very slight. The differences between monomers 1 and 2 are so subtle that two-fold disorder with respect to the inhibitor is frequently observed. For each inhibitor where 2-fold disorder is observed, there is always one dominant orientation in the population of PR-inhibitor complexes within the lattice. This dominant orientation is opposite for the statine-containing inhibitors relative to the others. The statine-containing inhibitors also appear to have lower ratios of one orientation to another than the other inhibitors. This may be a reflection of their lower inhibitory capacity.

3.3. P_1/P_1'

The protease side chains comprising these pockets (Tables 1 and 2, Fig. 2), with a notable exception of the active site aspartates, are mostly hydrophobic. The side chains of the active site aspartates and the main chain hydroxyl of those inhibitors containing such a central group are involved in polar contacts. All of the inhibitors compared here have hydrophobic moieties at P_1 and P_1', with the exception of the statine-containing inhibitors that have no groups occupying the subsite S_1. This may be one of the reasons for the relatively low inhibitory potency of statine-containing transition state analogs. In all the non-statine inhibitors, except the 2-fold symmetric A-74704, the side chain which extends furthest from the main chain of the inhibitor resides in the S_1 pocket. In MVT-101, though it also has the same side chain type in P_1 as P_1', the Nle in the S_1' pocket is compressed. The different conformations of the Nle side chains within the binding cleft making this bound inhibitor topologically asymmetric even at the P_1/P_1' site.

As can be seen from Fig. 2, the main chain and side chains of P_1/P_1' superim-

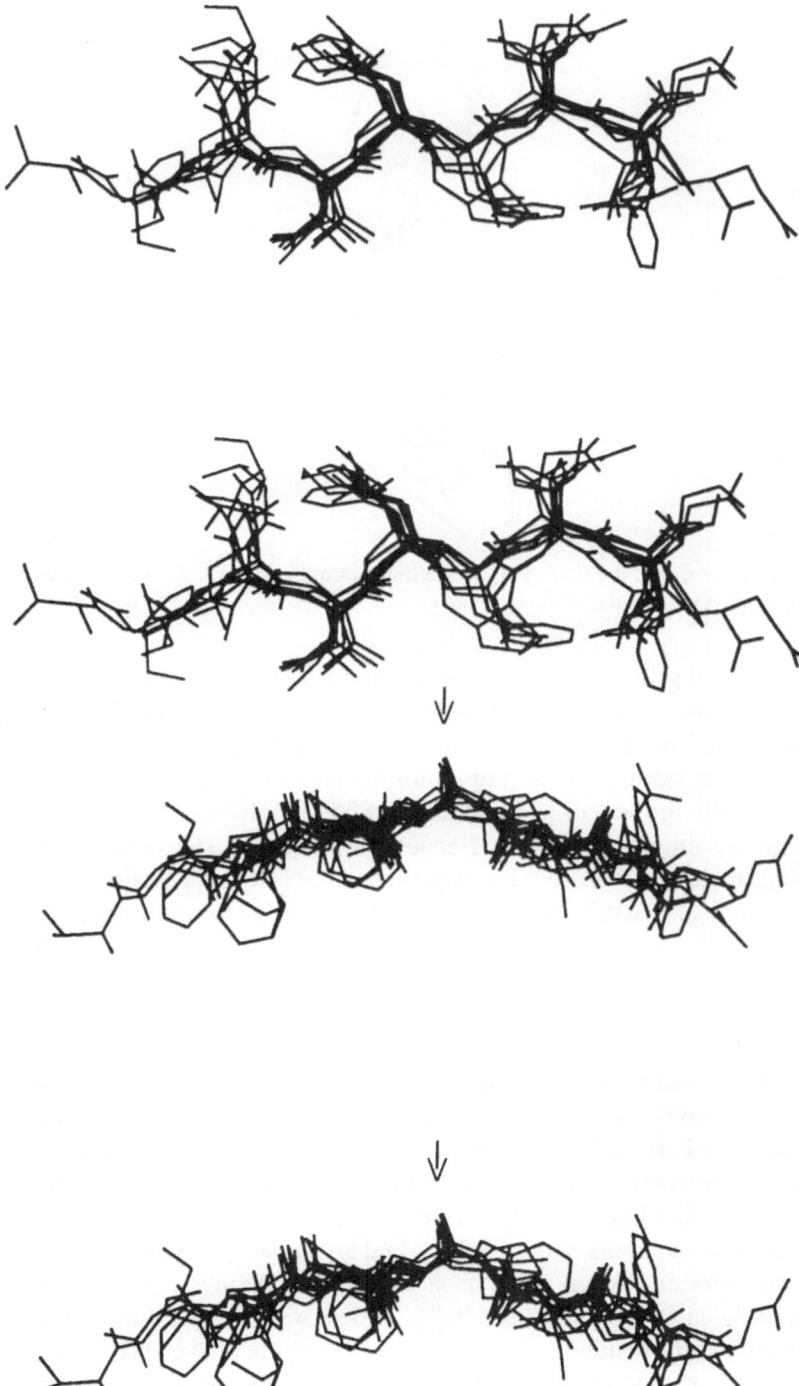

Fig. 1. Left: Superposition of 9 inhibitors in the conformation they assume within the protease-inhibitor complex crystals. *N*-termini are toward top; *C*-termini are toward bottom. The nonscissile hydroxyl group contained in 8 of the 9 inhibitors is noted with an arrow. **Right:** Same as **Left**, rotated 90° to illustrate superposition of side chains

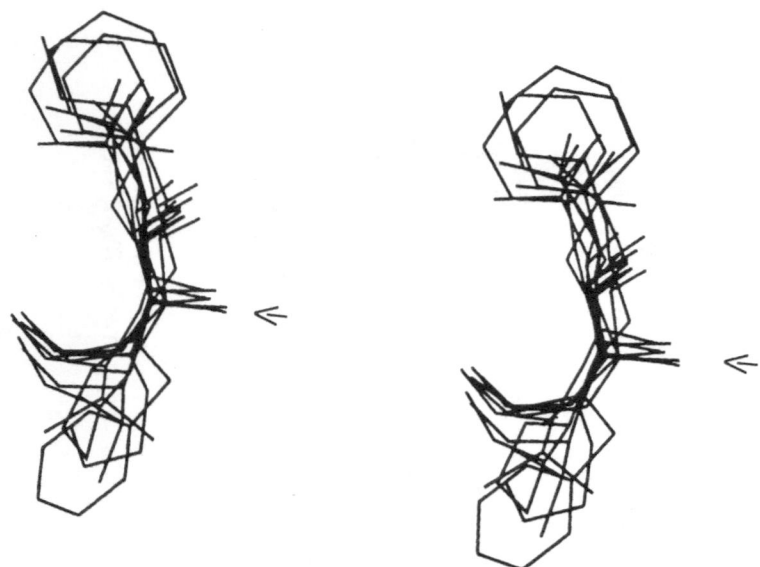

Fig. 2. P_1 and P_1' elements of the 9 inhibitors. The orientation is as in Fig. 1(left). An arrow points toward the hydroxyl group contained in 8 of the inhibitors

pose neatly. The hydroxyl groups of each inhibitor superimpose, except for MVT-101 which has no hydroxyl group. The main chain at P_1/P_1' of this inhibitor is positioned somewhat differently due to the reduced peptide bond. Of course, the nonscissile moiety of the statine-containing inhibitors superimpose most closely among themselves, while in the Ro-31-8558 and U-85548e inhibitor complexes, hydroxyethylene groups align very closely. Because of the extra CH_2 group of the hydroxyethylamine linkage of the Phe–Pro dipeptide of JG-365, the main chain bulges at this junction, causing this inhibitor to deviate slightly from the others in the superposition.

3.4. P_2/P_2'

Though the S_2 and S_2' pockets are hydrophobic (Tables 1 and 2, Fig. 3), both hydrophilic and hydrophobic residues can occupy these sites. The hydrophobic side chains are observed in different orientations for the different inhibitors, forming contacts with different groups in the binding pocket. In the case of the amides, however, the side chains point toward the extremity of the inhibitor, avoiding the hydrophobic contacts of the pockets, and are apparently held in place by electrostatic interactions. When there is a polar side chain or end group occupying the S_4 or S_4' site adjacent to the P_2/P_2' amide, a stabilizing force may exist. For both the P_2 and P_2' residues, the amide side chains are also stabilized by polar contact with the carbonyl oxygen of the previous inhibitor residues. There

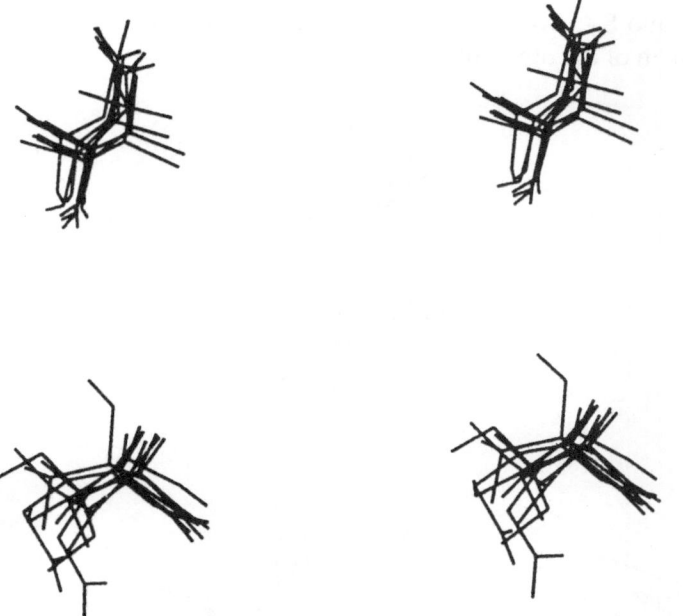

Fig. 3. P_2 and P_2' elements of the 9 inhibitors. The orientation is $\approx 180°$ from Fig. 1 (right). N-terminal groups are still toward top; C-terminal groups are toward bottom

are also polar contacts between some P_2/P_2' amide groups and polar side chains of the protease.

3.5. P_3/P_3'

Distal to S_2/S_2', the subsites become less defined (Figs. 1 (right) and 4). Therefore, all types of side chains are accomodated in the S_3 and S_3' subsites. However, neither Ro-31-8558 nor A-74704 have groups which occupy an S_3 subsite (Tables 1 and 2). The Ro-31-8558 N-terminal BOC group occupies the S_2 subsite, whereas the CBZ side chain of A-74704, which has the potential to fill S_3, is positioned along the path of the main chain of the other inhibitors instead. The other inhibitors have P_3 groups which extend into the S_3 subsite such that they superimpose on one another, though the orientations of the side chains are variable.

Only A-74704 does not have a group filling the S_3' subsite. The CBZ residue extends between the S_3' pocket and the usual direction of the main chain (Fig. 4 (bottom)). Pepstatin and AG4 are special cases. The P_3' position of pepstatin contains an isovaleryl residue. In this inhibitor, the main chain makes a turn such that the first valyl group of isovaleryl points toward the main chain path of the other inhibitors, while the second valyl group extends "sideways" into the S_3' subsite. In AG4, the main chain turns such that it, along with the N-terminal serine residue, extends into S_3', while the Gln side chain is positioned along the direction of the main chain of the other inhibitors.

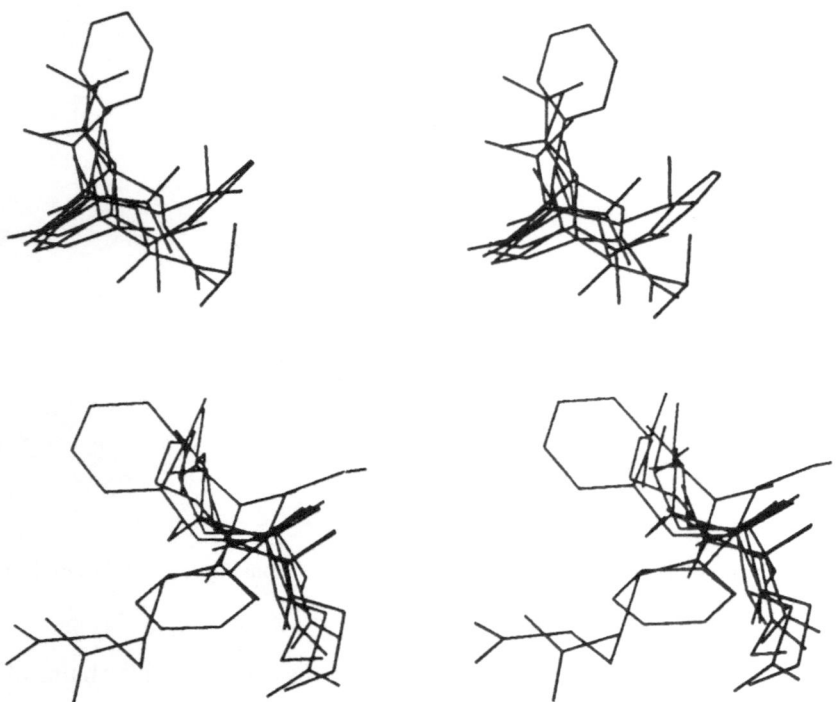

Fig. 4. Top: N-terminal P_3/P_3' elements of 9 inhibitors. The groups which fill the corresponding S_3/S_3' subsites are toward the right of the figure. **Bottom:** C-terminal P_3/P_3' elements of 9 inhibitors. The groups which fill the S_3/S_3' subsites are toward the left. The main chains of most inhibitors are toward the bottom, while the CBZ group of A-74704 and the C-terminal portion of AG4 are positioned in between

3.6. P_4/P_4'

Only JG-365 and U-85548e have a standard amino acid present in S_4. The hydroxyl groups of the N-terminal statines of pepstatin and acetyl-pepstatin reside at the P_4 site. In any case, S_4 can barely be referred to as a binding pocket because only the portion of the P_4 side chain which is oriented toward the carboxyl end of the inhibitor is surrounded by atoms of the protease. Though the Ser side chains are positioned differently in the models of different inhibitors, this polar side chain may play a role in stabilizing the amide residue in the adjacent P_2 site. The only inhibitor with a moiety at P_4' is AG4, which has one side of the Gln fork directed into the S_4' subsite.

3.7. P_5

U-85548e is the only inhibitor with a residue at P_5. There really is no complementary pocket, since this residue extends out of the protease binding groove. It makes no contact with the protease and its temperature factor is quite high, indicating probable motion. It is not expected that this residue contributes to the inhibitory capacity of U-85548e.

3.8. *Conserved Water Molecule*

Due to the limitations of resolution and differences in refinement strategies among these structures, a comparison of solvent cannot be done, though one intriguing water molecule is detected in all PR-inhibitor structures compared here. This special water molecule resides between each inhibitor and the tips of the flaps of the protease. This water molecule is tetrahedrally coordinated to main chain nitrogen atoms from residues 50 and 150 of the flaps, and to carbonyl oxygens from substituents P2 and P1' of the inhibitor. It has been postulated that by incorporating an equivalent atom covalently into the structure of the inhibitors, it might be possible to improve their specificity [4].

3.9. *Crystal Lattice Contacts*

The PR-inhibitor complexes grow in a number of different crystal forms, three of which are compared here. Two orthorhombic systems represented are $P2_12_12_1$ (Fig. 5 (top)) and $P2_12_12$ (Fig. 5 (bottom)). The higher symmetry class of these two, $P2_12_12_1$, relates the PR-inhibitor complexes in continuous, mutually perpendicular two-fold helices. The highest number of intermolecular contacts, 686, is observed in this crystal form. Replacement of one of the screw axes by a 2-fold axis reduces the number of contacts among molecules in the $P2_12_12$ crystal form to 553. The third crystal form is in the hexagonal crystal system with space group $P6_1$. This crystal form, characterized by a continuous 6-fold helix, has the fewest contacts, 542, among molecules. (Representing $P2_12_12_1$ is JG-365; $P2_12_12$ is acetyl pepstatin; $P6_1$ is A-74704).

4. Discussion

The structures of the protease in all the complexes are strikingly similar. The protease binds the inhibitors with a complementary conformational adjustment of both the protease and the inhibitor. The adjustment of the protease seems to be primarily the closing of the flaps over the protease along with a hinge motion at the interface of the two termini. An example of inhibitor adjustment is in the protease-MVT-101 complex where the P_1' Nle and the P_3' Arg side chains are bent to fit

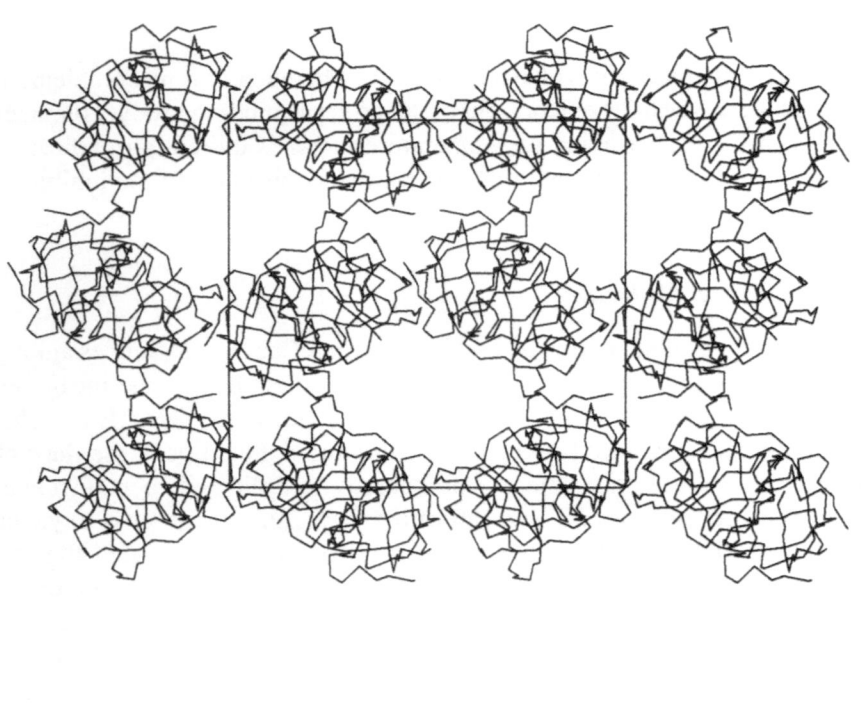

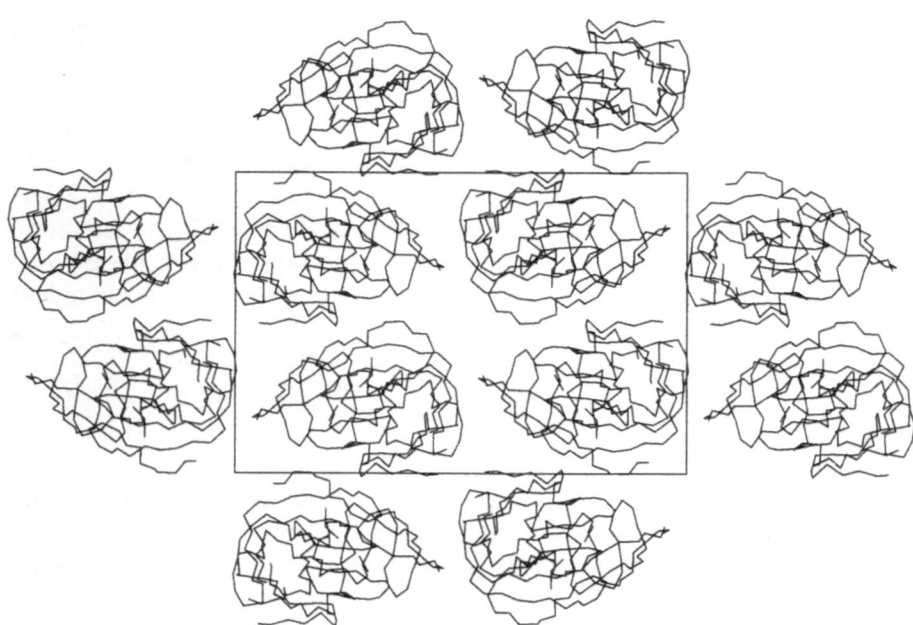

Fig. 5. Top: Packing diagram of PR-JG-365, space group P2$_1$2$_1$2$_1$, *x-z* plane. **Bottom**: Packing diagram of PR-acetyl-pepstatin, space group P2$_1$2$_1$2, *x-y* plane

into the pockets. The inhibitors also seem to aid in stabilization of their own conformation through an interaction between their own side chains of P_2 with P_4 or the N-terminus.

Most puzzling remains the difference between the two monomers in the orientation of the peptide bonds in the tips of the flaps. The orientation of this peptide bond has no bearing on contacts between the protease side chains and inhibitor. Evidence for this is the disorder of inhibitor observed in some protease-inhibitor complex structures [6]. The orientation of the flaps does make structural sense in providing an attractive force between the flaps to help maintain their closure around the inhibitor.

Acknowledgements. We are grateful to Drs. B. Graves and K. Appelt for sharing with us the results of their unpublished work. Research sponsored by the National Cancer Institute, DHHS, under contract N01-CO-74101 with ABL. The contents of this publication do not necessarily reflect the views or policies of the Department of Health and Human Services, nor does mention of trade names, commercial products, or organizations imply endorsement by the U. S. Government.

References

1. Navia MA, Fitzgerald PMD, McKeever BM, Leu C, Heimbach JC, Herber WK, Sigal IS, Darke PL and Springer JP, Three-dimensional structure of aspartyl protease from human immunodeficiency virus HIV-1. *Nature (London)* **337**: 615–620, 1989.
2. Wlodawer A, Miller M, Jaskólski M, Sathyanarayana BK, Baldwin ET, Weber IT, Selk LM, Clawson L, Schneider J and Kent SBH, Conserved folding in retroviral proteases: Crystal structure of a synthetic HIV-1 protease. *Science* **245**: 616–621, 1989.
3. Lapatto R, Blundell T, Hemmings A, Overington J, Wilderspin A, Wood S, Merson JR, Whittle PJ, Danley DE, Geoghegan KF, Hawrylik SJ, Lee SE, Scheld KG and Hobart PM, X-ray analysis of HIV-1 protease at 2.7 Å resolution confirms structural homology among retroviral enzymes. *Nature (London)* **342**: 299–302, 1989.
4. Miller M, Schneider J, Sathyanarayana BK, Toth MV, Marshall GR, Clawson L, Selk L, Kent SBH and Wlodawer A, Structure of a complex of synthetic HIV-1 protease with a substrate-based inhibitor at 2.3 Å resolution. *Science* **246**: 1149–1152, 1989.
5. Swain AL, Miller M, Green J, Rich DH, Schneider J, Kent SBH and Wlodawer A, X-ray crystallographic structure of a complex between a synthetic protease of human immunodeficiency virus 1 and a substrate-based hydroxyethylamine inhibitor. *Proc. Natl. Acad. Sci. USA* **87**: 8805–8809, 1990.
6. Fitzgerald PMD, McKeever BM, VanMiddlesworth JF, Springer JP, Heimbach JC, Leu C-T, Herber WK, Dixon AF and Darke PL, Crystallographic analysis of a complex between human immunodeficiency virus type 1 protease and acetyl-pepstatin at 2.0 Å resolution. *J. Biol. Chem.* **265**: 14209–14219, 1990.
7. Erickson J, Neidhart DJ, VanDrie J, Kempf DJ, Wang XC, Norbeck D, Plattner JJ, Rittenhouse J, Turon M, Wideburg N, Kohlbrenner WE, Simmer R, Helfrich R, Paul D and Knigg M, Design, activity and 2.8 Å crystal structure of a C2 symmetric inhibitor complexed to HIV-1 protease. *Science* **249**: 527–533, 1990.

8. Jaskólski M, Tomasselli AG, Sawyer TK, Staples DG, Heinrikson RL, Schneider J, Kent SBH and Wlodawer A, Structure at 2.5 Å resolution of chemically synthesized human immunodeficiency virus type 1 protease complexed with a hydroxyethylene-based inhibitor. *Biochemistry*, **30**: 1600–1609, 1991.
9. Gustchina A and Weber IT, Comparison of inhibitor binding in HIV-1 protease and in nonviral aspartic proteases: the role of the flap. *FEBS Lett.* **269**: 269–272, 1990.
10. Fitzgerald PMD, *MERLOT*, an integrated package of computer programs for the determination of crystal structures by molecular replacement. *J. Appl. Cryst.* **21**: 273–278.
11. Hendrickson WA, Stereochemically restrained refinement of macromolecular structures. *Methods Enzymol.* **115**: 252–270, 1985.
12. Rao JKM, Erickson JW and Wlodawer A, Structural and evolutionary relationships between retroviral and eucaryotic aspartic proteinases. *Biochemistry,* **30**: 4663–4671, 1991.
13. Satow Y, Cohen GH, Padlan EA and Davies DR, Phosphocholine binding immunoglobulin Fab McPC603, an X-ray diffraction study at 2.7 Å. *J. Mol. Biol.* **190**: 593–604, 1986.
14. Jones A, Interactive computer graphics: FRODO. *Methods Enzymol.* **115**: 157–171, 1985.

Anti-HIV Agents Interfering with the Viral gp120-Cellular CD4 Interaction

D. Schols and E. De Clercq

Rega Institute for Medical Research, Katholieke Universiteit Leuven, Minderbroedersstraat 10, B-3000 Leuven, Belgium

Abstract. The high affinity interaction of the viral envelope glycoprotein gp120 with the cellular CD4 receptor accounts for the tropism of human immunodeficiency virus (HIV) for CD4$^+$ cells [1] and represents an important target for therapeutic intervention. Any molecules that inhibit this interaction may be expected to block HIV infection at the earliest possible event of the viral replicative cycle, that is the virus adsorption step. Several classes of substances suppress HIV infection through interference with the virus adsorption/fusion step: i.e. sulfated polymers and polyanionic compounds in general, neoglycoproteins, mannose-specific binding lectins, glycosylation inhibitors and soluble CD4 derivatives.

1. Sulfated Polysaccharides, Polymers and Cyclodextrins

The inhibitory effects of sulfated polysaccharides (i.e. dextran sulfate, pentosan polysulfate and heparin) (Fig. 1) [2–16], sulfated polymers [sulfated polyvinylalcohol (PVAS) and the co-polymer of acrylic acid with sulfated vinylalcohol (PAVAS)] (Fig. 2) [17, 18] and sulfated cyclodextrins (Fig. 3) [19, 20] on the replication of HIV and HIV-induced syncytium formation have been well established.

PVAS and PAVAS are as efficient as dextran sulfate in inhibiting HIV replication, but more efficient than dextran sulfate in inhibiting HIV-induced syncytium formation [17, 18]. The mechanism of action of these compounds can be attributed to an inhibition of HIV attachment to CD4$^+$ cells, as has been shown by monitoring virus-cell binding with radiolabeled virus particles [8, 9, 11], flow cytometry [15, 18, 20, 21] and a radioimmunoassay [5].

Dextran sulfate and other sulfated polysaccharides do not interfere with the binding of OKT4A/Leu3a mAbs to the CD4$^+$ target cells [7, 8, 11, 21, 22], although dextran sulfate has been shown to interact with the CD4 molecule at an epitope distinct from the OKT4A epitope [22]. Furthermore, high-molecular-weight (500,000) dextran sulfate, but not low-molecular-weight (8,000) dextran sulfate or pentosan polysulfate, seems to modulate expression of CD4 receptors (also the OKT4A epitope), but only at very high concentrations [23]. It is not clear to what extent this effect on the CD4 receptor is correlated with the anti-HIV activity of dextran sulfate, since a molecular weight of 5,000 to 10,000 appears to

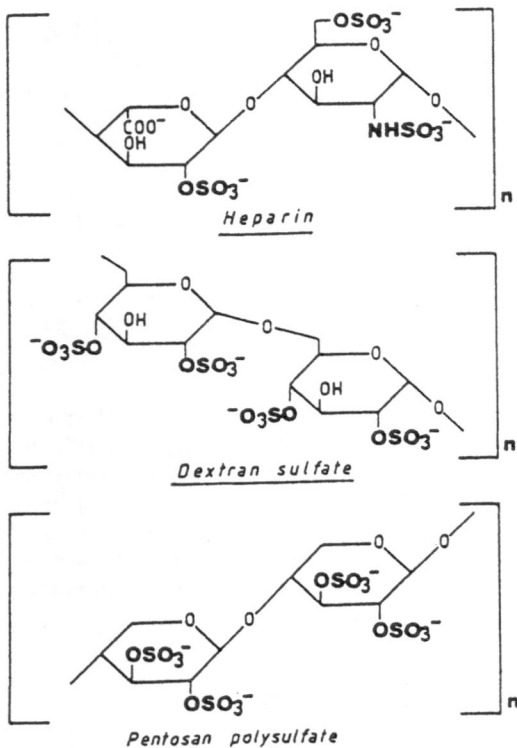

Fig. 1. Structural formulae of heparin and two sulfated polysaccharides: dextran sulfate and pentosan polysulfate. Heparin (MW: 6,000–20,000) consist of repeating units of L-iduronic acid (or D-glucuronic acid)-(1–4)-D-glucosamine, with sulfamide at C-2 of D-glucosamine and sulfate esters at C-6 of D-glucosamine and C-2 of L-iduronic acid. Dextran sulfate (MW: 1,000–500,000) contains a backbone of D-glucose units linked predominantly α-D-(1–6), with at an average two sulfate groups per glucose units. Pentosan polysulfate (MW: 3,100) can be considered as an oligomer of xylopyranose with at an average 1.8 sulfate groups per monomer

be optimal for the inhibitory effects of dextran sulfate on HIV-1 replication and giant cell formation [10] and such low-molecular-weight dextran sulfate samples do not directly interact with the CD4 receptor.

The susceptibility of HIV to the inhibitory effect of dextran sulfate varies among different HIV isolates and strains [15, 16, 24]. This suggests that dextran sulfate may interact differently with these different HIV isolates and strains depending on the nature of their envelope glycoproteins. A phase I/II trial of orally administered dextran sulfate has indicated little clinical efficacy [25]. This is not surprising since the oral bioavailibility of dextran sulfate is less than 1% [26]. Additional studies have been planned whereby dextran sulfate and pentosan polysulfate are administered intravenously [27]. Although the *in vitro* toxicity of the sulfated polymers for CD4$^+$ cells is extremely low, these compounds are

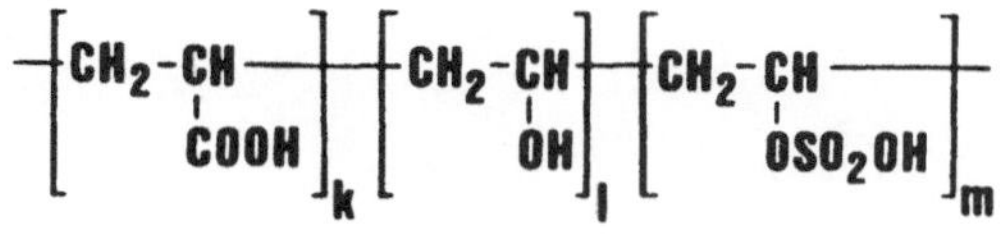

$$\left[CH_2-\underset{\underset{COOH}{|}}{CH}\right]_k\left[CH_2-\underset{\underset{OH}{|}}{CH}\right]_l\left[CH_2-\underset{\underset{OSO_2OH}{|}}{CH}\right]_m$$

PAVAS

$$\left[CH_2-\underset{\underset{OH}{|}}{CH}\right]_l\left[CH_2-\underset{\underset{OSO_2OH}{|}}{CH}\right]_m$$

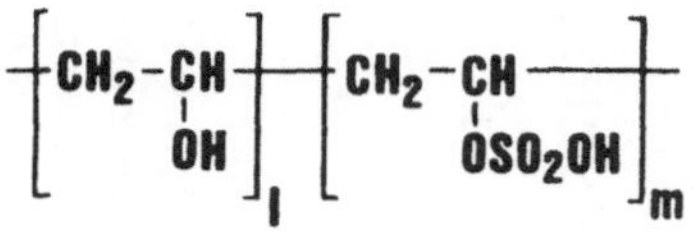

PVAS

Fig. 2. Structural formulae of the co-polymer of acrylic acid with sulfated vinylalcohol (PAVAS) and sulfated polyvinylalcohol (PVAS). For PAVAS, $k/(l+m) = 1/9$ and for PVAS, $m/l = 1$. Their molecular weight is 5,000–40,000, and the percentage of sulfation is 31 to 84%

known to interfere with the blood coagulation process. This anticoagulant activity might hamper their practical usefulness as anti-HIV drugs. Yet, dextran sulfate, pentosan polysulfate, PVAS and PAVAS are inhibitory to HIV *in vitro* at concentrations that are far below their anticoagulant threshold [17, 18], and as demonstrated with chemically modified heparin derivatives, anti-HIV activity can be dissociated from antithrombin activity [28].

That pentosan polysulfate and aurintricarboxylic acid (ATA) (see Section 2)

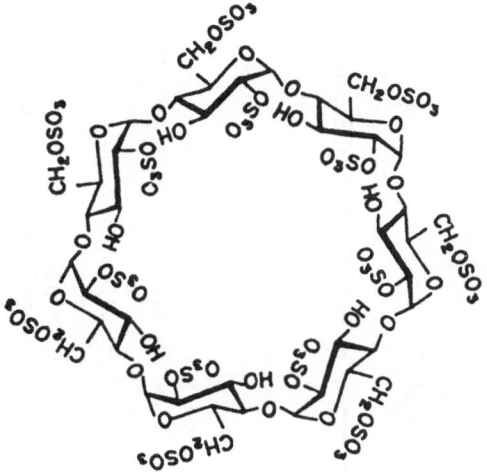

Fig. 3. Structural formula of β-cyclodextrin tetradecasulfate

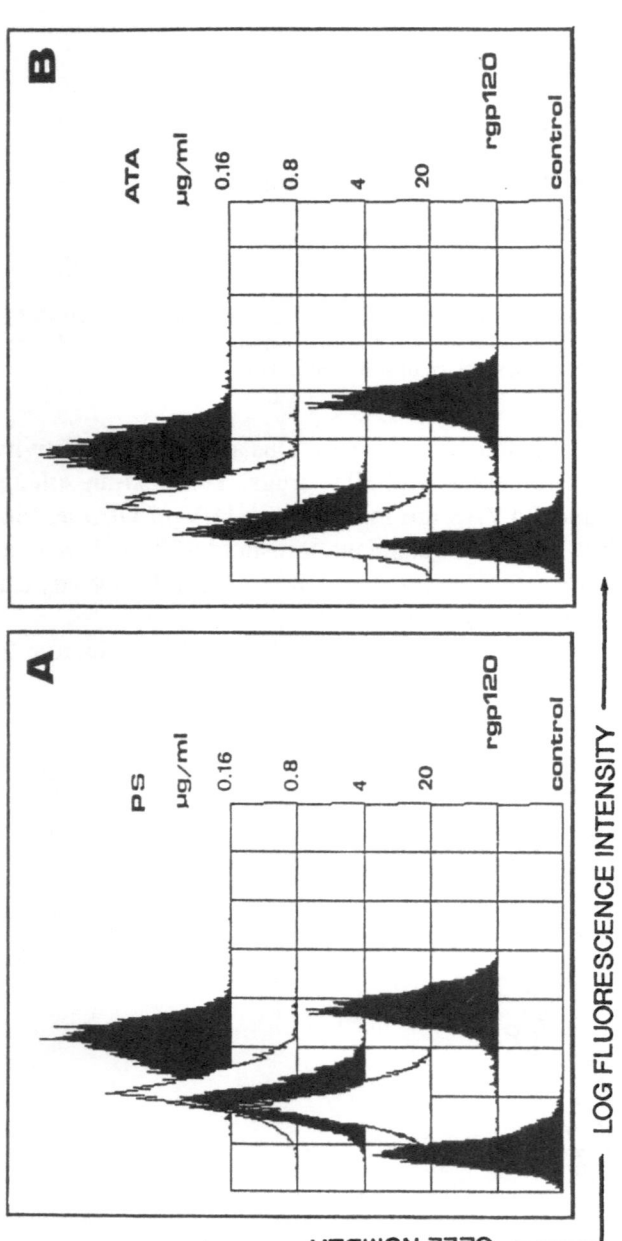

Fig. 4 A, B. Inhibitory effects of pentosan polysulfate (PS) (panel **A**) and aurintricarboxylic acid (ATA) (panel **B**) on the binding of rgp120 on CD4+ MT-4 cells. The first histogram (front) in both sets of histograms represents the control fluorescence of MT-4 cells incubated with anti-gp120 mAb and RaM-IgG-F(ab′)₂-FITC. The second histogram in each set of histograms represents the specific binding of rgp120 (10 µg/ml for 1 h at room temperature) (in the absence of PS or ATA) to the MT-4 cells. The third, fourth, fifth and sixth histograms represent the binding of rgp120 to the cells in the presence of PS or ATA concentrations varying from 0.16 µg/ml (back) to 20 µg/ml (front). The mean fluorescence intensity values were as follows: for set **A** of histograms (back to front): 39, 15, 12, 9, 87 and 3, respectively; and for set **B** of histograms (back to front): 29, 18, 17, 5, 87 and 3, respectively

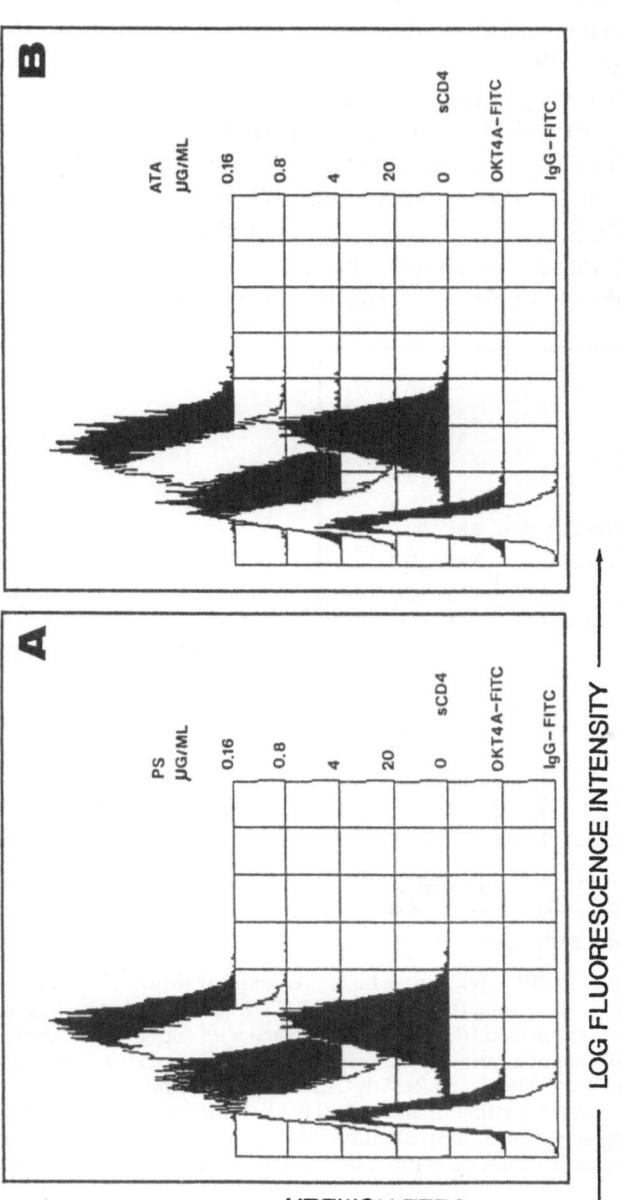

Fig. 5 A, B. Inhibitory effects of pentosan polysulfate (PS) (panel **A**) and aurintricarboxylic acid (ATA) (panel **B**) on the binding of sCD4 to the *env* glycoproteins expressed on persistently HIV-2(strain ROD)-infected HUT-78 cells. The first histogram (front) in both sets of histograms represents the control fluorescence of the HIV-2-infected HUT-78 cells incubated with IgG-FITC. The second histogram in each set of histograms represents the specific fluorescence of the HIV-2-infected HUT-78 cells incubated with OKT4-FITC (note the complete absence of CD4 on the cell surface of the infected cells). The third histogram represents the specific fluorescence of the infected cells incubated with sCD4 (10 μg/ml for 1 h at room temperature) and OKT4-FITC (in the absence of PS or ATA). The fourth, fifth, sixth and seventh histograms represent the binding of sCD4 to the infected cells in the presence of PS or ATA concentrations varying from 0.16 μg/ml (back) to 20 μg/ml (front). The mean fluorescence intensity values were as follows: for panel A of histograms (back to front) 26, 20, 9, 5, 31, 3 and 3, respectively; and for panel **B** of histograms (back to front) 25, 14, 8, 5, 31, 3 and 3, respectively

specifically interfere with the CD4-gp120 interaction is illustrated in Figs. 4 and 5. In Fig. 4 it is demonstrated that the binding of recombinant gp120 (rgp120) to CD4$^+$ target cells is inhibited by pentosan polysulfate (PS) and aurintricarboxylic acid (ATA) at very low concentrations. In Fig. 5 it is demonstrated that PS and ATA inhibit the binding of soluble recombinant CD4 (sCD4) to chronically HIV-2(strain ROD)-infected HUT-78 cells. Apparently, PS and ATA block the interaction of sCD4 with the viral (gp120) glycoproteins expressed on the surface of the chronically HIV-infected cells.

Dextran sulfate and pentosan polysulfate interfere with the binding of specific mAbs to the V3 fusion region of gp120 of HIV-1 (strain III-B) (Fig. 6) [29, 30]. These mAbs prevent a productive infection, although they do not prevent HIV binding to the CD4 receptor. This implies that dextran sulfate and its congeners not only interfere with the binding of HIV to the cells, but also interfere with a subsequent early event (i.e. fusion). It should also be noted that the IC$_{50}$ of pentosan polysulfate for inhibition of HIV-2 (strain ROD) cytopathicity in MT-4

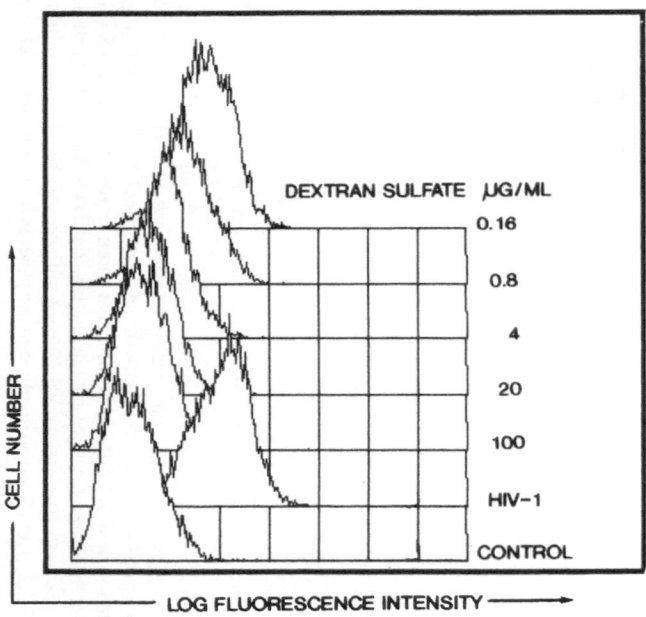

Fig. 6. Inhibitory effect of dextran sulfate (MW 5,000) on the binding of anti-gp120 mAb, recognizing the V3 loop of gp120, to persistently HIV-1(strain III-B)-infected HUT-78 cells. The first histogram represents the control fluorescence of uninfected HUT-78 cells incubated with anti-gp120 mAb and RaM-IgG-(Fab')$_2$-PE. The second histogram represents the specific fluorescence of HIV-1-infected HUT-78 cells incubated with the anti-gp120 mAb and RaM-IgG-(Fab')$_2$-PE. The third, fourth, fifth, sixth and seventh histograms represent the binding of anti-gp120 mAb to the infected cells which were pretreated for 15 min with dextran sulfate at concentrations varying from 0.16 µg/ml (back) to 100 µg/ml (front). The mean fluorescence values were (back to front) 81, 74, 63, 52, 47, 96 and 40, respectively

cells (0.005 µg/ml) is about 100-fold lower than the IC_{50} for inhibition of HIV-2 (strain ROD) binding to the cells, whereas the IC_{50} of pentosan polysulfate for inhibition of HIV-1 (strain III-B) binding to the cells (0.5 µg/ml) coincides with the IC_{50} for inhibition of HIV-1 (strain III-B) cytopathicity in MT-4 cells [15].

Recently, new sulfated polysaccharide derivatives have been described that, while retaining full anti-HIV activity, virtually lost all anticoagulant activity [28]. These compounds deserve to be further pursued for their therapeutic potential. Combination therapy of sulfated polymers with anti-HIV dideoxynucleoside analogues (i.e. AZT, DDI) can be considered, because of their different pharmacokinetics and different mechanism of anti-HIV activity. *In vitro,* combinations of these compounds have proved synergistic, or at least additive, in their anti-HIV activity [6, 19, 20, 31].

Sulfated polysaccharides, and sulfated polymers in general, are not only inhibitory to HIV but also to herpes simplex virus (HSV), cytomegalovirus (CMV) and various other enveloped viruses [18, 20, 32]. This increases their theurapeutic potential for the treatment of AIDS patients, where intercurrent HSV and in particular CMV infections have been found to accelerate disease progression and cause death [33, 34].

Recently, *N*-carbomethoxycarbonyl-prolyl-phenylalanyl benzyl esters (CPFs), have been reported to inhibit the binding of HIV-1 to CD4+ target cells. CPFs interact with gp120 and do not interfere with the binding of OKT4A/Leu3a mAbs to the CD4 epitope [35]. This result contrasts with that obtained for aurintricarboxylic acid (see Section 2), which can interact directly with the CD4 receptor [36].

2. Polyanionic Compounds

The polysulfonate suramin was the first compound found to inhibit HIV replication *in vitro* (Fig. 7A) [37–39]. Suramin had previously been shown to suppress the reverse transcriptase activity associated with murine and avian retroviruses [40]. Suramin also proved efficacious in suppressing retrovirus diseases in mice [41]. However, no significant clinical or immunological improvement was observed in AIDS or ARC patients treated with suramin [42–44]. As suramin is a polyanionic compound (containing six sulfonate groups), it was suggested that suramin may also interfere with the virus adsorption process [39]. We then demonstrated that suramin not only inhibits the reverse transcriptase of retroviruses, but also interferes with the binding of HIV virions to CD4+ cells [21]. In addition to suramin, several other sulfonated compounds, i.e. Evans Blue and naphthalenedisulfonic acid derivatives, have been found effective in inhibiting HIV cytopathicity [38, 45–48]. These compounds may also owe their anti-HIV activity to inhibition of virus adsorption [21; data not shown].

Another polyanionic compound that has been identified as a selective anti-HIV

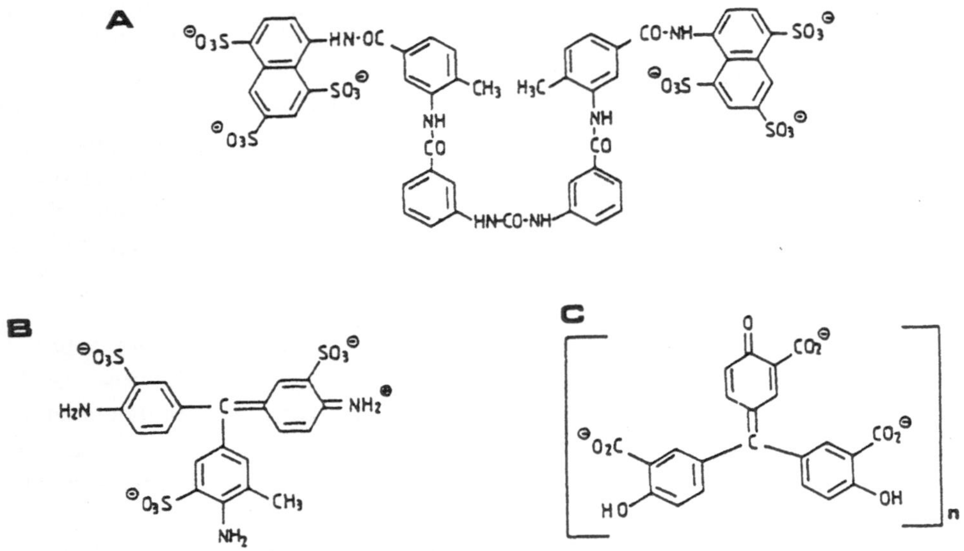

Fig. 7 A–C. Structural formulae of polyanionic compounds: polysulfonates (suramin (**A**), fuchsin acid (**B**)) and polycarboxylates (aurintricarboxylic acid (ATA) (**C**)). ATA is represented in its monomeric form, but actually occurs as a mixture of polymers

agent is aurintricarboxylic acid (ATA) [45]. Its IC_{50}, based on the inhibition of the HIV-1 cytopathicity in MT-4 cells, is 5.0 µM, and this concentration coincides with the concentration required to inhibit the HIV-1 reverse transcriptase (IC_{50}: 5.1 µM). Based on this coincidence, one may be tempted to attribute the anti-HIV-1 activity of ATA to an inhibitory effect on the reverse transcriptase [45].

However, ATA specifically modulates the expression of CD4 in peripheral blood lymphocytes (PBL) as demonstrated by a series of flow cytometry experiments [36]. In Fig. 8 dual fluorescence is used to demonstrate the specific interaction of ATA with anti-OKT4A mAb binding to PBL (green fluorescence). This specific interaction with OKT4A was observed at ATA concentrations ranging from 50 µM to 2 µM. At 50 µM, ATA also interfered with OKT4 expression, but to a lesser extent than with OKT4A expression. Aurin, lacking the carboxylic acid groups, had no effect on OKT4A expression in PBL. The specific effect of ATA on OKT4A expression was shown to be concentration-dependent, reversible, and cell type-independent [36]. The specificity of ATA as a modulator of OKT4A expression in PBL has been confirmed by others [49].

At those concentrations at which ATA was found to suppress anti-OKT4A binding to the cells, it also suppressed HIV adsorption to the cells [36]. Furthermore, at 50 µM ATA totally blocked HIV-induced cell fusion in co-cultures of

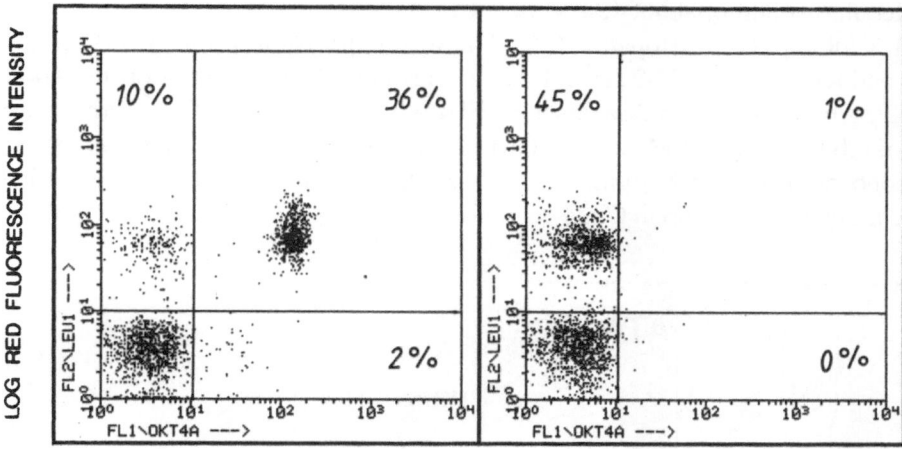

Fig. 8. Use of dual fluorescence to demonstrate that CD4 expression is specifically modulated by aurintricarboxylic acid (ATA). PBL (200,000 cells/100 µl PBS) were incubated with PBS (panel A) or 50 µM ATA (panel B) for 10–20 s, washed with PBS, incubated with anti-Leu1-PE and anti-OKT4A-FITC for 20 min at 4°C, washed with PBS, fixed and analyzed

persistently HIV-1- or HIV-2-infected cells with CD4$^+$ target cells [50, 51]. If ATA was added 3 hours after exposure of the MT-4 cells to HIV-1, it was no longer effective in inhibiting the replicative cycle of HIV-1. Thus, ATA is a unique compound that specifically modulates the expression of the cellular CD4 receptor without affecting the expression of other cell surface markers. Although ATA interacts specifically with the OKT4A epitope, this effect is noted only in the absence of serum. Also, ATA interacts with the viral gp120, even in the presence of serum [36, 52].

ATA is generally depicted as a triphenylmethane dye in its monomeric form (Fig. 7C) (MW: 422), although it occurs as an heterogenous mixture of polymers of the phenol formaldehyde type [52–54]. The polymer ATA preparations have been fractionated by dialysis, ultrafiltration, and gel permeation chromatography. All ATA fractions (MW: 138–>4,600) were evaluated for inhibition of HIV cytopathicity, inhibition of HIV antigen expression, inhibition of binding of OKT4A mAb to the CD4 receptor, inhibition of binding of anti-gp120 mAb to gp120, inhibition of attachment of HIV virions to CD4$^+$ cells, inhibition of HIV-1 reverse transcriptase activity, and cytotoxicity. In all assays, except for cytotoxicity, there was a direct correlation between activity and molecular weight. The higher the molecular weight, the higher the activity. The results obtained with the different ATA fractions indicate that the binding of ATA to gp120 suffices to explain its anti-HIV activity. The commercially available ATA behaved as if it had a molecular weight of about 4,600 [53, 54].

ATA inhibits viral cytopathicity and antigen expression of the HIV-1 strains III-B and RF and the HIV-2 strain ROD in MT-4 or CEM cells at a concentration of 1–2 μg/ml [15]. However, against the EHO strain of HIV-2, ATA is active at an IC_{50} of 0.004 μg/ml (CEM cells) or 0.01 μg/ml (MT-4 cells). As shown in Fig. 9, at a concentration of 0.03 μg/ml ATA completely blocks HIV-2(strain EHO)-induced antigen expression in CEM cells.

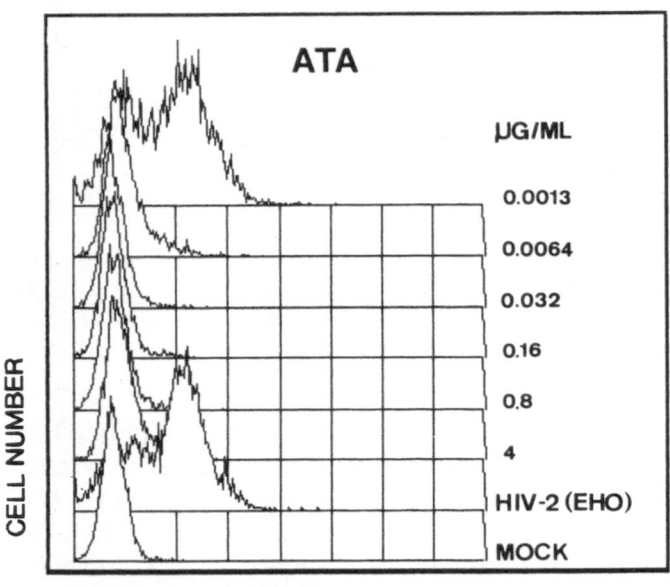

LOG FLUORESCENCE INTENSITY

Fig. 9. Effect of varying concentrations of aurintricarboxylic acid (ATA) on HIV-2 (strain EHO) antigen expression in CEM cells at 4 days post infection. The first histogram represents the control fluorescence of uninfected CEM cells incubated with specific HIV-2 polyclonal antibody and RaH-IgG-(Fab')$_2$-FITC. The second histogram represents the specific fluorescence of HIV-2-infected CEM cells incubated with the specific polyclonal antibody and RaH-IgG-(Fab')$_2$-FITC. The third, fourth, fifth, sixth, seventh and eighth histograms represent the binding of the polyclonal antibody to HIV-2(EHO)-infected cells incubated in the presence of ATA concentrations varying from 0.0013 μg/ml (back) to 4 μg/ml (front). The mean fluorescence values were (back to front) 57, 33, 26, 26, 27, 26, 57 and 25, respectively

Recently, various polyhydroxycarboxylates (MW: 3,800—14,000) derived from phenolic compounds have been found to selectively inhibit HIV replication *in vitro* [52]. The anti-HIV activity of this series of compounds can be attributed to the presence of the anionic carboxylate groups. The mechanism of action of these polyhydroxycarboxylates is similar to that of ATA. In this respect, ATA could be viewed as a prototype compound for probing the interaction between gp120 and CD4, with the aim to delineate the structural determinants that are needed to block this interaction and thus prevent virus binding to the cells.

Fuchsin acid, which is structurally related to ATA (Fig. 7B), also behaves as a selective inhibitor of HIV replication *in vitro* [55]. Unlike ATA, fuchsin acid does not interfere with the binding of HIV to the cells, does not inhibit the binding of mAbs to CD4 or gp120 and is only weakly inhibitory to the reverse transcriptase. The mechanism of anti-HIV activity of fuchsin acid remains subject of further study.

3. Neoglycoproteins

Neoglycoproteins, obtained by coupling sugars to human serum albumin, have also proved to inhibit HIV cytopathicity in MT-4 cells and syncytium formation between HIV-infected and uninfected CD4$^+$ cells [56]. Neoglycoproteins with the highest sugar content are the most effective. Their antiviral activity is determined primarily by their net negative charge. The compounds only partially inhibit the binding of free virus particles to the cells. Presumably they interact with the fusion of HIV virions of HIV-infected cells with the CD4$^+$ target cells (see also Section 4).

4. Mannose-Specific Plant Lectins

Lectins with specificity for D-mannose (i.e. Concanavalin A) inhibit HIV-induced cell fusion and infectivity, as originally shown by Lifson et al. [57]. Later, various other mannose-specific lectins have been found to inhibit HIV-induced syncytium formation [58–61]

In our study, we demonstrated that D-mannose-specific plant lectins achieve their anti-HIV activity by a mechanism that is clearly different from that of the sulfated polysaccharides. In contrast with the sulfated polysaccharides, the plant lectins do not inhibit HIV binding to the cells (Fig. 10), although they are able to inhibit HIV cytopathicity even if added 4 to 8 hours post infection. These data suggest that the mannose-specific lectins may be targeted at the fusion process, which corresponds to the earliest possible event following attachment of HIV to the cells. Also, combination of these plant lectins with sulfated polysaccharides leads to a synergistic inhibition of HIV-induced cytopathicity and syncytium formation [61].

5. Glycosylation Inhibitors

Since the envelope glycoprotein of HIV-1 is heavily glycosylated, inhibitors of the oligosaccharide transfer to protein or inhibitors of the processing of the protein

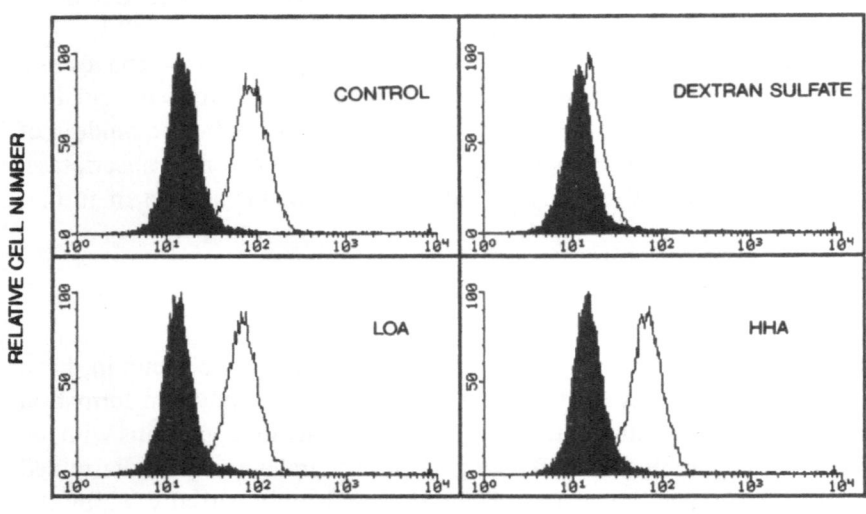

Fig. 10. Effects of mannose-specific plant lectins on HIV-1 binding to MT-4 cells. The dark histograms represent cellular fluorescence resulting from nonspecific binding of HIV-1 antibody to MT-4 cells (which were not exposed to HIV-1 virions) in the absence of any test compound (control), or in the presence of dextran sulfate (MW 5,000) (25 µg/ml), or α-mannose-specific lectin from *Listera ovata* (LOA) (100 µg/ml), or α-mannose-specific lectin from *Hippeastrum hybrid* (HHA) (100 µg/ml). The open histograms represent cellular fluorescence resulting from specific binding of HIV-1 antibody to MT-4 cells (which had adsorbed HIV-1 virions) in the absence of test compound (control) or in the presence of dextran sulfate (25 µg/ml), LOA (100 µg/ml) or HHA (100 µg/ml)

N-linked oligosaccharides may be expected to affect HIV replication, virus infectivity or virus-induced syncytium formation. These were examined for their inhibitory effects on HIV replication: inhibitors of the transfer of *N*-linked oligosaccharides to protein (tunicamycin) and inhibitors of the processing of *N*-linked oligosaccharides, e.g. inhibitors of the trimming enzymes, glucosidases I and II [castanospermine (1,6,7,8-tetrahydroxyoctahydroindolizine), 1-deoxy-nojirimycin and its methyl, ethyl, butyl and other derivatives], mannosidase I (1-deoxymannojirimycin), mannosidase II (swainsonine), and inhibitors of the protein transport from the endoplasmatic reticulum to the Golgi apparatus (monensin). Various effects were observed: reduction of HIV replication, infectivity and cytopathicity in the presence of tunicamycin and inhibitors of glucosidases I and II and mannosidase I, but not mannosidase II [62–70]; diminished cell surface expression of the envelope glycoproteins in the presence of castanospermine [64], monensin [71] and *N*-butyldeoxynojirimycin [69]; inhibition of cleavage of gp160 to gp120 and gp41 in the presence of glucosidase inhibitors [64, 68, 69, 73] or monensin [71]; no interference with viral envelope glycoprotein binding to CD4 [64, 68, 69], except for tunicamycin which abolished binding [73]. Among the glucosidase inhibitors, *N*-butyldeoxynojirimycin and 6-*O*-butanoylcastanospermine were found to have rather strong anti-HIV activity [65, 69, 74]. Castanospermine and its 6-*O*-butanoyl derivative also proved efficacious in mice infected with the Rauscher murine leukemia virus [75]. Johnson et al. [76] demonstrated synergistic action of castanospermine with AZT *in vitro*. From our own studies it appeared that the susceptibility of HIV to castanospermine varies considerably from one virus strain to another and from one cell type to another [15]. At present, *N*-butyldeoxynojirimycin is the subject of a phase I clinical trial in the United States [77].

6. Soluble CD4 and Derivatives

Various forms of soluble recombinant CD4 molecules (referred to as sCD4, rCD4 or rsCD4) have been obtained by recombinant DNA technology. Such rCD4 molecules inhibit HIV binding to the cells, HIV replication and HIV-induced syncytium formation [78–82]. A fragment containing only the two amino terminal Ig-like domains of rCD4 is almost as inhibitory as the whole soluble rCD4 [82, 83]. Also, synthetic CD4 peptide derivatives have been evaluated for their ability to inhibit HIV infection and HIV-induced syncytium formation [84–86]. Their mode of action can obviously be attributed to inhibition of the binding of HIV gp120 to intact CD4 molecules on the surface of the CD4$^+$ target cells. *In vitro*, soluble rCD4 molecules do not interfere with major MHC class II interactions [80, 87, 88].

Soluble rCD4 has a terminal half-life of about 15 min (in rabbits). To prolong this half-life, hybrid molecules have been constructed by genetically combining CD4 with the constant heavy chain domains of immunoglobulins (rCD4-Ig) [89–91]. Such constructs have been termed CD4-immunoadhesins. These hybrid molecules show anti-HIV-1 activity, comparable to that of soluble rCD4. Because of the presence of the Ig heavy chain domains, the CD4-immunoadhesins possess a much longer plasma half-life (7 to 48 h). They can also cross the placental barriers. The rCD4 portion of the CD4-immunoadhesins inhibits HIV infection by interfering with virion binding to the CD4$^+$ target cells.

Soluble rCD4 or mAbs specific for the viral envelope glycoproteins can be linked to toxins, such as ricin or *Pseudomonas exotoxin*, and thus may be expected to kill HIV-infected cells expressing gp120 [90, 92–96]. Such toxin-rCD4 hybrid molecules can be considered as rCD4-immunotoxins. They are able to destroy and thus eliminate persistently HIV-infected cells (i.e. macrophages), which otherwise may go on for ever in producing HIV virions. *In vitro*, rCD4-immunotoxins have no effect on major MHC class II interactions [96].

Fresh HIV-1 isolates appear to be less susceptible to sCD4 *in vitro* than the III-B strain of HIV-1 [97]. HIV-2 strains are also less susceptible to the inhibitory

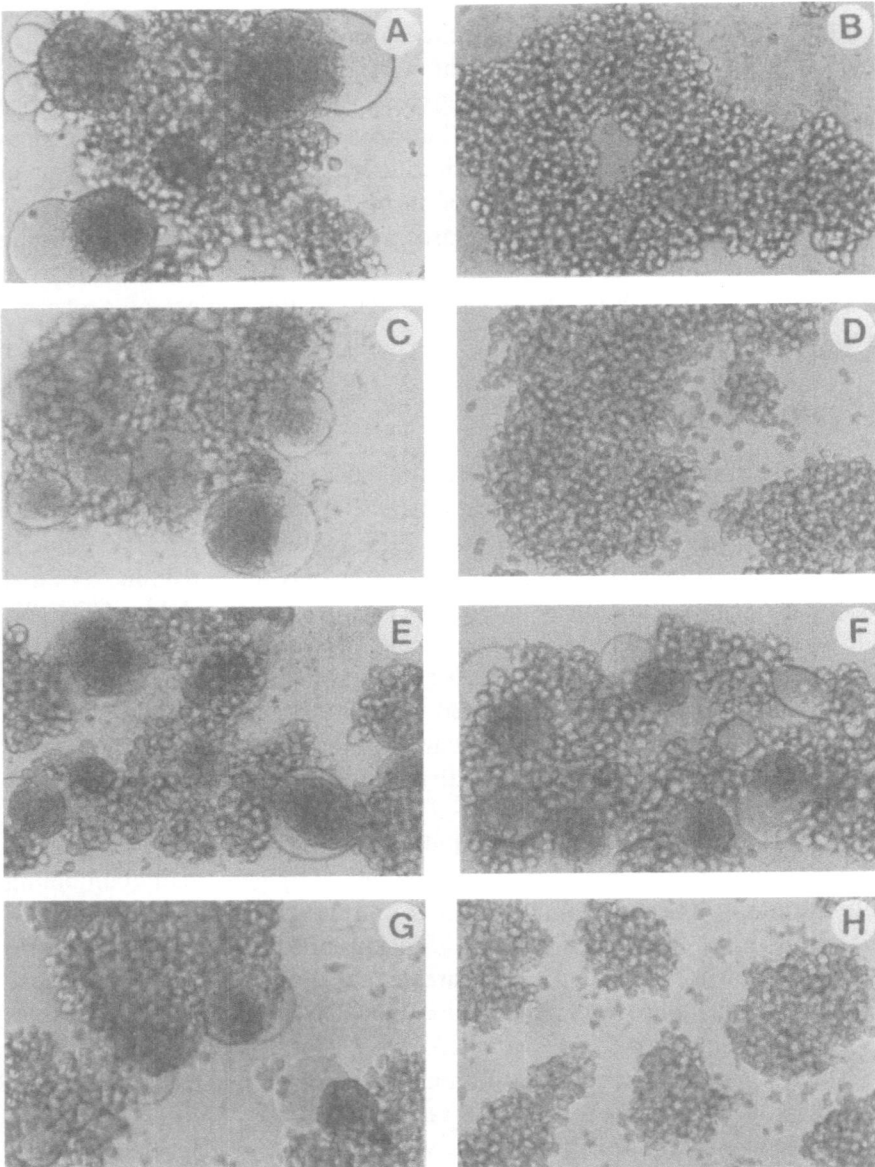

Fig. 11 A–H. Effect of sCD4 on syncytium formation between HIV-1(strain III-B)-infected or HIV-2(strain ROD)-infected HUT-78 cells and MOLT-4 or CEM cells. Equal numbers of HIV-1-infected HUT-78 cells and MOLT-4 cells (panels A and B), HIV-1-infected HUT-78 cells and CEM cells (panels C and D), HIV-2-infected HUT-78 cells and MOLT-4 cells (panels E and F) and HIV-2-infected HUT-78 cells and CEM cells (panels G and H) were cultured for 24 h in the presence of control medium (panels A, C, E and G) or in the presence of sCD4 (20 µg/ml) (panels B, D, F and H). The cells were cultured under the same conditions and photographed at the same time. Magnification ×100

effects of soluble rCD4 than HIV-1 strains [98]. Recently, it has been shown that sCD4 may even enhance SIV infection *in vitro* [99].

When sCD4 was evaluated for its inhibitory effect on syncytium formation, HIV-1(strain III-B)-induced giant cell formation was inhibited by sCD4 at an IC_{50} of 4 μg/ml (see also Figs. 11B and 11D). For HIV-1(strain RF)-induced syncytium formation with MOLT-4 as target cells, the IC_{50} of sCD4 was 100 μg/ml, whereas for HIV-1(strain RF)-induced syncytium formation with CEM as target cells, the IC_{50} was 2 μg/ml. The IC_{50} of sCD4 in the HIV-2 (strain ROD) syncytium assay with CEM as target cells was 2 μg/ml (Fig. 11H), but if MOLT-4 cells were used as the target cells, the IC_{50} was > 100 μg/ml (Fig. 11F) [15]. Sekigawa et al. [100] reported that when sCD4 was added to mixtures of HIV-2(strain ROD)-infected H9 cells with C8166 cells, it did not inhibit syncytium formation. A lower affinity of CD4 for HIV-2 than HIV-1 envelope glycoproteins [101] has been implicated as the reason for the failure of sCD4 to inhibit HIV-2-induced syncytium formation [100]. However, the inhibitory effect of sCD4 on HIV-induced formation is not only dependent on the virus strain [15, 100] but also on the target cell type [15].

In a recently completed phase I/II study, a decline in serum p24 antigen was observed in individuals receiving 30 mg sCD4 per day [102]. However, in another clinical trial with sCD4 no clear-cut evidence of anti-HIV activity in patients with AIDS or ARC was observed [103]. It may prove necessary to use rCD4, rCD4-Ig or rCD4-toxins in combination with other anti-retrovirus agents such as AZT [104–106].

In conclusion, compounds interfering with the cellular CD4 receptor and/or viral envelope gp120 glycoproteins could be considered as new markers for probing the CD4-gp120 interaction. They can also represent important leads in the development of anti-HIV agents that block HIV at the earliest possible stage of its replicative cycle, that is attachment to its cellular receptor.

References

1. Dalgleish AG, Beverley PCL, Clapham PR, Crawford DH, Greaves MF and Weiss RA, The CD4 (T4) antigen is an essential component of the receptor for the AIDS retrovirus. *Nature* **312**: 763–767, 1984.
2. Ito M, Baba M, Sato R, Pauwels R, De Clercq E and Shigeta S, Inhibitory effect of dextran sulfate and heparin on the replication of human immunodeficiency virus (HIV) *in vitro*. *Antiviral Res.* **7**: 361–367, 1987.
3. Ito M, Baba, M, Hirabayshi K, Matsumoto T, Suzuki S, Shigeta S and De Clercq E, *In vitro* activity of mannan sulfate, a novel sulfated polysaccharide, against human immunodeficiency virus type 1 and other enveloped viruses. *Eur. J. Clin. Microbiol. Infect. Dis.* **8**: 171–173, 1989.
4. Nakashima H, Yoshida O, Tochikura TS, Yoshida T, Mimura T, Kido Y, Motoki Y, Kaneko Y, Uryu T and Yamamoto N, Sulfation of polysaccharides generates potent and selective inhibitors

of human immunodeficiency virus infection and replication *in vitro*. *Jpn J. Cancer. Res.* **78**: 1164–1168, 1987.

5. Nakashima H, Yoshida O, Baba M, De Clercq E and Yamamoto N, Anti-HIV activity of dextran sulfate as determined under different experimental conditions. *Antiviral Res.* **11**: 233–246, 1989.

6. Ueno R and Kuno S, Dextran sulphate, a potent anti-HIV agent *in vitro* having synergism with zidovudine. *Lancet* i: 1379, 1987.

7. Bagasra O and Lischner HW, Activity of dextran sulfate and other polyanionic polysaccharides against human immunodeficiency virus. *J. Infect. Dis.* **158**: 1084–1087, 1988.

8. Baba M, Pauwels R, Balzarini J, Arnout J, Desmyter J and De Clercq E, Mechanism of inhibitory effect of dextran sulfate and heparin on replication of human immunodeficiency virus *in vitro*. *Proc. Natl. Acad. Sci. USA.* **85**: 6132–6136, 1988.

9. Baba M, Nakajima M, Schols D, Pauwels R, Balzarini J and De Clercq E, Pentosan polysulfate, a sulfated oligosaccharide, is a potent and selective anti-HIV agent *in vitro*. *Antiviral Res.* **9**: 335–343, 1988.

10. Baba M, Schols D, Pauwels R, Nakashima H and De Clercq E, Sulfated polysaccharides as potent inhibitors of HIV-induced syncytium formation: a new strategy towards AIDS chemotherapy. *J. Acquir. Immun. Defic. Syndromes* **3**: 493–499, 1990.

11. Mitsuya H, Looney DJ, Kuno S, Ueno R, Wong-Staal F and Broder S, Dextran sulfate suppression of viruses in the HIV family: inhibition of virion binding to CD4$^+$ cells. *Science* **240**: 646–649, 1988.

12. Hartman NR, Johns DG and Mitsuya H, Pharmacokinetic analysis of dextran sulfate in rats as pertains to its clinical usefulness for therapy of HIV infection. *AIDS Res. Hum. Retroviruses* **6**: 805–812, 1990.

13. von Briesen H, Meichsner C, Andreesen R, Esser R, Schrinner E and Rübsamen-Waigmann H, The polysulphated polyxylan Hoe/Bay-946 inhibits HIV replication on human monocytes/macrophages. *Res. Virol.* **141**: 251–257, 1990.

14. Yoshida T, Hatanaka K, Uryu T, Kaneko Y, Suzuki E, Miyano H, Mimura T, Yoshida O and Yamamoto N, Synthesis and structural analysis of curdlan sulfate with a potent inhibitory effect *in vitro* of AIDS virus infection. *Macromolecules* **23**: 3717–3722, 1990.

15. Schols D, Pauwels R, Desmyter J and De Clercq E, Differential activity of polyanionic compounds and castanospermine against HIV replication and HIV-induced syncytium formation depending on virus strain and cell type. *Antiviral Chem. Chemother.*, in press.

16. Witvrouw M, Schols D, Andrei G, Snoeck R, Pauwels R, Balzarini J and De Clercq E, Differential antiviral effects of low-molecular-weight dextran sulphate (derived from dextran MW 1,000) as compared to dextran sulphate samples of higher molecular weight. *Antiviral Chem. Chemother.*, in press.

17. Baba M, Schols D, De Clercq E, Pauwels R, Nagy M, Györgyi-Edelényi J, Löw M and Görög S, Novel sulfated polymers as highly potent and selective inhibitors of human immunodeficiency virus replication and giant cell formation. *Antimicrob. Agents Chemother.* **34**: 134–138, 1990.

18. Schols D, De Clercq E, Balzarini J, Baba M, Witvrouw M, Hosoya M, Andrei G, Snoeck R, Neyts J, Pauwels R, Nagy M, Györgyi-Edelényi J, Machovich R, Horváth I, Löw M and Görög S, Sulfated polymers are potent and selective inhibitors of various enveloped viruses, including herpes simplex virus, cytomegalovirus, vesicular stomatitis virus, respiratory syncytial virus, and toga-, arena- and retroviruses. *Antiviral Chem. Chemother.* **1**: 233–240, 1990.

19. Anand R, Nayyar S, Pitha J and Merril CR, Sulphated sugar alpha-cyclodextrin sulphate, a uniquely potent anti-HIV agent, also exhibits marked synergism with AZT, and lymphoproliferative activity. *Antiviral Chem. Chemother.* **1**: 41–46, 1990.

20. Schols D, De Clercq E, Witvrouw M, Nakashima H, Snoeck R, Pauwels R, Van Schepdael A and Claes P, Sulphated cyclodextrins are potent anti-HIV agents acting synergistically with 2′, 3′-dideoxynucleoside analogues. *Antiviral Chem. Chemother.* **2**: 45–53, 1991.

21. Schols D, Baba M, Pauwels R and De Clercq E, Flow cytometric method to demonstrate whether anti-HIV-1 agents inhibit virion binding to T4⁺ cells. *J. Acquir. Immun. Defic. Syndromes* **2** 10–15, 1989.

22. Lederman S, Gulick R and Chess L, Dextran sulfate and heparin interact with CD4 molecules to inhibit the binding of coat protein (gp120) of HIV. *J. Immunol.* **143**: 1149–1154, 1989.

23. Parish CR, Low L, Warren HS and Cunningham AL, A polyanion binding site on the CD4 molecule. Proximity to the HIV-gp120 binding region. *J. Immunol.* **145**: 1188–1195, 1990.

24. Busso ME and Resnick L, Anti-human immunodeficiency virus effects of dextran sulfate are strain dependent and synergistic or antagonistic when dextran sulfate is given in combination with dideoxynucleosides. *Antimicrob. Agents Chemother.* **34**: 1991–1995, 1990.

25. Abrams DI, Kuno S, Wong R, Jeffords K, Nash M, Molaghan JB, Gorter R and Ueno R, Oral dextran sulfate (UA001) in the treatment of the acquired immunodeficiency syndrome (AIDS) and AIDS-related complex. *Ann. Intern. Med.* **110**: 183–188, 1989.

26. Lorentsen KJ, Hendrix CW, Collins JM, Kornhauser DM, Petty BG, Klecker RW, Flexner C, Eckel RH and Lietman PS, Dextran sulfate is poorly absorbed after oral administration. *Ann. Intern. Med.* **111**: 561–566, 1989.

27. Biesert L, Suhartono H, Winkler I, Meichsner C, Helsberg M, Hewlett G, Klimetzek V, Mölling K, Schlumberger HD, Schrinner E, Brede HD and Rübsamen-Waigmann H, Inhibition of HIV and virus replication by polysulphated polyxylan: HOE/BAY 946, a new antiviral compound. *AIDS* **2**: 449–457, 1988.

28. Baba M, De Clercq E, Schols D, Pauwels R, Snoeck R, Van Boeckel C, Van Dedem G, Kraaijeveld N, Hobbelen P, Ottenheijm H and Den Hollander F, Novel sulfated polysaccharides: dissociation of anti-human immunodeficiency virus activity from antithrombin activity. *J. Infect. Dis.* **161**: 208–213, 1990.

29. Schols D, Pauwels R, Desmyter J and De Clercq E, Dextran sulfate and other polyanionic anti-HIV compounds specifically interact with the viral gp120 glycoprotein expressed by T-cells persistently infected with HIV-1. *Virology* **175**: 556–561, 1990.

30. Callahan LN, Phelan M, Mallinson M, and Norcross MA, Dextran sulfate blocks antibody binding to the principal neutralizing domain of human immunodeficiency virus type 1 without interfering with gp120-CD4 interactions. *J. Virol.* **65**: 1543–1550, 1991.

31. Sugawara I, Itoh W, Kimura S, Mori S and Shimada K, Further characterization of sulfated homopolysaccharides as anti-HIV agents. *Experientia* **45**: 996–998, 1989.

32. Baba M, Snoeck R, Pauwels R and De Clercq E, Sulphated polysaccharides are potent and selective inhibitors of various enveloped viruses, including herpes simplex virus, cytomegalovirus, vesicular stomatitis virus and human immunodeficiency virus. *Antimicrob. Agents Chemother.* **32**: 1742–1745, 1988.

33. Macher AM, Reichert CM, Stauss SE, Longo DL, Parrillo J, Lane HC, Fauci AS, Rook AH, Manischewitz JF and Quinnan GV, Death in the AIDS patient: role of cytomegalovirus. *New. Engl. J. Med.* **309**: 1454, 1983.

34. Webster A, Cytomegalovirus as a possible cofactor in HIV disease progression. *J. Acquir. Immun. Defic. Syndromes* **4**: S47–S52, 1991.

35. Finberg RW, Diamond DC, Mitchell DB, Rosenstein Y, Soman G, Norman TC, Schreiber SL and Burakoff SJ, Prevention of HIV-1 infection and preservation of CD4 function by the binding of CPFs to gp120. *Science* **249**: 287–291, 1990.

36. Schols D, Baba M, Pauwels R, Desmyter J and De Clercq E, Specific interaction of aurintricarboxylic acid with the human immunodeficiency virus/CD4 cell receptor. *Proc. Natl. Acad. Sci. USA* **86**: 3322–3326, 1989.

37. Mitsuya H, Popovic M, Yarchoan R, Matsushita S, Gallo RC and Broder S, Suramin protection of T cells *in vitro* against infectivity and cytopathic effect of HTLV-III. *Science* **222**: 172–174, 1984.

38. Balzarini J, Mitsuya H, De Clercq E and Broder S, Comparative inhibitory effects of suramin and other selected compounds on the infectivity and replication of human T-cell lymphotropic virus (HTLV-III)/lymphadenopathy-associated virus (LAV). *Int. J. Cancer* **37**: 451–457, 1986.

39. De Clercq E, Suramin in the treatment of AIDS: mechanism of action. *Antiviral Res.* **7**: 1–10, 1987.

40. De Clercq E, Suramin: a potent inhibitor of the reverse transcriptase of RNA tumor viruses. *Cancer Lett.* **8**: 9–22, 1979.

41. Ruprecht RM, Rossoni LD, Haseltine WA and Broder S, Suppression of retroviral propagation and disease by suramin in murine systems. *Proc. Natl. Acad. Sci. USA* **82**: 7733–7737, 1985.

42. Broder S, Yarchoan R, Collins JM, Lane HC, Markham PD, Klecker RW, Redfield RR, Mitsuya H, Hoth DF, Gelmann E, Groopman JE, Resnick L, Gallo RC, Myers CE and Fauci AS, Effects of suramin on HTLV-III/LAV infection presenting as Kaposi's sarcoma or AIDS-related complex: clinical pharmacology and suppression of virus replication *in vivo*. *Lancet* **ii:** 627–630, 1987.

43. Kaplan LD, Wolfe PR, Volberding PA, Feorino P, Levy JA, Abrams DI, Kiprov D, Wong R, Kaufman L and Gottlieb MS, Lack of response to suramin in patients with AIDS and AIDS-related complex. *Am. J. Med.* **82**: 615–620, 1987.

44. Rasheed S, Gowda S, Gill PS, Meyer PR and Levine AM, Antiviral effects of suramin in patients with the acquired immunodeficiency syndrome. *Int J. Immunother.* **3**: 81–88, 1987.

45. Balzarini J, Mitsuya H, De Clercq E and Broder S, Aurintricarboxylic acid and Evans Blue represent two different classes of anionic compounds which selectively inhibit the cytopathogenicity of human T-cell lymphotropic virus type III/lymphadenopathy-associated virus. *Biochem. Biophys. Res. Commun.* **136**: 64–71, 1986.

46. Mohan P, Singh R, Wepsiec J, Gonzalez I, Sun DK and Sarin PS, Inhibition of HIV replication by naphthalenemonosulfonic acid derivatives and a bis naphtalenedisulfonic acid compound. *Life Sci.* **47**: 993–999, 1990.

47. Mohan P, Singh R and Baba M, Potential anti-AIDS agents. Synthesis and antiviral activity of naphthalenesulfonic acid derivatives against HIV-1 and HIV-2. *J. Med. Chem.* **34**: 212–217, 1991.

48. Mohan P, Singh R and Baba M, Anti-HIV-1 and HIV-2 activity of naphthalenedisulfonic acid derivatives. Inhibition of cytopathogenesis, giant cell formation, and reverse transcriptase activity. *Biochem. Pharmacol.* **41**: 642–646, 1991.

49. Weaver JL, Gergely P, Pine PS, Patzer E and Aszalos A, Polyionic compounds selectively alter availability of CD4 receptors for HIV coat protein rgp120. *AIDS Res. Hum. Retroviruses* **6**: 1125–1130, 1990.

50. Schols D, Pauwels R, Baba M, Desmyter J and De Clercq E, Syncytium formation and destruction of bystander CD4$^+$ cells cocultured with T cells persistently infected with human immunodeficiency virus as demonstrated by flow cytometry. *J. Gen. Virol.* **70**: 2397–2408, 1989.

51. Schols D, Pauwels R, Desmyter J and De Clercq E, Flow cytometric method to monitor the destruction of CD4$^+$ cells following their fusion with HIV-infected cells. *Cytometry* **11**: 736–743, 1990.

52. Schols D, Wutzler P, Klöcking R, Helbig B and De Clercq E, Selective inhibitory activity of polyhydroxycarboxylates derived from phenolic compounds against human immunodeficiency virus replication. *J. Acquir. Immun. Defic. Syndromes* **4**: 677–685, 1991.

53. Cushman M, Wang P, Chang SH, Wild C, De Clercq E, Schols D, Goldman ME and Bowen JA, Preparation and anti-HIV activities of aurintricarboxylic acid fractions and analogues: direct correlation of antiviral potency with molecular weight. *J. Med. Chem.* **34**: 329–337, 1991.

54. Cushman M, Kanamathareddy S, De Clercq E, Schols D, Goldman ME and Bowen JA, Synthesis and anti-HIV activities of low molecular weight aurintricarboxylic acid fragments and related compounds. *J. Med. Chem.* **34**: 337–342, 1991.

55. Baba M, Schols D, Pauwels R, Balzarini J and De Clercq E, Fuchsin acid selectively inhibits human immunodeficiency virus (HIV) replication *in vitro*. *Biochem. Biophys. Res. Commun.* **155**: 1404–1411, 1988.

56. Jansen RW, Molema G, Pauwels R, Schols D, De Clercq E and Meijer DKF, Potent *in vitro* anti-HIV-1 activity of modified human serum albumins. *Molec. Pharmacol.* **39**: 818–823, 1991.

57. Lifson JD, Coutré S, Huang E and Engleman E, Role of envelope glycoprotein carbohydrate in human immunodeficiency virus (HIV) infectivity and virus-induced cell fusion. *J. Exp. Med.* **164**: 2101–2106, 1986.

58. Müller WEG, Renneisen K, Kreuter MH, Schröder HC and Winkler I, The D-mannose-specific lectin from *Gerardia savaglia* blocks binding of human immunodeficiency virus type I to H9 cells and human lymphocytes *in vitro*. *J. Acquir. Immun. Defic. Syndromes* **1**: 453–458, 1988.

59. Hansen JES, Nielsen CM, Nielsen C, Heegaard P, Mathiesen LR and Nielsen JO, Correlation between carbohydrate structures on the envelope glycoprotein gp120 of HIV-1 and HIV-2 and syncytium inhibition with lectins. *AIDS* **3**: 635–641, 1989.

60. Weiler BE, Schröder HC, Stefanovich V, Stewart D, Forrest JMS, Allen LB, Bowden BJ, Kreuter MH, Voth R and Müller WEG, Sulphoevernan, a polyanionic polysaccharide, and the narcissus lectin potently inhibit human immunodeficiency virus infection by binding to viral envelope protein. *J. Gen. Virol.* **71**: 1957–1963, 1990.

61. Balzarini J, Schols D, Neyts J, Van Damme E, Peumans W and De Clercq E, α-(1–3)- and α-(1–6)-D-mannose-specific plant lectins are markedly inhibitory to human immunodeficiency virus and cytomegalovirus infection *in vitro*. *Antimicrob. Agents Chemother.* **35**: 410–416, 1991.

62. Gruters RA, Neefjes JJ, Tersmette M, de Goede REY, Tulp A, Huisman HG, Miedema F and Ploegh HL, Interference with HIV-induced syncytium formation and viral infectivity by inhibitors of trimming glucosidase. *Nature* **330**: 74–77, 1987.

63. Tyms AS, Berrie EM, Ryder TA, Nash RJ, Hegarty MP, Taylor DL, Mobberley MA, Davis JM, Bell EA, Jeffries DJ, Taylor-Robinson D and Fellows LE, Castanospermine and other plant alkaloid inhibitors of glucosidase activity block the growth of HIV. *Lancet* **ii**: 1025–1026, 1987.

64. Walker BD, Kowalski M, Goh WC, Kozarsky K, Krieger M, Rosen C, Rohrschneider L, Haseltine WA and Sodroski J, Inhibition of human immunodeficiency virus syncytium formation and virus replication by castanospermine. *Proc. Natl. Acad. Sci. USA* **84**: 8120–8124, 1987.

65. Karpas A, Fleet GWJ, Dwek RA, Petursson S, Namgoong SK, Ramsden NG, Jacob GS and Rademacher TW, Aminosugar derivatives as potential anti-human immunodeficiency virus agents. *Proc. Natl. Acad. Sci. USA* **85**: 9229–9233, 1988.

66. Montefiori DC, Robinson WE and Mitchell WM, Role of protein *N*-glycosylation in pathogenesis of human immunodeficiency virus type 1. *Proc. Natl. Acad. Sci. USA* **85**: 9248–9252, 1988.

67. Pal R, Hoke GM and Sarngadharan MG, Role of oligosaccharides in the processing and maturation of envelope glycoproteins of human immunodeficiency virus type 1. *Proc. Natl. Acad. Sci. USA* **86**: 3384–3388, 1989.

68. Pal R, Kalyanaramen VS, Hoke GM and Sarngadharan MG, Processing and secretion of envelope glycoproteins of human immunodeficiency virus type 1 in the presence of trimming glucosidase inhibitor deoxynojirimycin. *Intervirology* **30**: 27–35, 1989.

69. Dedera D, Vander Heyden N and Ratner L, Attenuation of HIV-1 infectivity by an inhibitor of oligosaccharide processing. *AIDS Res. Hum. Retroviruses* **6**: 785–793, 1990.

70. Shimizu H, Tsuchie H, Yoshida K, Morikawa S, Tsuruoka T, Yamamoto H, Ushijima H and Kitamura T, Inhibitory effect of novel 1-deoxynojirimycin derivatives on HIV-1 replication. *AIDS* **4**: 975–979, 1990.

71. Dewar RL, Vasudevachari MB, Natarajan V and Salzman NP, Biosynthesis and processing of human immunodeficiency virus type 1 envelope glycoproteins: effects of monensin on glycosylation and transport. *J. Virol.* **63**: 2452–2456, 1989.

72. Tyms AS, Taylor DL, Sunkara PS and Kang MS, Glycoprotein synthesis and human immunodeficiency viruses: a target for chemotherapy, pp. 257–318. In: De Clercq E (ed), *Design of Anti-AIDS Drugs*. Elsevier Science Publishers, Amsterdam, 1990.

73. Fennie C and Lasky LA, Model for intracellular folding of the human immunodeficiency virus type 1 gp120. *J. Virol.* **63**: 639–646, 1989.

74. Sunkara PS, Taylor DL, Kang MS, Bowlin TL, Liu PS, Tyms AS and Sjoerdsma A, Anti-HIV activity of castanospermine analogues. *Lancet* **i**: 1206, 1989.

75. Ruprecht RM, Bernard LD, Bronson R, Sosa MAG and Mullaney S, Castanospermine vs its 6-*O*-butanoyl analog: a comparison of toxicity and antiviral activity *in vitro* and *in vivo*. *J. Acquir. Immun. Defic. Syndromes* **4**: 48–55, 1991.

76. Johnson VA, Walker BD, Barlow MA, Paradis TJ, Chou TC and Hirsch MS, Synergistic inhibition of human immunodeficiency virus type 1 and type 2 replication in vitro by castanospermine and 3′-azido-3′-deoxythymidine. *Antimicrob. Agents Chemother.* **33**: 53–57, 1989.

77. Mitsuya H, Yarchoan R and Broder S, Molecular targets for AIDS therapy. *Science* **249**: 1533–1544, 1990.

78. Smith DH, Byrn RA, Marsters SA, Gregory T, Groopman JE and Capon DJ, Blocking of HIV-1 infectivity by a soluble, secreted form of the CD4 antigen. *Science* **238**: 1704–1707, 1987.

79. Fisher RA, Bertonis JM, Meier W, Johnson VA, Costopoulos DS, Liu T, Tizard R, Walker BD, Hirsch MS, Schooley RT and Flavell RA, HIV infection is blocked *in vitro* by recombinant soluble CD4. *Nature* **331**: 76–78, 1988.

80. Hussey RE, Richardson NE, Kowalski M, Brown NR, Chang HC, Siliciano RF, Dorfman T, Walker B, Sodroski J and Reinherz EL, A soluble CD4 protein selectively inhibits HIV replication and syncytium formation. *Nature* **331**: 78–81, 1988.

81. Deen KC, McDougal JS, Inacker R, Folena-Wasserman G, Arthos J, Rosenberg J, Maddon PJ, Axel R and Sweet RW, A soluble form of CD4 (T4) protein inhibits AIDS virus production. *Nature* **331**: 82–84, 1988.

82. Traunecker A, Lucke W and Karjalainen K, Soluble CD4 molecules neutralize human immunodeficiency virus type 1. *Nature* **331**: 84–86, 1988.

83. Wang J, Yan Y, Garrett TPJ, Liu J, Rodgers DW, Garlick RL, Tarr GE, Husain Y, Reinherz EL and Harrison SC, Atomic structure of a fragment of human CD4 containing two immunoglobulin-like domains. *Nature* **348**: 411–418, 1990.

84. Lifson JD, Hwang KM, Nara PL, Fraser B, Padgett M, Dunlop NM and Eiden LE, Synthetic CD4 peptide derivatives that inhibit HIV infection and cytopathicity. *Science* **241**: 712–716, 1988.

85. Hayashi Y, Ikuta K, Fujii N, Ezawa K and Kato S, Inhibition of HIV-1 replication and syncytium formation by synthetic CD4 peptides. *Arch. Virol.* **105**: 129–135, 1989.

86. Nara PL, Hwang KM, Rausch DM, Lifson JD and Eiden LE, CD4 antigen-based antireceptor peptides inhibit infectivity of human immunodeficiency virus *in vitro* at multiple stages of the viral life cycle. *Proc. Natl. Acad. Sci. USA* **86**: 7139–7143, 1989.

87. Liu MA and Liu T, Effect of recombinant soluble CD4 on human peripheral blood lymphocyte responses *in vitro*. *J. Clin. Invest.* **82**: 2176–2180, 1988.

88. Lamarre D, Ashkenazi A, Fleury S, Smith DH, Sekaly RP and Capon DJ, The MHC-binding and gp120-binding functions of CD4 are separable. *Science* **245**: 743–746, 1989.

89. Byrn RA, Sekigawa I, Chamow SM, Johnson JS, Gregory TJ, Capon DJ and Groopman JE, Characterization of *in vitro* inhibition of human immunodeficiency virus by purified recombinant CD4. *J. Virol.* **63**: 4370–4375, 1989.

90. Capon DJ, Chamow SM, Mordenti J, Marsters SA, Gregory T, Mitsuya H, Byrn RA, Lucas C, Wurm FM, Groopman JE, Broder S and Smith DH, Designing CD4 immunoadhesins for AIDS therapy. *Nature* **337**: 525–531, 1989.

91. Traunecker A, Schneider J, Kiefer H and Karjalainen K, Highly efficient neutralization of HIV with recombinant CD4-immunoglobulin molecules. *Nature* **339**: 68–73, 1989.

92. Chaudhary VK, Mizukami T, Fuerst TR, FitzGerald DJ, Moss B, Pastan I and Berger EA, Selective killing of HIV-infected cells by recombinant human CD4-*Pseudomonas exotoxin* hybrid protein. *Nature* **335**: 369–372, 1988.

93. Till MA, Ghetie V, Gregory T, Patzer EJ, Porter JP, Uhr JW, Capon DJ and Vitetta ES, HIV-infected cells are killed by rCD4-ricin A chain. *Science* **242**: 1166–1168, 1988.

94. Till MA, Zolla-Pazner S, Gorny MK, Patton JS, Uhr JW and Vitetta ES, Human immunodeficiency virus-infected T cells and monocytes are killed by monoclonal human anti-gp41 antibodies coupled to ricin A chain. *Proc. Natl. Acad. Sci. USA* **86**: 1987–1991, 1989.

95. Berger EA, Clouse KA, Chaudhary VK, Chakrabarti S, FitzGerald DJ, Pastan I and Moss B, CD4-*Pseudomonas exotoxin* hybrid protein blocks the spread of human immunodeficiency virus infection *in vitro* and is active against cells expressing the envelope glycoproteins from diverse primate immunodeficiency retroviruses. *Proc. Natl. Acad. Sci. USA* **86**: 9539–9543, 1989.

96. Berger EA, Chaudhary VK, Clouse KA, Jaraquemada D, Nicholas JA, Rubino KL, FitzGerald DJ, Pastan I and Moss B, Recombinant CD4-*Pseudomonas exotoxin* hybrid protein displays HIV-specific cytotoxicity without affecting MHC class II-dependent functions. *AIDS Res. Hum. Retroviruses* **6**: 795–803, 1990.

97. Daar ES, Li XL, Moudgil T and Ho DD, High concentrations of recombinant soluble CD4 are required to neutralize primary human immunodeficiency virus type 1 isolates. *Proc. Natl. Acad. Sci. USA* **87**: 6574–6578, 1990.

98. Looney DJ, Hayashi S, Nicklas M, Redfield RR, Broder S, Wong-Staal F and Mitsuya H, Differences in the interaction of HIV-1 and HIV-2 with CD4. *J. Acquir. Immun. Defic. Syndromes* **3**: 649–657, 1990.

99. Werner A, Winskowsky G and Kurth, Soluble CD4 enhances simian immunodeficiency virus SIV$_{agm}$ infection. *J. Virol.* **64**: 6252–6256, 1990.

100. Sekigawa I, Chamow SM, Groopman JE and Byrn RA, CD4 Immunoadhesin, but not recombinant soluble CD4, blocks syncytium formation by human immunodeficiency virus type 2-infected lymphoid cells. *J. Virol.* **64**: 5194–5198, 1990.

101. Moore JP, Simple methods for monitoring HIV-1 and HIV-2 gp120 binding to soluble CD4 by enzyme-linked immunosorbent assay: HIV-2 has a 25-fold lower affinity than HIV-1 for soluble CD4. *AIDS* **4**: 297–305, 1990.

102. Schooley RT, Merigan TC, Gaut P, Hirsch MS, Holodniy M, Flynn T, Liu S, Byington RE, Henochowicz S, Gubish E, Spriggs D, Kufe D, Schindler J, Dawson A, Thomas D, Hanson DG, Letwin B, Liu T, Gulinello J, Kennedy S, Fisher R and Ho DD, Recombinant soluble CD4 therapy in patients with the acquired immunodeficiency syndrome (AIDS) and AIDS-related complex. A phase I–II escalating dosage trial. *Ann. Intern. Med.* **112**: 247–253, 1990.

103. Kahn JO, Allan JD, Hodges TL, Kaplan LD, Arri CJ, Fitch HF, Izu AE, Mordenti J, Sherwin SA, Groopman JE and Volberding PA, The safety and pharmacokinetics of recombinant soluble CD4 (rCD4) in subjects with the acquired immunodeficiency syndrome (AIDS) and AIDS-related complex. A phase 1 study. *Ann. Intern. Med.* **112**: 254–261, 1990.

104. Johnson VA, Barlow MA, Merrill DP, Chou TC and Hirsch MS, Three-drug synergistic inhibition of HIV-1 replication *in vitro* by zidovudine, recombinant soluble CD4, and recombinant interferon-alpha A. *J. Infect. Dis.* **161**: 1059–1067, 1990.

105. Ashorn P, Moss B, Weinstein JN, Chaudhary VK, FitzGerald DJ, Pastan I and Berger EA, Elimination of infectious human immunodeficiency virus from human T-cell cultures by

synergistic action of CD4-*Pseudomonas exotoxin* and reverse transcriptase inhibitors. *Proc. Natl. Acad. Sci. USA* **87**: 8889–8893, 1990.

106. Tsubota H, Winkler G, Meade HM, Jakubowski A, Thomas DW and Letvin NL, CD4-*Pseudomonas exotoxin* conjugates delay but do not fully inhibit human immunodeficiency virus replication in lymphocytes *in vitro*. *J. Clin. Invest.* **86**: 1684–1689, 1990.

Polyionic Compounds Selectively Alter Availability of CD4 Receptors for HIV Coat Protein rgp 120

A. Aszalos, P. S. Pine and J. Weaver

Food and Drug Administration, Washington DC, USA

Abstract. We studied the ability of several polyionic compounds, previously shown to have anti-HIV activity *in vitro*, to block binding of anti-CD4 and recombinant HIV gp120 to the CD4 receptor on human lymphocytes. We found that Evans Blue and aurin tricarboxylic acid (ATA) could completely inhibit binding of anti-CD4 (Leu 3a) and rgp120 and have selectivity for the CD4 receptor. A number of other compounds including dextran sulfate and heparin had no effect on binding of rgp120 and were shown to be nonspecific for inhibition of binding of monoclonal antibodies to different T-cell receptors. Studies using a number of membrane-active drugs showed that changes in membrane potential or ion fluxes were not involved in the inhibition of binding of rgp120 by Evans Blue or aurin tricarboxylic acid. Using our developed flow cytometric method we could detect novel compounds which can prevent binding of rgp120 specifically and have good anti-HIV activity.

We have also found that ATA prevents the binding of interferon-α to its receptor in a dose dependent manner (12–50 μM range). Membrane potential shift, associated with the binding of interferon-α to its receptor, was also blocked by ATA in a dose dependent fashion. Several other agents, able to prevent the binding of rgp120 to the CD4 molecule, did not block the binding of interferon-α to its receptor. Our results indicate that potential anti-AIDS drugs should be screened for such undesired side effects.

1. Introduction

One avenue for possible therapeutic compounds for the treatment of HIV infections are compounds that interfere with the binding of the virus through its gp120 coat protein to the cellular CD4 receptor. A variety of molecules have been studied that may act at this site. These include soluble CD4 and CD4 peptides [1, 2], large polysulfated molecules, such as dextran sulfate (DS) and relatives [3–5], and smaller polyionic molecules, such as aurin tricarboxylic acid (ATA) [6, 7], and dyes, such as Evans Blue (EB) [8, 9]. The exact mechanism of action of these compounds is not well known. DS has been reported in prevent the binding of HIV to CD4 and possibly fusion related events [10]. However, DS is unable to prevent binding of recombinant HIV gp120 coat protein (rgp120) or anti-CD4 (αCD4) monoclonal antibodies to CD4 [10, 11]. ATA has been reported to be able to specifically prevent binding of HIV-1 viral particles to CD4, whereas DS acts in a more nonspecific fashion [6].

In this research we have studied the mechanism of action of several of these compounds in defined systems. We have found that dextran sulfate does not interfere with the binding of recombinant HIV gp120 (rgp120) to the CD4 molecule on cell surfaces. We have found that Evans Blue and ATA inhibit this binding. In addition, we have found that ATA inhibits the binding of interferon-α to its receptor. This inhibition is not seen with the other anti-HIV agents we have studied.

2. Materials and Methods

Reagents. Bis(1,3-dibutylbarbituric acid)trimethine oxonol (DiBaC$_4$(3)), a membrane potential sensing dye, was obtained from Molecular Probes (Eugene, OR). Aurin tricarboxylic acid was purchased from Sigma Chemical Co. (St. Louis, MO). Recombinant interferon-α (rIFN-α2b), specific activity 1.6×10^8 U/mg protein, was obtained from Schering Corporation (Kenilworth, NJ). Monoclonal anti-human CD3, CD4, and CD8 (FITC-labeled) were obtained from Becton-Dickinson (Mountain View, CA). 463 anti-CD4, which binds near the OKT4 epitope, recombinant HIV coat protein gp120 (rgp120), and FITC-labeled anti-gp120 were generous gifts from Gonentech Inc. (San Francisco, CA).

Cell preparation. Daudi cells, originated from a Burkitt lymphoma (B-cell), were obtained from the laboratory of Dr. P. Grimley, USUHS, Bethesda, MD. These cells were grown in suspension culture using RPMI-1640 with 20 μM HEPES supplemented with penicillin/streptomycin (Gibco, Grand Island, NY) and 10% NuSerum (Collaborative Research, Bedford, MA). Human peripheral blood lymphocytes (PBL) were prepared as previously described [12].

Binding of ligands to PBL. Binding of ligands to human PBL was performed as previously described [12]. Briefly, human PBL were prepared as described [12] and resuspended in PBS at 1×10^8 cells/ml. Ligands were added as indicated and the cells were incubated for 30 min at 4°C. With directly labeled ligands (αCD3, αCD4a, αCD8) cells were washed, resuspended, and analyzed by flow cytometry. Binding of rgp120 was detected by binding of FITC-labeled anti-gp120.

Membrane potential measurement. Flow cytometric determination of membrane potential was performed as previously described [13]. Briefly, a cell suspension of 1×10^8 cells/ml was equilibrated for 1 min in PBS and ligands were added 1 min later; oxonol (140 nM) was added 8 min after addition of ligands and histograms were collected 2 min after oxonol addition (10 min after cells were equilibrated). This timing showed the maximum changes in mean fluorescence intensity for compounds known to affect membrane potential. All measurements were made on a FACScan flow cytometer (Becton-Dickinson).

Interferon-α receptor assay. IFN-α binding and receptor number were determined in Daudi cells using ^{125}I-labeled IFN-α as previously described [14].

3. Results

Effect of drugs on binding of monoclonal antibodies to PBL. Table 1 shows that dextran and protamine had no effect on the binding of the monoclonal antibodies to various cell surface molecules. Larger polyanions dextran sulfate (DS) and heparin, partially decreased the binding of αCD4 (Leu3a) and αCD8 while having a smaller effect on the binding of αCD3. The smaller polyanions Evans Blue (EB) and aurin tricarboxylic acid (ATA) completely inhibited binding of αCD4, while producing no inhibition of the binding of αCD3 and only a partial inhibition of

Table 1. The effect of various polyions on the binding of monoclonal antibodies to the CD3, CD4, and CD8 receptors of human peripheral blood T-cells[1]

Drug	Dose	αCD3	αCD4[2]	αCD8	463 αCD4[3]
Dextran	20 μM	94.6	117.0	105.0	ND[4]
	40 μM	100.0	105.7	100.0	ND
Dextran SO₄	20 μM	78.7	69.2	59.8	ND
	40 μM	75.9	64.7	44.2	
Protamine	20 μM	103.0	95.0	100.0	ND
	40 μM	111.5	96.0	105.0	ND
Heparin SO₄	2 U/ml	96.3	60.7	66.0	ND
	10 U/ml	92.0	58.1	44.2	ND
ATA	3 μM	106.0	0.0	47.8	61.1
Evans Blue	1 μM	99.0	0.0	47.5	62.4
rgp120	0.3 μM	100.0	0.0	ND	96.0

[1] Data from [12], results in % logarithmic fluorescence intensity of control.
[2] Anti-Leu 3a.
[3] Binds near the Leu3/OKT4 epitope.
[4] ND = Not Done.

αCD8 binding at concentrations that completely inhibit αCD4 binding. Treatment with rgp120 also totally blocked the binding of αCD4 to CD4. This is consistent with previous observations that Leu 3a inhibits HIV binding and infectivity [15].

Effect of drugs on the binding of gp120 to PBL. Table 2 shows the ability of various drugs to inhibit binding of rgp120. Comparison of Tables 1 and 2 shows that the ability to completely inhibit binding of αCD4 correlates with the ability to block binding of gp120. The larger polyionic compounds (DS, heparin), which have a limited capability to inhibit binding of αCD4, are unable to inhibit the binding of rgp120 to cellular CD4. These results were similar when the cells were washed after the drug binding step or when rgp120 was added directly to the cell/drug incubation mixture.

Table 2. The effect of drugs on the binding of rgp120 to PBL[1]

Drug	Dose	% Control[2]
Dextran	40 µM	99
Dextran SO₄	20 µM	118
Protamine Cl	40 µM	106
Heparin	10 U/ml	118
Evans Blue	1 µM	0
ATA	3 µM	3
Amphotericin B	0.8 µM	119
Amantadine	1 µM	85
Cyclosporin A	4 µM	91
Ascorbic acid	60 µM	84
Mycophenolic acid	30 µM	80
Poly-L-aspartic acid	10 µM	78
Poly-L-glutamic acid	100 µM	86

[1]Data from [12].
[2]Variations of ±20% were observed with repeated experiments with the same drug.

Dose response of the effect of compounds on the binding of αCD4 or rgp120 to PBL. Figure 1 indicates the dose response curves for the inhibition of the binding of αCD4 or rgp120 by DS, EB, ATA, and rgp120. EB produces complete inhibition at concentrations of 1 µM and ATA at 3 µM. The dose response of EB inhibition of rgp120 binding is not changed after cells are activated by co-culture with αCD3 for three days (not shown). Activation was confirmed by an increase in IL-2 receptor positive cells from <5 to 66% at three days of stimulation. DS

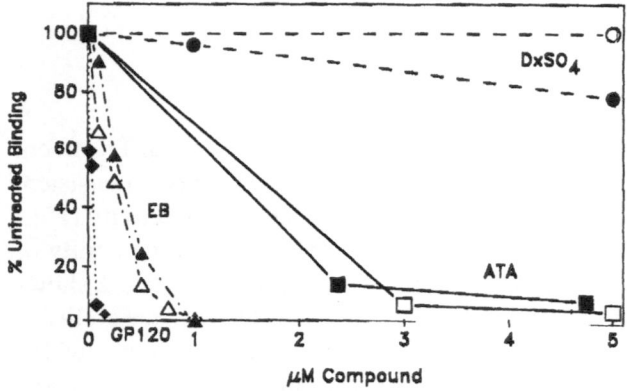

Fig. 1. Inhibition of binding of anti-CD4-FITC or rgp120 by polyionic compounds. Data from [12]. Solid symbols — anti-CD4 binding, open symbols — rgp120 binding; squares — aurin tricarboxylic acid, circles — dextran sulfate, triangles — Evans Blue, diamonds — rgp120

shows only partial inhibition of αCD4 binding and no inhibition of rgp120 binding under any conditions (not shown). The most effective reagent for blocking αCD4 binding is rgp120 with inhibition complete at 0.15 μM.

The inhibition of αCD4 binding by EB or ATA is not due to loss of cell surface CD4, since the degree of inhibition is similar when the cells are precooled to 4°C for 15 min before the drug is added and are maintained at 4°C for the entire assay (not shown). In addition, ATA or EB reduce the binding of 463 αCD4 to only 61

Table 3. Membrane-active drugs do not affect the binding of αCD4 to PBL[1]

Drug	Dose[2]	Site of action	Oxonol shift[3]	αCD4 shift[3]
Ionomycin	2 μg/ml	Ca^{++} flux	−8.55	−0.01
A23187	2 μg/ml	Ca^{++} flux	−7.52	−0.19
Verapamil	40 μM	Ca^{++} &K$^+$	+10.49	−0.59
Nifedipine	20 μM	Ca^{++} &K$^+$	−5.29	−0.41
Tamoxifen	1 μM	PKC/Calmodulin	+7.14	−0.25
R24571	0.3 μM	Calmodulin	−10.87	+0.25
H7	10 μg/ml	PKC/Calmodulin	−1.38	0.00
Valinomycin	4 μg/ml	K$^+$ flux	−12.15	−1.14
Nigericin	1.9 μg/ml	K$^+$ flux	−7.96	+0.14
Quinine	40 μM	K$^+$ flux	−3.45	−0.32
Monensin	20 μg/ml	Na$^+$ flux	?[4]	−0.72
Amiloride	2 μg/ml	Na$^+$ flux	−3.75	−0.40
Ouabain	100 μM	Na$^+$/K$^+$ ATP-ase	+2.73	+0.02
CCCP	2 μg/ml	Na$^+$ flux	+2.00	+0.51
Nystatin	10 μg/ml	Intercalates	+20.02	−0.19
Amphotericin B	0.2 μg/ml	Intercalates	+6.01	0.00
Amantadine	1 mM	Intercalates	+22.26	−0.20
Colchicine	5 μg/ml	Microtubules	ND	−0.29
Cytochalasin B	10 μM	Microfilaments	ND	−0.26
Indocin	1 μM	Cyclooxygenase	−5.82	−0.65
Gramicidin S	0.5 μg/ml	Pore former	−7.28	−0.19
Propranolol	30 μM	β-adrenergic	+7.09	−0.42
ATA	3 μM	?	−7.19	−6.12
Evans Blue	1 μM	?	−10.68	−6.40

[1]Data from [12].
[2]All compounds were dissolved in DMSO except: propranolol, amantadine, TP-5, ouabain, and cytochalasin B in PBS; monensin in ethanol.
[3]Shift indicates the change in mean logarithmic peak fluorescence channel number for the given compound as compared to an equal volume solvent control.
[4]Monensin caused complex changes in the uptake of bis-oxonol.

and 62% of control levels respectively (Table 1). This confirms that the CD4 molecule is still accessible on the cell surface.

Failure of 23 drugs that affect membrane potential to affect the binding of αCD4. We have shown [12] that the compounds that inhibit αCD4 and/or rgp120 binding also affect membrane potential. We tested the effect on the binding of αCD4 of 23 different drugs known to affect cell physiology at or near the plasma membrane. Concentrations were chosen either to give a small but significant shift in membrane potential as indicated by bisoxonol uptake, or at a level known to be biologically effective in other systems for those compounds that did not affect membrane potential directly. As shown in Table 3, none of these compounds, at the concentrations indicated, had any effect whatsoever on the binding of αCD4 to PBL. Amantadine and amphotericin B, membrane-intercalating molecules, likewise have no effect on rgp120 binding. In contrast, ATA and EB, while showing comparable shifts in bis-oxonol uptake, could completely inhibit the binding of αCD4 and rgp120. Thus we can conclude that the ability to change membrane potential is not related to the ability to inhibit binding of αCD4 or rgp120.

Effect of ATA on binding of RIFN-α. Table 4 shows that pretreatment of cells with ATA (2 min, 22°C) inhibits the binding of RIFN-α to the cell surface receptor of Daudi cells in a dose-dependent manner. Similar decreases in specific binding were observed with both 2.0 nM and 0.2 nM concentrations of $[^{125}I]$-rIFN-α2b (Table 3). It is noteworthy that with increasing concentrations of radiolabeled

Table 4. Aurin tricarboxylic acid inhibits the binding of rIFN-α2b to cell surface receptors of Daudi cells[1]

[ATA][3]	% Control specific binding		% Nonspecific binding[2]	
	2.0 nM[4]	0.2 nM	2.0 nM	0.2 nM
0.0	100.0[5]	100.0	17.7	6.4
2.5	77.7	87.9	21.4	8.6
5.0	81.0	75.3	22.9	8.9
10.0	40.4	38.4	40.3	18.8
15.0	15.6	28.9	70.5	23.4

[1]Data from [14], total binding volume was 400 μl; binding temperature was 4°C; binding buffer was RPMI-1640, 2 mM glutamine, 0.1 mg/ml gentamicin.
[2]As the % of total binding.
[3]Concentration of ATA in μg/200 μl cell culture.
[4]Concentration of $[^{125}I]$-rIN-α2b.
[5]100% specific binding is 3074 dpm for 2 nM and 873 dpm for 0.2 nM $[^{125}I]$-rIFN-α2b.

rIFN-α2b, there is an increase in nonspecific binding as a function of increasing concentrations of ATA. However, the inhibition of binding by ATA was independent of temperature between 4 and 37°C.

Effect of ATA on membrane potential. As Fig. 2 demonstrates, ATA is able to inhibit the hyperpolarization of membrane potential induced by rIFN-α treatment in Daudi cells in a dose-dependent manner.

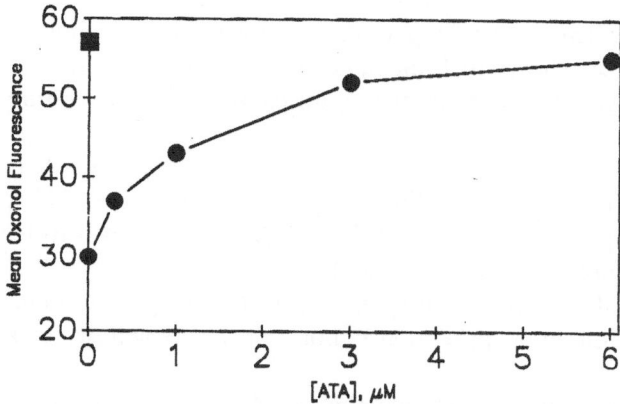

Fig. 2. The effect of ATA on membrane potential of rIFN-α treated Daudi cell. Data from [14]. Circles indicate the mean oxonol fluorescence intensities obtained from 10^6 Daudi cells treated with 400 U/ml doses of rIFN-α and various concentrations of ATA. ATA was added 1 min before rIFN-α and oxonol 8 min after rIFN-α. The square indicates cells not treated with rIFN-α or ATA. Fluorescence histograms were collected by flow cytometry 2 min after cells were stained with 140 nM oxonol

4. Discussion

We have investigated the mode of action of several compounds that have been shown to have anti-HIV activity. We have shown that the large polysulfated compounds, such as DS and heparin sulfate, do not affect the binding of αCD4 (Table 1) or rgp120 (Table 2) to CD4. In contrast, we have shown that the smaller polyionic compounds EB and ATA can completely inhibit the binding of rgp120 to CD4 in a dose dependent fashion. This inhibition is not a general property of polyions since a number of other polyionic molecules were not able to inhibit rgp120 binding (Table 2).

DS has been reported to be able to block the infection of cells in culture by HIV [3, 10, 16]. These studies have shown that DS is able to inhibit the binding of HIV to cells. The results of our studies suggest that DS does not act to prevent gp120-CD4 binding, but it may interfere at other steps of the infection process.

The inhibition of CD4 binding does not appear to be due to loss of cell surface CD4 since we found that the 463 anti-CD4 antibody bound at 61% of control levels [14]. In addition, ATA and EB can inhibit αCD4 binding at 0°C. Thiele et al. [17] have shown that CD4 is not modulated at 0°C. In addition, we can rule out resonance energy transfer from FITC emission at 520 nm to EB or ATA since the absorption maxima are at 605 nm for EB and < 400 nm for ATA.

We have previously shown that ATA and EB can affect membrane potential as measured by two potential sensitive fluorescent dyes [12]. We studied the effect of 23 membrane active compounds on αCD4 binding to determine if changes in membrane potential or other membrane related factors could affect HIV binding.

We have used anti-Leu3a (αCD4) as a surrogate marker for rgp120 binding to conserve our stock of rgp120. As shown in Table 3, none of these drugs were able to affect αCD4 binding. ATA, EB and DS do not significantly affect the resting Ca^{++} levels in PBL [12].

Additionally we have shown that ATA but not EB blocks the binding of IFN-α to its cell surface receptor (Table 4). In addition, ATA can inhibit the IFN-α mediated shift in membrane potential (Fig. 2). This shift has been shown to be associated with the anti-proliferative signal of IFN-α [13]. These data suggest that ATA could inhibit any part of the immune response requiring a signal through IFN-α. Additional potential immunosuppression is not a desirable activity of a potential anti-AIDS drug.

Cushman et al. [7, 18] have recently shown that ATA is a heterogeneous mixture of various sized polymers of the nominal ATA structure. They have demonstrated that the anti-HIV activity is found in the polymeric structures. Our experiments were carried out with the commercially available mixture. We have recently obtained some ATA fractions of defined size and experiments are underway to examine the size dependence of the rgp120 binding inhibition and IFN-α binding inhibition.

References

1. Lifson JD, Hwang KM, Nara PL, Fraser B, Padgett M, Dunlop NM and Eiden LE, Synthetic CD4 peptide derivatives that inhibit HIV infection and cytopathicity. *Science* **241**: 712–716, 1988.
2. Smith DH, Byrn RA, Marsters SA, Gregory T, Groopman JE and Capon D, Blocking of HIV-1 infectivity by a soluble secreted form of the CD4 antigen. *Science* **238**: 1704–1707, 1987.
3. Baba M, Pauwels R, Balzarini J, Arnout J and Desmyter J, Mechanism of inhibitory effect of dextran sulfate and heparin on replication of human immunodeficiency virus *in vitro*. *Proc. Natl. Acad. Sci. USA* **85**: 6132–6136, 1988.
4. Schols D, Baba M, Pauwels R and De Clercq E, Flow cytometric method to demonstrate whether anti-HIV-1 agents inhibit virion binding to T4+ cells. *J. Acq. Imm. Def. Synd.* **2**: 10–15, 1989.
5. Mizumoto K, Sugawara I, Ito W, Kodama T, Hayami M and Mori S, Sulfated homopolysaccharides with immunomodulating activities are more potent anti-HTLV-III agents than sulfated heteropolysaccharides, *Japan J. Exp. Med.* **58**: 145–151, 1988.
6. Schols D, Baba M, Pauwels R, Desmyter J and De Clercq E, Specific interaction of aurintricarboxylic acid with the human immunodeficiency virus/CD4 receptor. *Proc. Natl. Acad. Sci. USA* **86**: 3322–3326, 1989.
7. Cushman M, Kanamathareddy S, De Clercq E, Schols D, Goldman ME and Bowen JA, Synthesis and anti-HIV activities of low molecular weight aurintricarboxylic acid fragments and related compounds. *J. Med. Chem.* **34**: 337–342, 1991.
8. Balzarini J, Mitsuya H, De Clercq E and Broder S, Aurintricarboxylic acid and Evans Blue represent two different classes of anionic compounds which selectively inhibit the cytopathogenicity of human T-cell lymphotropic virus type III/lymphadenopathy-associated virus. *Biochem. Biophys. Res. Commun.* **136**: 64–71, 1986.
9. Balzarini J, Mitsuya H, De Clercq E and Broder S, Comparative inhibitory effects of suramin and other selected compounds on the infectivity and replication of human T-cell lymphotrophic virus (HTLV-III)/lymphadenopathy-associated virus (LAV). *Int. J. Cancer* **37**: 451–457, 1986.

10. Mitsuya H, Looney DJ, Kuno S, Ueno R, Wong-Stall F and Broder S, Dextran sulfate suppression of viruses in the HIV family: Inhibition of virion binding to CD4 + cells. *Science* **240**: 646–649, 1988.

11. Lederman S, Gullck R and Chess L, Dextran sulfate and heparin interact with CD4 molecules to inhibit the binding of coat protein (GP120) of HIV. *J. Immunol.* **143**: 1149–1154, 1989.

12. Weaver JL, Gergeley P, Pine PS, Patzer E and Aszalos A, Polyionic compounds selectively alter availability of CD4 receptors for HIV coat protein rgp120. *AIDS Res. Human Retrovir.* **6**: 1125–1130, 1990.

13. Grimley PM and Aszalos A, Early membrane depolarization by alpha interferon: Biologic correlation with antiproliferative signal. *Biochem. Biophys. Res. Commun.* **146**: 300–306, 1987.

14. Gan Y, Weaver JL, Pine PS, Zoon KC and Aszalos A, Aurintricarboxylic acid, the anti-AIDS compound, prevents the binding of interferon-alpha to its receptor. *Biochem. Biophys. Res. Commun.* **172**: 1298–1303, 1990.

15. McDougal JS, Nicholson JKA, Cross GD, Cort SP, Kennedy MS and Mawle AC, Binding of the human retrovirus HTLV-III/LAC/ARV/HIV the CD4 (T4) molecule; conformation dependence, epitope mapping, antibody inhibition, and potential for idiotype mimicry. *J. Immunol.* **137**: 2937–2944, 1986.

16. Chang RS, Tabba HD, He Y and Smith KM, Dextran sulfate as an inhibitor against the human immunodeficiency virus (42811). *Proc. Soc. Exp. Biol. Med.* **189**: 304–309, 1988.

17. Thiele B, Braig HR, Ehm I, Kunze R and Ruf B, Influence of sulfated carbohydrates on the accessibility of CD4 and other CD molecules on the cell surface and implications for human immunodeficiency virus infection. *Eur. J. Immunol.* **19**: 1161–1164, 1989.

18. Cushman M, Wang P, Chang SH, Wild C, De Clercq E, Schols D, Goldman ME and Bowen JA, Preparation and anti-HIV activities of aurintricarboxylic acid fractions and analogues: direct correlation of antiviral potency with molecular weight. *J. Med. Chem.* **34**: 329–337, 1991.

Penciclovir and Famciclovir, Selective Anti-Herpesvirus Agents

M. R. Harnden

SmithKline Beecham Pharmaceuticals, Biosciences Research Centre, Great Burgh, Epsom, Surrey KT18 5XQ, UK

Abstract. The synthesis of penciclovir and famciclovir and data leading to their progression to clinical trials against herpes simplex virus and varicella zoster virus infections are summarized. The spectra of antiviral activity and potency of penciclovir and acyclovir in cell culture are similar, but penciclovir treatment results in a much more persistent suppression of virus replication. This is attributable to the rapid formation and stability of its triphosphate derivative in infected cells. As a consequence of this persistent activity, penciclovir is more active than acyclovir when administered as a single daily dose in some animal infection models. The biotransformation of famciclovir to penciclovir following oral administration is discussed. Famciclovir is very well absorbed and efficiently converted to the antiviral acyclonucleoside in mice, rats and humans and it is therefore being clinically evaluated as an orally active, selective anti-herpesvirus agent.

In 1985 we reported the synthesis and anti-herpesvirus activity of 9-[4-hydroxy-3-(hydroxymethyl)butyl]guanine (BRL 39123, penciclovir) [1].

This acyclonucleoside, which is structurally related to acyclovir and ganciclovir, but lacks the ether oxygen in the acyclic substituent, is readily prepared by alkylation of 2-amino-6-chloropurine [2].

penciclovir (BRL 39123)

Penciclovir has a spectrum of antiviral activity that is similar to that of acyclovir [3, 4]. Thus, in plaque reduction tests carried out in MRC-5 (human fibroblast) cells against both laboratory strains (Table 1) and clinical isolates (Table 2) penciclovir was slightly less active than acyclovir against herpes simplex virus

Table 1. Activity of penciclovir and acyclovir against her- pesviruses, determined by plaque reduction assay

Virus strain	IC_{50} (μM)	
	Penciclovir	Acyclovir
HSV-1 (HFEM)	2.0	1.3
HSV-2 (MS)	3.2	1.8
VZV (Ellen)	9.5	15
CMV (AD-169)	205	111

Table 2. Activity of penciclovir and acyclovir against clinical isolates of HSV-1, HSV-2 and VZV, determined by plaque reduction assay

Virus	No. of isolates tested	Mean IC_{50} (μM) $\pm$ SD	
		Penciclovir	Acyclovir
HSV-1	17	1.6±0.4	0.9±0.9
HSV-2	13	5.9±1.6	2.7±0.9
VZV	5	12.0±3.2	17.0±3.1

type 1 (HSV-1), 2-fold less active against herpes simplex virus type 2 (HSV-2) and slightly more active against varicella zoster virus (VZV). Like acyclovir and unlike ganciclovir, it shows activity against cytomegalovirus (CMV) only at very high concentrations.

In a test which measured the inhibition of virus capsid antigen expression Epstein-Barr virus (EBV) was also susceptible to both penciclovir and acyclovir, activity being observed at 40 μM.

In contrast to the results obtained in plaque reduction assays, penciclovir was, however, appreciably more active than acyclovir against HSV-1 strains and slightly more active against HSV-2 strains in virus yield reduction assays in MRC-5 cells, in which a higher multiplicity of infection was used and infectious virus production measured (Table 3) [4, 5].

Table 3. Activity of penciclovir and acyclovir against clinical strains of HSV-1 and HSV-2, determined by virus yield reduction assay

| Virus | No. of isolates tested | Concn. (μM) for 99% redn. in infectious virus 24 h after infection | |
		Penciclovir	Acyclovir
HSV-1	3	1.6	4.4
HSV-2	4	2.8	4.0

At concentrations up to 400 μM, penciclovir caused no morphological abnormalities in control uninfected, established MRC-5 cell monolayers used in these antiviral tests, nor did it significantly affect the growth of subconfluent monolayers of MRC-5 cells over a 3 day period.

Thus, from these initial data, penciclovir appeared to be a highly selective anti-herpesvirus agent with properties rather similar to those of acyclovir. However, although both penciclovir and acyclovir were only slightly active against an

Table 4. Activity of penciclovir against acyclovir-resistant mutants of HSV-1, determined by plaque reduction assay in Vero cells

| Virus strain | Phenotype | IC_{50} (μM) | |
		Penciclovir	Acyclovir
Cl (101)	Wild type	4.3	1.8
Cl (101) TK$^-$	TK$^-$	343	258
Cl (101) P$_2$C$_5$	TK$^+$, DNA polymerase mutant (acyclovir resistant)	5.9	93

HSV-1 strain that does not specify a viral thymidine kinase, penciclovir was active against a DNA polymerase mutant selected for resistance to acyclovir (Table 4).

Additionally, interesting differences between the two compounds were encountered in experiments in which the time of contact between the acyclonucleoside and cell monolayers infected with HSV-1 and HSV-2 was varied.

In the first of these experiments (Fig. 1) MRC-5 cells infected with HSV-2

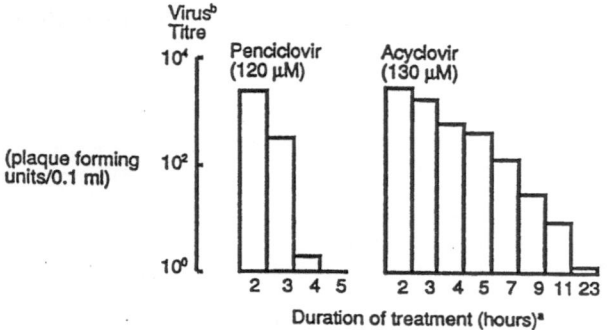

Fig. 1. Effect of duration of treatment with penciclovir and acyclovir on HSV-2 replication in MRC-5

(MS) were treated with either penciclovir (120 µM) or acyclovir (130 µM) for various lengths of time beginning 2 hours after infection, and cell-free virus titres were measured 25 hours after infection. Brief treatment with penciclovir resulted in considerably lower virus yield than did a similar period of treatment with acyclovir. Indeed, the continuous presence of acyclovir was required in order to achieve maximum reduction in virus yield, whereas this was obtained after 5 hours of incubation with penciclovir. Similar results were obtained with HSV-1 in MRC-5 cells.

In another experiment (Fig. 2), MRC-5 cells infected with HSV-2 were treated with penciclovir and acyclovir for 18 hours after infection, and thereafter incubation was continued after removal of extracellular compound. Infectivity titres of supernatants from these cultures were then measured for up to 6 days. The amount of virus produced by infected cells after treatment with acyclovir at 130 µM rapidly increased upon removal of the compound to give levels similar to those released by untreated cells. In contrast, after treatment with penciclovir at 12 µM, infectivity titres declined. This persistent antiviral activity with penciclovir has been observed at concentrations as low as 4 µM and is also apparent against HSV-1 and in other cell lines.

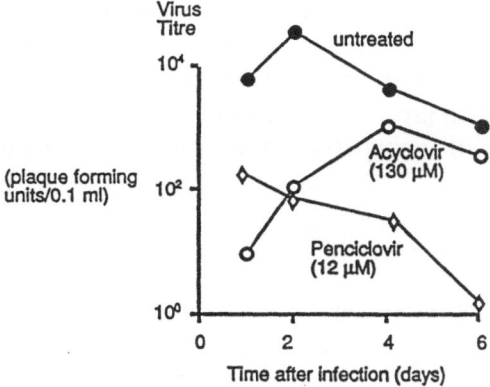

Fig. 2. Persistence of activity after 18 hour treatment of HSV-2 infected MRC-5 cells with penciclovir and acyclovir

The explanation for the persistent activity of penciclovir lies in its mode of action [5, 6]. Like acyclovir, following entry into virus infected cells penciclovir is phosphorylated initially by the virus induced thymidine kinase to give a monophosphate, which is further phosphorylated by cellular enzymes to give the triphosphate. The triphosphate is a competitive inhibitor of herpesvirus DNA polymerase and inhibits replication of viral DNA. However, penciclovir is phosphorylated far more quickly and to a greater extent than is acyclovir in cells infected with HSV and VZV (Fig. 3). Using ^{13}C labelling of one hydroxymethyl

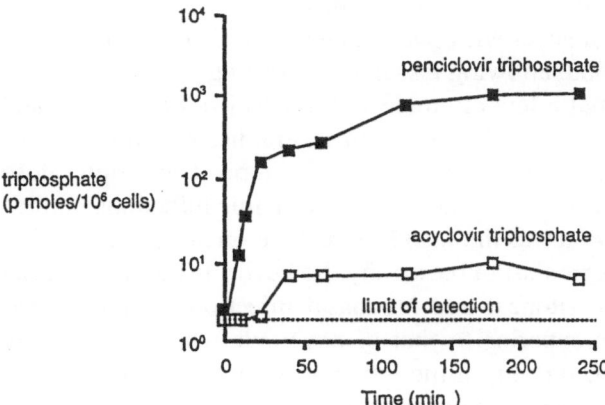

Fig. 3. Formation of the triphosphates of penciclovir and acyclovir in MRC-5 cells infected with HSV-1

group, it has been established that in HSV-1 infected cells the absolute configuration of the penciclovir triphosphate produced is (S), with an enantiomeric purity >95% [7].

Following removal of the acyclonucleoside from the cell culture medium, the stability of penciclovir triphosphate is much greater than that of acyclovir triphosphate, the half-lives in HSV-1 infected cells being 10 and 0.7 hours, respectively (Fig. 4).

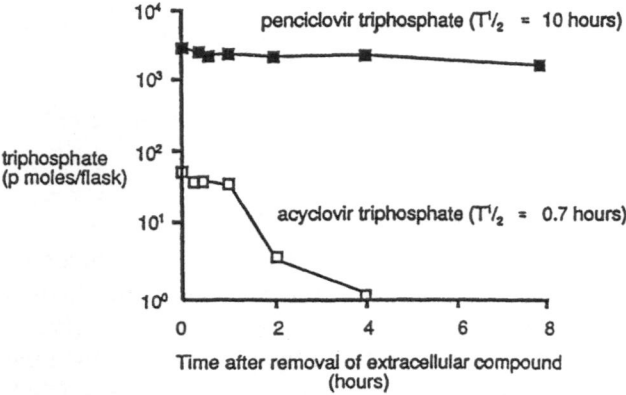

Fig. 4. Degradation of the triphosphates of penciclovir and acyclovir in MRC-5 cells infected with HSV-1

These data provided an explanation for the persistent activity of penciclovir in cell culture and suggested to us that the activity of penciclovir *in vivo* might be less dependent than is the activity of acyclovir upon maintenance of concentrations of the acyclonucleoside in the blood, allowing less frequent dosing.

This was confirmed in animal infection models (Table 5) [8]. Penciclovir and acyclovir had significant and comparable activity in several herpes simplex virus animal models when administered by parenteral or topical multiple dose schedules. These models included both cutaneous and systemic infections in mice and guinea pigs in which infection was initiated by scarification or by intranasal or intraperitoneal inoculation. Particularly noteworthy, however, is the observation that in the treatment of intraperitoneal and intranasal infections in mice, penciclovir had activity that was superior to that of acyclovir when single daily subcutaneous doses were used. Indeed, in the intraperitoneal infection model, a single subcutaneous dose of penciclovir gave greater reduction in virus replication than did three equivalent doses of acyclovir during a 24 hour period and a single subcutaneous penciclovir dose of 5.5 mg/kg was significantly more active than

50 mg/kg of acyclovir when treatment was initiated 24 hours after infection. When the compounds were administered semi-continuously in the drinking water, acyclovir was somewhat more active than penciclovir in mice infected by the intranasal route. This was attributed to the poor oral absorption of penciclovir and indicated the requirement for an orally active form.

In our studies on analogues of penciclovir many modifications of the acyclic substituent and the purine base were synthesized. These included compounds with acyclic moieties that were longer or shorter, or substituted at various positions with hydroxy, hydroxymethyl, methyl and fluoro groups and also compounds with different substituents at positions 2 and 6 in the purine [9–12].

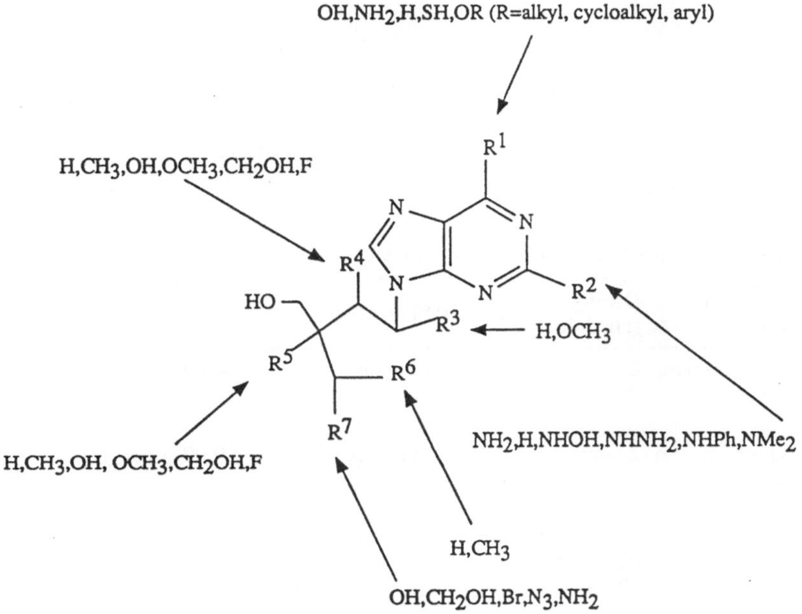

OH,NH$_2$,H,SH,OR (R=alkyl, cycloalkyl, aryl)

H,CH$_3$,OH,OCH$_3$,CH$_2$OH,F

R^1

R^4

HO

R^3 ← H,OCH$_3$

R^2

R^5

R^6

NH$_2$,H,NHOH,NHNH$_2$,NHPh,NMe$_2$

R^7

H,CH$_3$,OH, OCH$_3$,CH$_2$OH,F

H,CH$_3$

OH,CH$_2$OH,Br,N$_3$,NH$_2$

Although some of these compounds retained selective anti-herpesvirus activity, none was as potent or more selective than penciclovir.

However, several analogues and derivatives, for example 6-deoxy penciclovir, some 6-alkoxy analogues and their esters, were better absorbed than penciclovir after oral administration and were converted to penciclovir *in vivo* [5, 13]. After extensive studies in mice and rats and using human tissue preparations 2-amino-9-[4-acetoxy-3-(acetoxymethyl)butyl]purine (famciclovir) was selected for evaluation in humans [13, 14]. Famciclovir is conveniently prepared from the same synthetic precursor as is penciclovir.

Table 5. Summary of activity of penciclovir in herpesvirus infected animals

Model	Dose and route	Dosing schedule	Assessment	Penciclovir vs placebo	Penciclovir vs acyclovir
HSV-1 cutaneous lesions in male guinea pigs	1% aq. cream, topical	2× daily, days 1–6	lesion score	>(p < 0.001)	= (NS)
HSV-1 cutaneous lesions in male hairless mice	50 mg/kg, sc	2× daily, 5 days, starting time of infection	lesion score	>(p < 0.001)	= (NS)
HSV-1 infection of ear pinna of BALB/c mice	100 mg/kg sc	1× daily, 5 days, starting 5 h after infection	lesion score	>(p < 0.01)	= (NS)
HSV-1 intranasal infection of BALB/c mice	100 mg/kg, sc	1× daily, days 0–6	survival time	>(p < 0.001)	> (NS)
	250 mg/kg, sc	1× daily, days 0 and 1		>(p < 0.01)	>(NS)
	200 mg/kg, oral (drinking water)	continuous, days 1–5		>(p < 0.001)	<(NS)
HSV-1 intraperitoneal infection of DBA/2 mice	50 mg/kg, sc	1×, given 5 h after infection	infectivity titre of peritoneal wash, 24 and 48 h after infection	>(p < 0.01)	>(p < 0.01) [5.5 mg/kg PCV >50 mg/kg ACV (p < 0.05)]*

* Dosing 1×, given 24 h after infection, assessment at 48 h after infection.

Upon oral administration to rodents, famciclovir was well absorbed and efficiently converted to penciclovir. Thus, in mice, the highest concentrations of penciclovir detected in the blood following oral famciclovir were 15× higher than those seen after administration of penciclovir itself (Table 6).

Table 6. Concentrations of penciclovir in the blood following oral administration of penciclovir and famciclovir (0.2 mmol/kg) to mice

Compound administered	Concn. (μM) of penciclovir in blood at time after dosing		
	0.25 h	1 h	3 h
Penciclovir	4.9	4.0	1.1
Famciclovir	74	17	1.1

Similarly, in rats comparison of blood concentrations obtained after oral dosing with those seen following intravenous penciclovir (Fig. 5) indicated that the bioavailability of penciclovir is only 1.5% following oral penciclovir, but 41% following oral famciclovir.

The major pathway for conversion of famciclovir to penciclovir involves sequential removal of the acetyl groups by esterases, followed by oxidation at the purine 6-position by xanthine oxidase. Some oxidation can, however, occur prior to deesterification, but all metabolites generated are convertible to penciclovir.

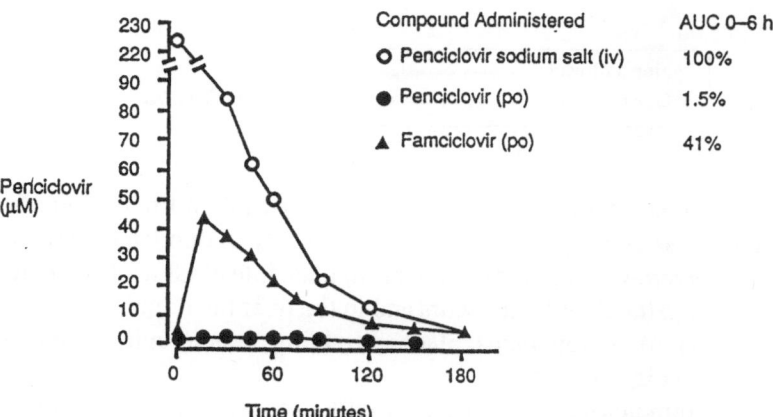

Fig. 5. Mean concentrations of penciclovir in blood after administration of penciclovir and famciclovir (0.2 mmol/kg) to rats

Biotransformation of famciclovir to penciclovir

Major pathway ——————▶
Minor pathway ——————▶   PENCICLOVIR

Studies in which human duodenal contents, intestinal wall and liver preparations were used indicated that famciclovir has high stability in duodenal contents and one acetyl group is removed upon passage through the intestinal wall. Transfer through the portal vein to the liver then occurs and in the liver the remaining acetyl group is removed and oxidation takes place providing penciclovir, which is released into the blood (Fig. 6).

After a single oral famciclovir dose of 20 mmol/kg (equivalent to 5 mg/kg of penciclovir) to healthy volunteers, more than half the dose was absorbed and converted rapidly to penciclovir. Peak plasma concentrations of penciclovir (mean

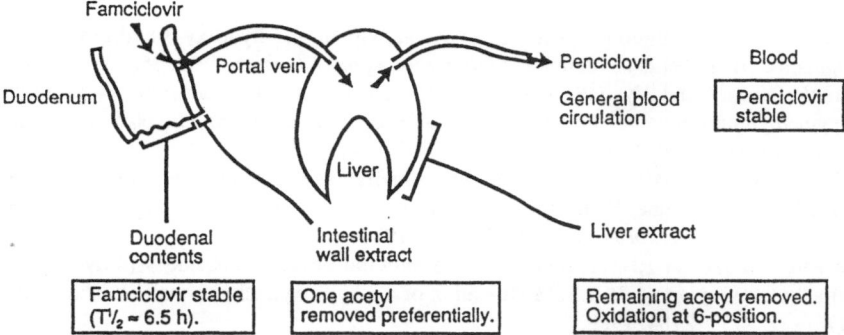

Fig. 6. Schematic representation of conversion of famciclovir to penciclovir by human tissues after oral administration

4.5 µg/ml) were observed within 1 hour and were over 10-fold higher than those detected following the equivalent oral dose of penciclovir. Both penciclovir and famciclovir are well tolerated in humans and clinical trials against herpes simplex virus and varicella zoster virus infections are now underway using an intravenous penciclovir formulation and oral famciclovir.

Acknowledgements. I am very grateful to many colleagues at SmithKline Beecham who have been responsible for generation of the data that has been reviewed, particularly to Dr. R. L. Jarvest for synthesis of penciclovir and famciclovir and to Mr. M. R. Boyd, Dr. R. A. Vere Hodge and their co-workers for the antiviral and biochemical studies.

References

1. Harnden MR and Jarvest RL, An improved synthesis of the antiviral acyclonucleoside 9-[4-hydroxy-3-(hydroxymethyl)but-1-yl]guanine. *Tetrahedron Lett.* **26**: 4265–4268, 1985.
2. Harnden MR and Jarvest RL (Beecham Group Plc) EP 141, 927.
3. Harnden MR, Jarvest RL, Bacon TH and Boyd MR, Synthesis and antiviral activity of 9-[4-hydroxy-3-(hydroxymethyl)but-1yl]purines. *J. Med. Chem.* **30**: 1636–1642, 1987.
4. Boyd MR, Bacon TH, Sutton D and Cole M, Anti-herpesvirus activity of 9-(4-hydroxy-3-hydroxymethylbut-1-yl)guanine (BRL 39123) in cell culture. *Antimicrob. Agents Chemother.* **31**: 1238–1242, 1987.
5. Harnden MR, Development of 9-[4-hydroxy-3-(hydroxymethyl)but-1-yl]purines as potential therapeutic agents for treatment of human herpesvirus infections. *Drugs of the Future* **14**: 347–358, 1989.
6. Vere Hodge RA and Perkins RM, Mode of action of 9-(4-hydroxy-3-hydroxymethylbut-1-yl)guanine (BRL 39123) against herpes simplex virus in MRC-5 cells. *Antimicrob. Agents Chemother.* **33**: 223–229, 1989.
7. Jarvest RL, Barnes RD, Earnshaw DL, O'Toole KJ, Sime JT and Vere Hodge RA, Synthesis of isotopically chiral [^{13}C]penciclovir (BRL 39123) and its use to determine the absolute configuration of penciclovir triphosphate formed in herpes virus infected cells. *J. Chem. Soc. Chem. Commun.* 555–556, 1990.

8. Boyd MR, Bacon TH and Sutton D, Antiherpesvirus activity of 9-(4-hydroxy-3-hydroxymethyl-but-1-yl)guanine (BRL 39123) in animals. *Antimicrob. Agents Chemother.* **32**: 358–363, 1988.

9. Harnden MR, Parkin A and Wyatt PG, Analogues of the antiviral acyclonucleoside 9-(4-hydroxy-3-hydroxymethylbutyl)guanine. Part 1. Substitutions on C-2' of the acyclic N-9 substituent. *J. Chem. Soc. Perkin Trans.* **1**: 2757–2765, 1988.

10. Bailey S and Harnden MR, Analogues of the antiviral acyclonucleoside 9-(4-hydroxy-3-hydroxymethylbutyl)guanine. Part. 2. Substitutions on C-1' and C-3' of the acyclic N-9 substituent. *J. Chem. Soc. Perkin Trans.* **1**: 2767–2775, 1988.

11. Harnden MR and Jarvest RL, Analogues of the antiviral acyclonucleoside 9-(4-hydroxy-3-hydroxymethylbutyl)guanine. Part 3. Modification of a 3-hydroxymethyl group. *J. Chem. Soc. Perkin Trans.* **1**: 2777–2784, 1988.

12. Harnden MR and Jarvest RL, Analogues of the antiviral acyclonucleoside 9-(4-hydroxy-3-hydroxymethylbutyl)guanine. Part. 4. Substitution on the 2-amino group. *J. Chem. Soc. Perkin Trans.* **1**: 2207–2213, 1989.

13. Harnden MR, Jarvest RL, Boyd MR, Sutton D and Vere Hodge RA, Prodrugs of the selective anti-herpesvirus agent 9-[4-hydroxy-3-(hydroxymethyl)but-1-yl]guanine (BRL 39123) with improved gastrointestinal absorption properties. *J. Med. Chem.* **32**: 1738–1743, 1989.

14. Vere Hodge RA, Sutton D, Boyd MR, Harnden MR and Jarvest RL, Selection of an oral prodrug (BRL 42810; famciclovir) for the antiherpesvirus agent BRL 39123 [9-(4-hydroxy-3-hydroxymethylbut-1-yl)guanine; penciclovir]. *Antimicrob. Agents Chemother.* **33**: 1765–1773, 1989.

2'3'Dideoxy-3'Thiacytidine (SddC) as an Anti-Human Hepatitis B Virus (HBV) and Anti-Human Immunodeficiency Virus (HIV) Agent

Y-C. Cheng*, C-N. Chang, S-L. Doong, G. E. Dutschman, C-H. Tsai, E. A. Murphy and J. H. Zhou

Department of Pharmacology, Yale University School of Medicine, 333 Cedar Street, New Haven, CT, 06510, USA

Abstract. Dideoxycytidine (ddC) was found to be a potent inhibitor of HIV and HBV growth in culture. The activity of ddC against HIV was also demonstrated in patients with acquired immunodeficiency syndrome (AIDS). It was noted at the dosage employed to be effective against HIV, but patients developed peripheral neuropathy after a few months of treatment. The mechanism responsible for the delayed toxicity observed was proposed by us to be the inhibition of mitochondrial DNA synthesis in the target tissue. SddC (2',3'dideoxy-3'thiacytidine) was discovered by others to have potent anti-HIV activities. This was confirmed by us. However, this compound is much less active than ddC in its inhibition of mitochondrial DNA synthesis. It is also less toxic to cell growth in culture. When SddC and 5-fluoro SddC were examined against HBV using an HBV-producing cell line (2.2.15) in culture, it was found to be at least 40-fold more potent than ddC in inhibiting HBV. The mechanism is due to the inhibition of episomal HBV DNA synthesis. HBV RNA synthesis and the integrated form of HBV DNA were not affected. The antiviral action is reversible, suggesting that continuous usage of SddC may be needed in order to achieve a therapeutic effect against AIDS and hepatitis. The SddC examined is a racemic mixture. Subsequent studies indicated that (–)SddC was more potent than (+)SddC against HBV and HIV in culture, whereas the cytotoxicity of (+)SddC is more than that of (–)SddC. The potential use of (–)SddC for the treatment of HIV and HBV is indicated. It is anticipated that (–)SddC will be less toxic than ddC in patients.

Human hepatitis B virus (HBV) is recognized as a major worldwide health problem. Not only does it cause the infectious disease, but it is also closely associated with primary hepatocellular carcinoma [1–3]. No effective therapy which can control the replication of this virus without substantial toxicity has been found.

HBV is a double-stranded DNA virus and codes a unique DNA polymerase which catalyzes both DNA-dependent as well as RNA-dependent DNA polymerization reactions [4, 5]. Unlike other DNA viruses, the reverse transcriptase reaction is a critical step involved in HBV DNA replication [6]. Given the unique nature of this viral DNA polymerase, and the important role this enzyme

---

* Corresponding author.

plays in viral replication, HBV DNA polymerase is an obvious target for developing selective anti-HBV drugs. The search for compounds which target on HBV DNA polymerase has been pursued by many laboratories around the world [7–10]. Although there is still difficulty in preparing this enzyme for study, the development of cell lines transfected by HBV DNA and capable of secreting HBV in cell culture systems [11–15] has facilitated the development of selective anti-HBV agents targeting on viral specified proteins including DNA polymerase. Using these HBV-transfected hepatoma cell lines, several laboratories have indentified a number of nucleoside analogs which are selective against HBV [7–10, 16–18]. Furthermore, a number of animal hepatitis B virus equivalents, such as duck and woodchuck hepatitis virus, were also characterized and shown to share some similar molecular and pathological properties with human HBV [19–20]. The animals infected with these hepatitis viruses are also useful models for testing anti-HBV compounds. It should be pointed out that each testing system has its limitations. The ultimate test of the effectiveness of drugs has to be in patients.

Dideoxycytidine (ddC) and other dideoxynucleoside analogs were found to be potent and selective inhibitors [21] of human immunodeficiency virus (HIV) which is thought to be the etiological agent responsible for acquired immunodeficiency syndrome (AIDS) [22, 23]. These compounds are phosphorylated inside the cell and, as triphosphate nucleotides, they interact with HIV reverse transcriptase preferentially and are incorporated into viral DNA to terminate DNA chain elongation [24–26]. This is the key mechanism responsible for their antiviral action. It has been noted that these analogs, as triphosphates, could also interact well with DNA polymerase gamma which is responsible for mitochondrial DNA synthesis [24, 27, 28]. In view of their lack of cytotoxicity when the cells were exposed to the drugs for less than 4 days, ddC and dideoxyinosine (ddI) were introduced into the clinic for testing their efficacy as therapeutic agents for the treatment of patients with AIDS. Since the action of these compounds is to suppress viral replication, and they are incapable of getting rid of the cells which harbor HIV DNA in their chromosomal DNA, continued usage of these compounds is required and used to demonstrate their clinical effectiveness. Indeed, both compounds were shown to have anti-HIV activity in the clinic and to retard the progress of AIDS in patients. However, delayed toxicity such as peripheral neuropathy was also observed [29]. The mechanism responsible for this delayed toxicity by ddC, ddI or other deoxynucleoside analogs was attributed by us to their inhibitory activities against mitochondrial DNA synthesis in target organs [30, 31]. There continues to be a positive correlation between the potency of inhibition of mitochondrial DNA synthesis and development of delayed toxicity [32]. Anti-HIV nucleoside analogs which have the same inhibitory effects on HIV without inhibiting mitochondrial DNA synthesis may not have these delayed toxic effects upon continuous usage in patients.

ddC and other deoxynucleoside analogs have also been shown to have potent activity against HBV replication in cell culture and anti-hepatitis virus activity in duck and other animal systems [8, 33–36]. Although their mechanism of action is not clear, it is conceivable that their selectivity is also due to the potent interaction of their nucleoside triphosphate metabolites and HBV DNA polymerase. Since these nucleoside analogs could only suppress HBV replication, and HBV DNA could exist in a latent episomal form or be incorporated into chromosomal DNA, long-term usage of these anti-HBV nucleosides is again required in order to demonstrate clinical efficacy against HBV in patients. Thus, nucleoside analogs with anti-HBV activity without potent activity against human DNA polymerases, including DNA polymerase gamma, will be more desirable.

2′3′dideoxy-3′thiacytidine (SddC) was recently synthesized and shown to have good anti-HIV activity [37–39]. Its 5-fluoro analog (5-fluoro SddC) was also shown to have comparable activity [40]. Those results were confirmed by us as shown in Table 1. The compounds which we examined have two stereo isomers in

Table 1. Comparative potencies (ID_{50} values) of ddC analogs as monitored by anti-HBV, anti-HIV, cytotoxicity and mtDNA effects

Compound	HBV[a]	HIV[b]	Cytotoxicity		anti-mtDNA[a]
			CEM[a]	MT2[b]	
			ID_{50} (μM)		
(±)S-ddC	0.05[e]	3	37[e]	> 20	47[e]
(−)S-ddC	0.01	2	50	> 20	> 50
(+)S-ddC[c]	0.5	10	2	13	4
(±)5-F-SddC[d]	0.1[e]	—	> 200[e]	—	> 200[e]
(−)5-F-SddC	0.02	—	> 200	—	> 200
ddC	2.8[e]	1	10[e]	—	0.022[e]

[a] The detailed procedure was the same as the one published by Doong et al. [42]. [b] The detailed procedure was according to the methods of Larder et al. [43]. [c] The compound was kindly provided by C. K. Chu, J. W. Beach, L. S. Jeong and A. Alves (The University of Georgia, Athens, GA). [d] The compound was kindly provided by R. Schinazi and D. Liotta (Emory University, Atlanta, GA). [e] Previously published results [42].

each composition, as shown in Fig. 1. These compounds were examined for their activity against mitochondrial DNA synthesis in a human cell line, CEM. In comparison with ddC, their potency against either cell growth or against mitochondrial DNA replication are much lower. In view of the fact that several anti-HIV compounds were also shown to have anti-HBV activity, we examined the potency of SddC and 5-fluoro SddC against HBV replication in a HBV-transfected Hep-G2 hepatoma cell line [4, 41] kindly provided by Dr. Acs from Mount Sinai Medical School. Both compounds are shown to have more potent activity against HBV than ddC. No viral DNA replication could be detected at 0.5 μM as monitored by the absence of both intracellular episomal and secreted viral DNA.

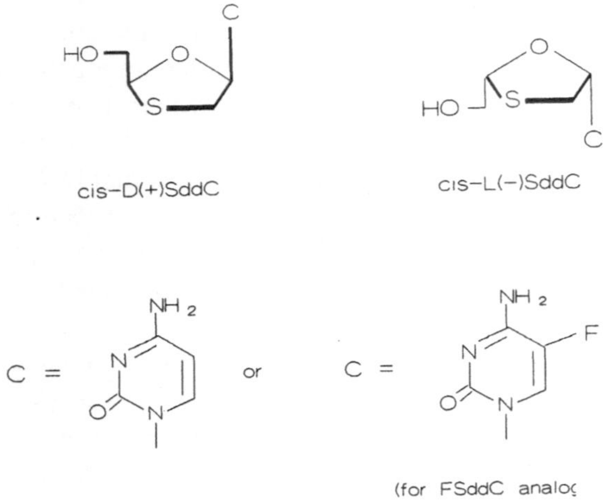

cis–D(+)SddC

cis–L(–)SddC

C =

or

C =

(for FSddC analog

Fig. 1. Structures of SddC analogues and their stereo isomers used in this study

The mechanism for their antiviral activity is likely due to their selective action against HBV DNA synthesis, since the viral RNA synthesis presumably made from integrated HBV DNA in this cell line was not inhibited at this concentration. The antiviral action of SddC and 5-fluoro SddC is reversible [42]. This is not surprising since there are multiple copies of episomal HBV DNA, and integrated HBV DNA is present in this cell line. This result did suggest that, if SddC or 5-fluoro SddC is introduced into the clinic for treatment of HBV, continued usage of this drug is required to obtain better clinical results. Currently one form of the SddC is in Phase 1 trials as an anti-HIV compound. Its potential for the treatment of HBV infection in patients should also be explored as soon as possible.

As indicated, there are two stereo isomers, the (+) and (–) forms, in the compounds we examined; which of these two forms of SddC or 5-fluoro SddC is the active species was explored. Upon prolonged incubation of (±)SddC or the 5-fluoro (±)SddC with human deoxycytidine deaminase, only about 50% of the nucleosides were found to be deaminated. Subsequently, it was shown that the (+)SddC and (+)5-fluoro SddC were substrates, while the (–) forms of both compounds were not substrates of human deoxycytidine deaminase. The cytotoxicity of the (+)SddC is much more pronounced than that of (–)SddC, and the (–) form seems to decrease the toxicity of the (+) form of SddC. It was surprising to learn that the anti-HBV activity of the (–) form is much more pronounced than that of the (+) form. The ID_{50} was determined to be approximately 0.01 μM and 0.02 μM for (–)SddC and (–)5-fluoro SddC, respectively (Table 1). The (–)SddC can be phosphorylated to (–)SddCMP, (–)SddCDP, and (–)SddCTP inside the cell. Two additional metabolites were also detected. The

(–)SddCTP and (+)SddCTP were shown to be inhibitors of HBV-associated DNA polymerase with (–)SddCTP more potent than (+)SddCTP (Table 2). This inhibition of HBV polymerase by SddCTP is consistent with the hypothesis that the mechanism of action of SddC against HBV replication is due to the inhibition of HBV DNA synthesis, but may not be sufficient to explain why (+)SddC is less active than (–)SddC in inhibiting HBV (unpublished results). It was also of interest to note that the (–)SddC is also more potent than the (+)SddC against HIV (Table 1). The detailed information will be published later.

Table 2. Effect of S-ddCTP on HBV-associated DNA polymerase[a]

Additive	Concentration (µM)	Activity (%)
Polymerase alone	—	100
(–)S-ddCTP	0.175	35
(+)S-ddCTP	0.175	50

[a] The methodology is very similar to that of Chang et al. [44] with some modifications; detailed procedure will be published elsewhere.

The (–)SddC or 5-fluoro(–)SddC are two interesting nucleoside analogs. Both have activity against HBV and HIV at a concentration much lower than that required to inhibit either nuclear or mitochondrial DNA synthesis. Both compounds should have the potential to exert good activity against HIV- and HBV-associated infection in patients. The potential use of these two analogs for the prevention and the treatment of hepatocellular carcinoma is also worthwhile exploring in view of the fact that HBV may play an important role in the pathogenesis of hepatocellular carcinoma. (Supported by NIH Grant CA44358).

References

1. Ayoola EA, Balayan MS, Deinhardt F, Gust I, Kureshi AW, Maynard JE, Nayak NC, Brodley DW, Ferguson M, Melnick J, Purcell RH and Zuckerman AJ, Progress in the control of viral hepatitis: Memorandum from a WHO Meeting. *Bull. World Health Org.* **66**: 443–455, 1988.
2. Beasley RP, Hwang LY, Lin CC and Chien CS, Hepatocellular carcinoma and hepatitis B virus: A prospective study of 22,707 men in Taiwan. *Lancet* **ii:** 1129–1133, 1981.
3. Di Bisceglei AM, Rustgi VK, Hoofnagle JH, Dusheik GM and Lotze MT, Hepatocellular carcinoma. *Ann. Intern. Med.* **108**: 390–401, 1988.
4. Ganem D and Varmus HE, The molecular biology of the hepatitis-B virus. *Ann. Rev. Biochem.* **56**: 651–693, 1987.
5. Miller RH and Robinson WS, Common evolutionary origin of hepatitis B virus and retroviruses. *Proc. Natl. Acad. Sci. USA* **83**: 2531–2535, 1986.
6. Summes J and Mason WS, Replication of the genome of a hepatitis B-like virus by reverse transcription of an RNA intermediate. *Cell* **29**: 403–415, 1982.

236

7. Matthes E, Langen P, von Janta-Lipinski M, Will H, Schroder HC, Merz H, Weiler BE and Muller WEG, Potent inhibition of hepatitis B virus production *in vitro* by modified pyrimidine nucleosides. *Antimicrob. Agents Chemother.* **34**: 1986–1990, 1990.
8. Lee B, Luo W, Suzuk S, Robins M and Tyrrell DLJ, *In vitro* and *in vivo* comparison of the abilities of purine and pyrimidine 2′,3′-dideoxynucleosides to inhibit duck hepadnavirus. *Antimicrob. Agents Chemother.* **33**: 336–339, 1989.
9. Meisel H, Reimer K, von Janta-Lipinski M, Barwolf D and Matthes E, Inhibition of hepatitis B virus DNA polymerase by 3′-fluorothymidine triphosphate and other modified nucleoside triphosphate analogs. *J. Med. Virol.* **30**: 137–141, 1990.
10. Matthes E, Reimer K, von Janta-Lipinski M, Meisel H and Lehmann C, Comparative inhibition of hepatitis B virus DNA polymerase and cellular DNA polymerase by triphosphate of sugar-modified 5-methyldeoxycytidine, and of other nucleoside analogs. *Antimicrob. Agents Chemother.* **35**: 1254–1257, 1991.
11. Sureau C, Romet-Lomonne J-L, Mullins JI and Essex M, Production of hepatitis B virus by a differentiated human hepatoma cell line after transfection with cloned circular HBV DNA. *Cell* **47**: 37–47, 1986.
12. Chang C, Jeng K-S, Hu C-P, Lo SJ, Su T-S, Ting L-P, Chou C-K, Han S-H, Pfaff E, Salfeld J and Schaller H, Production of hepatitis B virus *in vitro* by transient expression of cloned HBV DNA in hepatoma. *EMBO J.* **6**: 675–680, 1987.
13. Tsurimoto T, Fujiyama A and Matsubara K, Stable expression and replication of hepatitis B virus genome in an integrated state in a human hepatoma cell line transfected with the cloned viral DNA. *Proc. Natl. Acad. Sci. USA* **84**: 444–448, 1987.
14. Sells MA, Chen M-L and Acs G, Production of hepatitis B virus particles in HepG2 cells transfected with cloned hepatitis B virus DNA. *Proc. Natl. Acad. Sci. USA* **84**: 1005–1009, 1987.
15. Shih C, Li L-S, Roychoudhury S and Ho M-H, *In vitro* propagation of human hepatitis B virus in a rat hepatoma cell line. *Proc. Natl. Acad. Sci. USA* **86**: 6323–6327, 1989.
16. Yokota T, Konno K, Chonan E, Mochizuki S, Kojima K, Shigeta S and de Clercq E, Comparative activities of several nucleoside analogs against duck hepatitis B virus *in vitro*. *Antimicrob. Agents Chemother.* **34**: 1326–1330, 1990.
17. Price PM, Banerjee R and Acs G, Inhibition of the replication of hepatitis B virus by the carbocyclic analogue of 2′-deoxyguanosine. *Proc. Natl. Acad. Sci. USA* **86**: 8541–8544, 1989.
18. Tyrrell DLJ, Inhibition of duck hepatitis B virus replication by purine 2′,3′-dideoxynucleosides. *Biochem. Biophys. Res. Commun.* **156**: 1144–1151, 1988.
19. Galle P, Schlicht R, Fischer M and Schaller H, Production of infectious duck hepatitis B virus in a human hepatoma cell line. *J. Virol.* **62**: 1736–1740, 1988.
20. Horwich AL, Furtak K, Pugh J and Summers J, Synthesis of hepadnavirus particles that contain replication-defective duck hepatitis B virus genomes in cultured HuH7 cells. *J. Virol.* **64**: 642–650, 1990.
21. Mitsuya H, Yarchoan R and Broder S, Molecular targets for AIDS therapy. *Science* **249**: 1533–1544, 1990.
22. Barre-Sinoussi F, Chermann JC, Rey F, Nugeyre MT, Chamarel S, Gruest T, Dauguet CD, Axler-Blin C, Vezin-Brun F, Rouzious C, Rozenbaum W and Montagnier L, Isolation of a T lymphocyte retrovirus from a patient at risk for acquired immunodeficiency syndrome (AIDS). *Science* **220**: 868–870, 1983.
23. Gallo RC, Salahuddin SZ, Popovic M, Shearer GM, Kaplan M, Haynes BF, Paulker TJ, Redfield R, Oleske J, Sasai B, White G, Foster P and Markham PD, Frequent detection and isolation of cytopathic retroviruses (HTLV-III) from patients with AIDS and at risk of AIDS. *Science* **224**: 500–503, 1983.
24. Starnes MC and Cheng Y-C, Cellular metabolism of 2′,3′-dideoxycytidine, a compound active against human immunodeficiency virus *in vitro*. *J. Biol. Chem.* **262**: 988–991, 1987.

25. Cheng Y-C, Dutschman GE, Bastow KF, Sarngadharan MG and Ting RYC, Human immunodeficiency virus reverse transcriptase. General properties and its interactions with nucleoside triphosphate analogs. *J. Biol. Chem.* **262**: 2187–2189, 1987.
26. Starnes MC and Cheng Y-C, Inhibition of human immunodeficiency virus reverse transcriptase by 2′,3′-dideoxynucleoside triphosphates: Template dependence, and combination with phosphonoformate. *Virus Genes* **2**: 241–251, 1989.
27. Meyer RR and Simpson MV, Deoxyribonucleic acid biosynthesis in mitochondria: Purification and general properties of rat liver mitochondrial deoxyribonucleic acid polymerase. *J. Biol. Chem.* **245**: 3426–3435, 1970.
28. Lestienne P, Evidence for a direct role of the DNA polymerase gamma in the replication of the human mitochondrial DNA *in vitro. Biochem. Biophys. Res. Commun.* **146**: 1146–1153, 1987.
29. Dubinsky RM, Yarchoan R, Dalakas M and Broder S, Reversible axonal neuropathy from the treatment of AIDS and related disorders with 2′,3′-dideoxycytidine (ddC). *Muscle and Nerve* **12**: 856–860, 1989.
30. Chen C-H and Cheng Y-C, Delayed cytotoxicity and selective loss of mitochondrial DNA in cells treated with the anti-human immunodeficiency virus compound 2′,3′-dideoxycytidine. *J. Biol. Chem.* **264**: 11934–11937, 1989.
31. Cheng Y-C, Gao W-Y, Chen C-H, Vazquez-Padua M and Starnes MC, DNA polymerases versus HIV reverse transcriptase in AIDS therapy. In : *AIDS: Anti-HIV Agents, Therapies, and Vaccines.* **616**: 217–223, *Ann. N.Y. A.S.,* 1990.
32. Chen C-H, Vazquez-Padua M and Cheng Y-C, The effect of anti-HIV nucleoside analogs on mitochondrial DNA and its implication on delayed toxicity. *Mol. Pharmacol.* **39**: 27–33, 1991.
33. Yokota T, Mochizuki S, Konno K, Mori S, Shigeta S and De Clercq E, Inhibitory effects of selected antiviral compounds on human hepatitis B virus DNA synthesis. *Antimicrob. Agents Chemother.* **35**: 394–397, 1991.
34. Ueda K, Tsurimoto T, Nagahata T, Chisaka O and Matsubara K, An *in vitro* system for screening anti-hepatitis B virus. *Virology* **169**: 213–216, 1989.
35. Suzuki S, Lee B, Luo W, Tovell D, Robins MJ and Tyrrell LJ, Inhibition of duck hepatitis B virus replication by purine 2′,3′-deoxynucleosides. *Biochem. Biophys. Research Comm.* **156**: 1144–1151, 1988.
36. Kassianides L, Hoofnagle JH, Miller RH, Doo E, Ford H, Broder S and Mitsuya H, *Gastroenterology* **97**: 1275–1280, 1989.
37. Bellau B, Dixit D, Nguyen-Ga N and Kraus JL, *V Internat. Conf. on AIDS.* Montreal, Canada, June 4–9, 1989, Abstract No. T.C.O.I.
38. Greenberg ML, Allaudeen HS and Hershfield MS, Metabolism, toxicity, and anti-HIV activity of 2′-deoxy-3′-thia-cytidine (BCH-189) in T and B cell lines. *New York Acad. Sci.* **616**: 517–518, 1990.
39. Wainberg MA, Tremblay M, Rooke R, Blain N, Soudeyns H, Parniak MA, Yao X-J, Li X-G, Fanning M, Montaner JSG, O'Shaughnessy M, Tsoukas C, Falutz J, Dionne G, Belleau B and Reudy J, Characterization of reverse transcriptase activity and susceptibility to other nucleosides of AZT-resistant variants of HIV-1. *Ann. N.Y. A.S.* **616**: 346–355, 1990.
40. Schinazzi R and Liotta D, unpublished results.
41. Sells MA, Zelent AZ, Shvartsman and Acs G, *J. Virol.* **62**: 2836–2844, 1988.
42. Doong SL, Tsai CH, Schinazi RF, Liotta DC and Cheng Y-C, Inhibition of the replication of hepatitis B virus *in vitro* by 2′,3′-dideoxy-3′thiacytidine and related analogues. *Proc. Natl. Acad. Sci. USA* **88**: 8495–8499, 1991.
43. Larder BA, Chesebro B and Richman DD, Susceptibilities of zidovudine-susceptible and -resistant human immunodeficiency virus isolates to antiviral agents determined by using a quantitative plaque reduction assay. *Antimicrob. Agents Chemother.* **34**: 436–441, 1990.

44. Chang L-J, Hirsch RC, Ganem D and Varmus HE, Effects of insertional and point mutations on the functions of the duck hepatitis B virus polymerase. *J. Virol.* **64**: 5553-5558, 1990.

Phosphorylating Enzymes Involved in Activation of Chemotherapeutic Nucleosides and Nucleotides

D. Shugar

Institute of Biochemistry & Biophysics, Polish Academy of Sciences, 36 Rakowiecka St., 02-532 Warszawa, and Department of Biophysics, Institute of Experimental Physics, University of Warsaw, 93 Żwirki Wigury St., 02-089 Warszawa, Poland

Abstract. A brief survey is presented of the enzymes responsible for the activation (by phosphorylation) of chemotherapeutically active nucleoside analogues, including acyclonucleosides, dideoxynucleosides, and the recently reported oxetanocin and cyclobut (carbocyclic oxetanocin) nucleosides, and relevance to the latter of 3′-branched 2′-deoxynucleosides. Properties of the known cellular and viral-encoded nucleoside kinases, and nucleoside phosphotransferases (cytosolic 5′-nucleotidases) are described. Phosphates and cyclic phosphates of acyclonucleosides with more than one hydroxyl in the acyclic chain, and their behaviour towards a variety of enzymes, are examined in relation to the possible mechanism of antiviral activity of the cyclic phosphate of Ganciclovir, which is an analogue of cGMP. Reported observations on the role of cAMP and cGMP on viral replication are briefly reviewed.

1. Introduction

The majority of nucleoside analogues which exhibit antitumour, antiviral and/or antiparasitic activities are dependent for their activity on phosphorylation by intracellular enzymes, so that the nucleotides are really the active chemotherapeutic agents. In the case of antiviral agents, the active species is frequently the nucleoside 5′-triphosphate, which may selectively inhibit the viral DNA or RNA polymerase relative to the host cell polymerase. Furthermore, in viral infected cells, initial monophosphorylation of a nucleoside is often mediated preferentially or exclusively by a viral encoded kinase, so that such a nucleoside may

---

Abbreviations employed: TK, thymidine kinase; TK1, cytosolic TK; TK2, mitochondrial TK; AK, adenosine kinase; dCK, deoxycytidine kinase; dGK, deoxyguanosine kinase; GK, guanosine kinase; UK, uridine kinase (actually uridine/cytidine kinase);PDase, phosphodiesterase; cPDase, cyclic nucleotide phosphodiesterase; HSV, herpes simplex virus; CMV, cytomegalovirus; HCMV, human CMV; ACV, Acyclovir or 9-(2-hydroxyethoxymethyl)guanine; DHPG, 9-(1,3-dihydroxy-2-propoxymethyl)guanine; DHP-Ade, adenine analogue of DHPG; DHBG, 9-(3,4-dihydroxy-butyl)guanine; –MP, monophosphate; –TP, triphosphate; –cMP, cyclic monophosphate; N, nucleoside.

be inactive against viruses or viral mutants which do not code for a viral-specific kinase.

Although this brief overview is limited largely to the initial monophosphorylation step, obvious advantages can be envisaged if this could be by-passed by the use of the corresponding nucleotide. However, nucleotides do not readily traverse the cell membrane as such, but are hydrolyzed to the nucleoside before entry. Efforts have been made in the past to circumvent this problem by use of phosphate esters of nucleotides, or nucleoside phosphonates [1], but with little success.

A promising new development is the synthesis of a series of nucleotide analogues, characterized as phosphonyl methyl ethers of acyclonucleosides (see Scheme 1). These compounds are apparently readily taken up as such by cells via

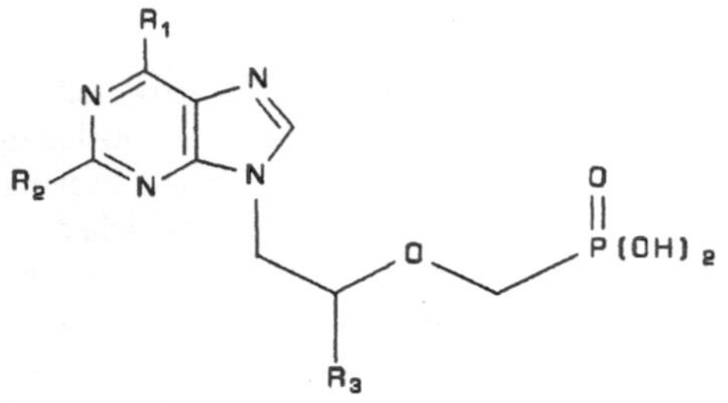

Compound	R₁	R₂	R₃
PMEG	OH	NH$_2$	H
PMEA	NH$_2$	H	H
HPMPG	OH	NH$_2$	CH$_2$OH
HPMPA	NH$_2$	H	CH$_2$OH

Scheme 1. Examples of phosphonyl methyl ethers of acyclonucleosides with potent antiviral activities, but which do not require "activation" by cellular or viral-encoded nucleoside kinases

a process resembling endocytosis and exhibit broad-spectrum activities against a variety of viruses, including TK⁻ strains, and retroviruses, including HIV. An international symposium has already been devoted to this field [2], also extensively reviewed by De Clercq [3]. It has also been shown that these phosphate analogues can undergo intracellular phosphorylation to the "triphosphates" in a single step by transfer of the pyrophosphate group of 5-phosphoribosyl 1-pyrophosphate (PRPP) mediated by PRPP synthetase [4]; but there is evidence that other phosphorylation routes exist. The triphosphates also turn out to be selective inhibitors of viral polymerases or reverse transcriptases; but it is probably too soon to determine whether this is their only mechanism of action. In the interim it has been reported that the four compounds shown in Scheme 1 are, under defined conditions, also good cytotoxic agents, with activities against tumours such as P388 leukemia and B16 melanoma [5].

There are also instances of nucleoside analogues the activities of which are not dependent on prior metabolism, such as phosphorylation. A very recent, and indeed striking, example is that reported by Tanaka et al. [6] of two acyclonucleoside analogues of 5-ethyl-6-thiophenyluracil, with the terminal hydroxyl replaced by a methyl or phenyl group (see Scheme 2). There is, consequently, no site for phosphorylation. Nonetheless these compounds are potent inhibitors of HIV-1, but not of HIV-2; and, in accord with this, they are likewise potent specific inhibitors of the reverse transcriptase activity of HIV-1, but not of HIV-2.

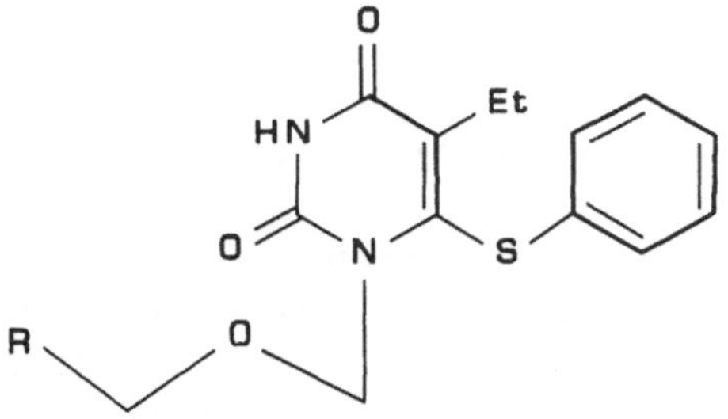

R = Me or phenyl

Scheme 2. Two of a series of acyclonucleosides with anti-HIV-1 activity, which cannot be phosphorylated, but specifically inhibit the reverse transcriptase activity of HIV-1, but not HIV-2

No details have as yet appeared regarding the mechanism of action of these compounds. However, a cursory examination shows that the substituted uracil residue is still capable of Watson–Crick pairing with an adenine residue in the growing fork of the RNA template, or reverse Watson–Crick pairing with another base, which may in either case inhibit binding of the natural NTP substrates. If so, then methylation of the ring N(3) of the uracil moiety should abolish the inhibitory activity. These compounds are strikingly reminiscent of the 6-arylhydrazino derivaties of uracil and cytosine, which are excellent specific inhibitors of the DNA polymerase III of Gram-positive bacteria, and the mechanism of action of which has been extensively studied and well documented (see Ref. 7 for review). In the case of 6-phenylhydrazinouracil, for example, inhibition involves base pairing in an unusual manner (see Scheme 3) to a cytosine residue at the growing

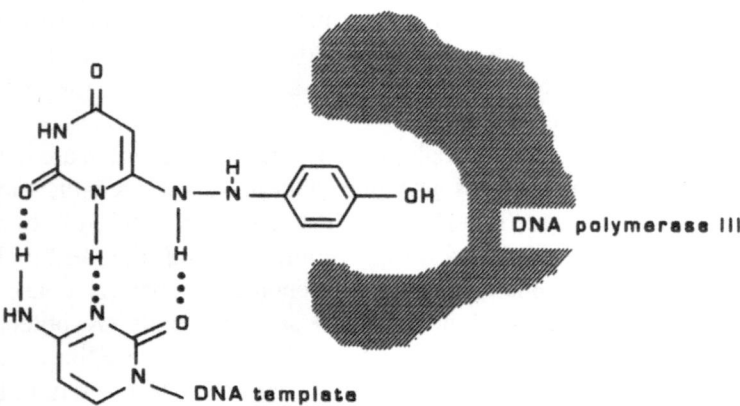

Scheme 3. Mechanism of specific inhibition of DNA polymerase III of Gram-positive bacteria by 6-phenylhydrazinouracil, which base pairs as shown to a cytosine residue in the growing fork of the DNA template. Simultaneously the phenyl group is sequestered by the polymerase, leading to a ternary stable complex which inhibits further replication

fork in the DNA template. Simultaneously the 6-aryl group interacts with a unique site on the enzyme, which is thus sequestered to an inactive complex. The resultant ternary DNA : Inhibitor : Pol III complex has been isolated and identified. The corresponding 6-phenylhydrazinoisocytosine inhibits in an analogous manner by hydrogen bonding to a thymine residue in the growing fork of the DNA template. Furthermore, replacement of the 6-phenylhydrazino substituent by 6-anilino leads to compounds which are good inhibitors of mammalian DNA polymerase α. The comprehensive review by Wright & Brown [7] should be consulted for a general overview of deoxynucleotide analogues as polymerase inhibitors.

2. Cyclic Phosphates of Acyclonucleosides

The first reported nucleotide with antiviral activity, synthesized by Tolman et al. [8], was the cyclic phosphate of Ganciclovir (DHPG-cMP, see Scheme 4). It is a potent broad-spectrum agent vs DNA viruses in cell culture, more active than DHPG vs murine CMV, and comparable to DHPG vs human CMV. It is 15- to 20-fold more effective than DHPG vs guinea-pig CMV and exhibited therapeutic efficacy in infected guinea-pigs (Ref. 9, and references cited).

The mode of action of DHPG-cMP differs from that of DHPG in that its activity is independent of viral and cellular TK. It undergoes uptake in cells essentially intact. A small proportion then undergoes opening of the cyclic phosphate ring to liberate the monophosphate, DHPG-MP, largely (but perhaps not exclusively) as the (S)-enantiomer, which is that formed by phosphorylation of

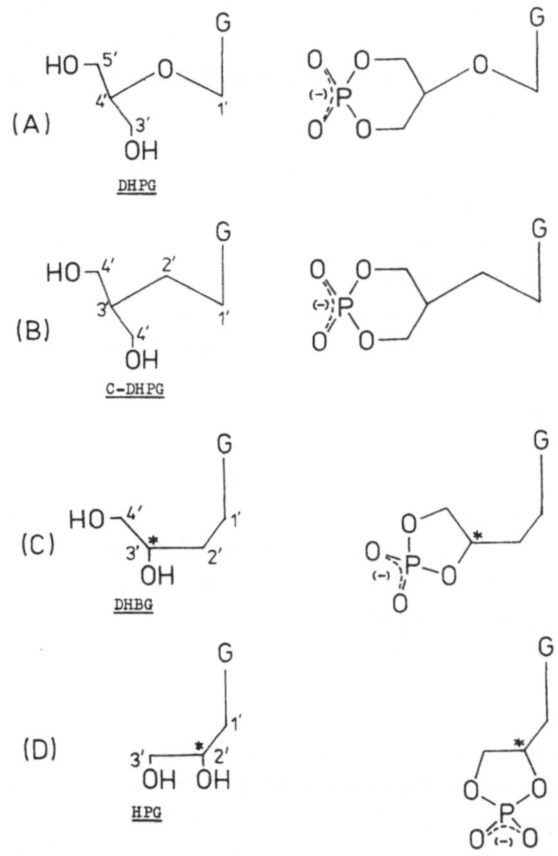

Scheme 4. Examples of simple acyclonucleoside analogues of guanosine (G) with two hydroxyl groups in the acyclic chain, hence capable of being converted to their cyclic phosphates. The guanine base may be replaced by other bases

DHPG by HSV-1 TK. This is further phosphorylated to the triphosphate, both in HSV-1 infected or non-infected cells. But the amount of triphosphate produced, which is an effective inhibitor of HSV-1 DNA polymerase, was much too low to account for the antiviral activity of DHPG-cMP, which itself was not an inhibitor of DNA polymerase. Furthermore, the cyclic phosphate ring was resistant to hydrolysis by PDases I and II, and by commercial preparations of cAMP and cGMP PDases. It was found to be opened on exposure of DHPG-cMP to mouse whole blood or extracts of mouse liver or HeLa cells [10].

The mechanism of action of DHPG-cMP remains to be elucidated. Since it is only minimally catabolized in intact cells, one may envisage the possibility that it acts as such as an analogue of cGMP. This prompted us to prepare the mono- and cyclic phosphates of several acyclonucleosides which contain more than one hydroxyl, and to examine their behaviour in several enzyme systems. It is perhaps pertinent to first briefly review research on the role of cyclic nucleotides on viral replication.

2.1. Cyclic nucleotides and viral replication

The ubiquity of nocleoside 3′,5′-cyclic phosphates, and their well-known regulatory roles in both eucaryotic and procaryotic systems, has prompted some limited studies of their potential effects on viral replication. For example, replication of adenovirus and herpes simplex virus has been reported to be inhibited, whereas that of mammary tumour virus-like particles is stimulated 30-fold by cAMP (Ref. 11, and references cited).

Infection of human fibroblasts and HEp-2 cells with HSV-1 led to a marked decrease in intracellur levels of cAMP and a concomitant 3-fold increase in the level of cGMP, pointing a close relationship between cyclic nucleotides and HSV-1 in productively infected cells. Early addition to infected fibroblasts of cAMP-enhancing compounds decreased the yield of HSV-1, whereas addition of cGMP-enhancing compounds increased the yield of virus. Similar effects have been observed with other viral systems. Furthermore, viral transformation of cell cultures has been variously reported to be inhibited by cAMP-enhancing compounds (Ref. 12, and references cited).

Measles virus (MV) replication has been most extensively investigated. Robbins and Rapp [13] first reported that exogenous cAMP inhibits MV (but not VSV or HSV-1) replication in AV_3 (but not Vero) cells, accompanied by blockage of synthesis of M and P viral polypeptides. Subsequently Miller and Carrigan [14] found that acute MV infection of neural cells could be converted to an indolent state by treatment of infected cells with agents that affect cyclic nucleotide metabolism. Cyclic nucleotides such as cAMP, dibutyryl-cAMP and 8-bromo-cAMP, at concentrations of 1 mM, specifically reduced MV replication in cells of neural origin, but not in CV-1 or Vero cultures, by a factor of about 10^2. Viral replication was also reduced by cyclic nucleotide phosphodiesterase inhibitors, such as isobutylmethylxanthine and papaverine, the latter at a concentration as low as 7 μM. In all instances inhibition of replication was reversible by removal of the inhibitor or, somewhat surprisingly, by addition of cGMP. Paradoxically, when cell cultures were exposed to cGMP alone for 3 days, and then infected, yields of infectious virus decreased by a factor of 10^2, as with cAMP.

By contrast, Robbins [15] found that cGMP (≥ 0.5 mM), but not 5′-GMP, added immediately after infection of AV_3 cells, led to acceleration of virus-mediated cell fusion, enhanced synthesis of virus-specified proteins, and a 50-fold increase in production of infectious particles. There was evidence of altered viral protein phosphorylation.

It is obvious from the foregoing that, in cell lines of neural origin, MV replication is dependent on certain cellular functions, in line with other observations on host control of replication of various viruses. It should, on the other hand, be noted that the effects elicited by cyclic nucleotides required rather high concentrations, ≥ 0.1 mM, with no measurements of the resulting intracellular concentrations, particularly in infected cells.

It is clear that further, and more in-depth, studies in this field may be rewarding. Relevant to this subject is a recent review by Cho-Chung et al. [16] on site-selective cAMP analogues which exhibit a potent growth inhibitory effect on a variety of human tumours at micromolar concentrations; one of these, 8-Cl-cAMP, has been selected by NCl for phase I trials.

3. Chemical Phosphorylation of Acyclonucleosides

Standard procedures for chemical phosphorylation of acyclonucleosides with a single primary hydroxyl on the acyclic chain, e.g. Acyclovir, lead, as expected, to a single monophosphate product. More interesting are the products when the acyclic chain contains more than one hydroxyl.

Tolman et al. [8] treated DHPG with $POCL_3$ in triethylphosphate at 0°C (Yoshikawa procedure) to obtain the 3',5'-*bis*monophosphate (73%) and the 3':5'-cyclic phosphate (DHPG-cP, see Scheme 4) as a minor product (7%). Application of the same procedure to DHP-Ade gave two major products, the (R,S)-monophosphate (30%) and the 3',5'-*bis*monophosphate (34%). Prisbe et al. [17] prepared the (R,S)-monophosphate of DHPG by tritylation, followed by phosphorylation with cyanoethylphosphate, in 30% yield. The racemate may also be obtained by alkaline hydrolysis of DHPG-cP [18]. Both these procedures are more arduous and less efficient than enzymatic phosphorylation (see below).

We have applied $POCl_3$ in trimethylphosphate for phosphorylation of 2',3'-*seco*nucleosides and acyclonucleosides, followed by isolation of products by column chromatography, as described in detail elsewhere [19–22].

Application of the foregoing to *seco*Ado led to isolation of products shown in Scheme 5. Note the high yield of the cyclic phosphate, which may also be obtained by cyclization with dicyclohexylcarbodiimide of the 5'-phosphate of the *seco*nucleoside (Scheme 6). Identification of products was based on the use of specific enzymes, chromatography, and ^1H and ^{31}P NMR spectroscopy, also employed to determine solution conformations of the various products.

The same procedure applied to the adenine analogue of DHPG, i.e. DHP-Ade, and to DHPG, led to the mixture of expected products, as shown in Scheme 7. Note the high yield of the cyclic phosphate, as well as the overall yield of products of phosphorylation. Quite unexpectedly, C-DHPG, the carbocyclic analogue of DHPG, yielded only the cyclic phosphate as the major product under these conditions.

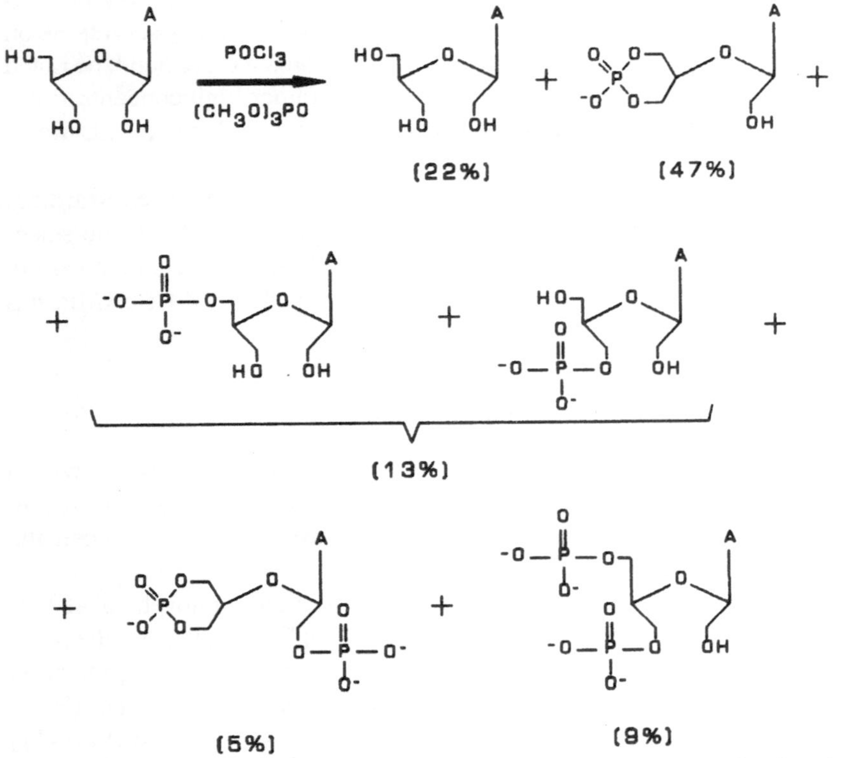

Scheme 5. Products of phosphorylation of 2′,3′-*seco*adenosine following fractionation by column chromatography

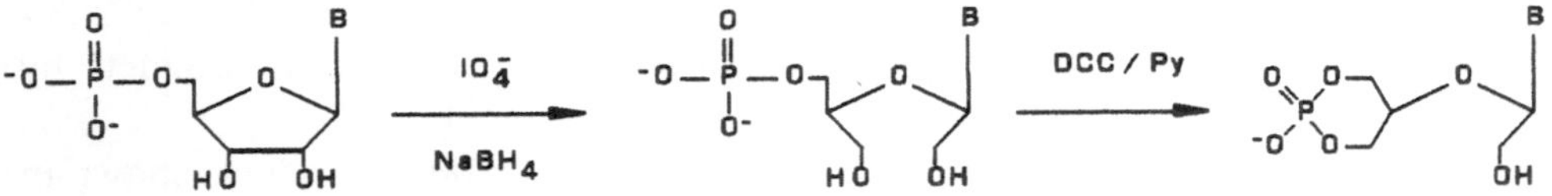

Scheme 6. Synthetic procedure for the cyclic phosphate of a 2′,3′-*seco*nucleoside (B = base). The nucleoside 5′-phosphate is converted to the *seco*nucleotide by periodate oxidation and borohydride reduction, and then to the cyclic phosphate with dicyclohexylcarbodiimide (DCC) in pyridine (Py)

DHBG, which contains one primary, and one secondary, hydroxyl, also gave a single major product, the cyclic phosphate. Applied separately to each of its enantiomers, the cyclic phosphate of (S)-DHPG was isolated in 75% yield, and that of the (R)-enantiomer in 45% yield.

Reference was made in the previous section to the differences in phosphoryla-

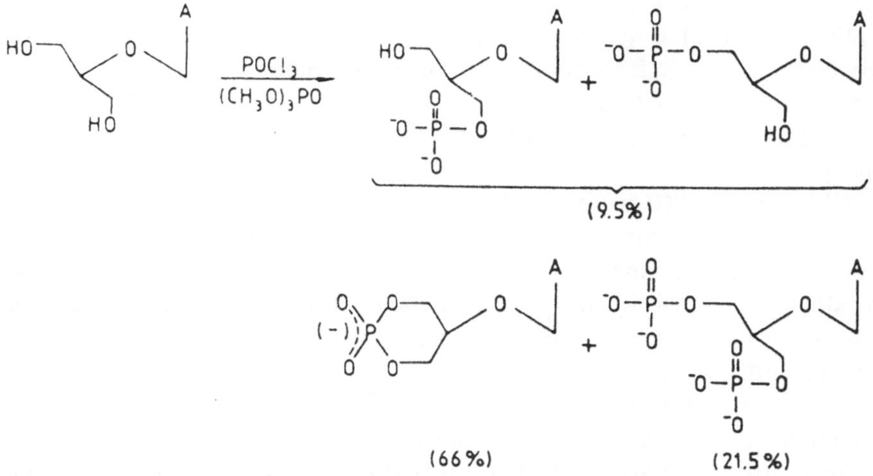

Scheme 7. Products of phosphorylation of DHP-Ade, the adenine analogue of DHPG, with POCl$_3$ in trimethylphosphate. The monophosphates are a racemate of the two enantiomers

tion products reported by Tolman et al. [8] and Stolarski et al. [22] following POCl$_3$ phosphorylation of acyclonucleosides. It now appears that this is due to differences in phosphorylating conditions (M. MacCoss, personal communication; see Ref. 22), and that these may be appropriately varied to amplify yields of desired products. Furthermore, other phosphorylating agents such as cyanoethylphosphate may more effectively yield the mono- and *bis*monophosphates.

3.1. Enzymatic monophosphorylation

We have, however, repeatedly found that, in most instances, the monophosphates of acyclonucleosides may be easily prepared in good yields with the use of the wheat shoot nucleoside phosphotransferase system, using *p*-nitrophenylphosphate as the phosphate donor, as described below.

3.2. Substrate/inhibitor properties

Our studies on the substrate/inhibitor properties of the cyclic phosphates of acyclonucleosides has profited from the use of a highly purified cyclic nucleotide phosphodiesterase (cPDase) from higher plants (see Ref. 23, and references cited). This is a multifunctional enzyme which hydrolyzes both 2':3'- and 3':5'-cyclic phosphates of nucleosides. We have also found that the activity vs 3':5'-cyclic phosphates can be selectively inactivated, without affecting activity vs 2':3'-cyclic phosphates.

Neither *seco*-5'-, nor 2'-, CMP was a substrate or inhibitor of snake venom 5'-nucleotidase (which has no phosphotransferase activity) or rye grass 3'-nucleotidase. By contrast *seco*-3',5'-*bis*monophosphate was quantitatively converted, by 3'-nucleotidase, at about half the rate for 3'-CMP, to *seco*-5'-CMP, consistent with the fact that this enzyme hydrolyzes ribonucleoside-3',5'-*bis*monophosphates to the 5'-phosphates.

Seco-cCMP and cAMP were resistant to the conventional mammalian beef heart cPDase, but were hydrolyzed to the monophosphates by purified plant cPDase at about 10% the rate for cAMP. Extracts of mammalian cells were also found to slowly open the cyclic phosphate rings. With a sonicated extract of Ehrlich ascites cells, *seco*-cCMP and *seco*-cAMP were hydrolyzed at about 15% the rate for cAMP.

Since the *seco*-cNMPs are resistant to the conventional ubiquitous beef heart cPdase, their hydrolysis by cell extracts must be due to some non-conventional cPDase. Such enzymes have, in fact, been partially purified and characterized from a mammalian source (pig liver) and from *Lactuca* cotyledons [24, 24a]. Both exhibit broad specificities strikingly similar to that of our plant (potato tuber) enzyme [23].

Selective inactivation of the activity of the plant enzyme vs cAMP was accompanied, as anticipated, by loss of activity vs the *seco* nucleoside cyclic phosphates.

DHP-Ade-cMP was hydrolyzed by the plant enzyme at a rate 5-fold higher than the *seco* analogue, with a K_m of 10 mM; but the rate for DHPG-cMP was only one-half that for the adenine analogue. Neither was a substrate for the selectively inactivated enzyme, consistent with the fact that, like the *seco* nucleoside cyclic phosphates, they contain 6-membered cyclic phosphate rings. Both also exhibited opening of the cyclic phosphate rings on exposure to extracts of Ehrlich ascites and L1210 cells. Somewhat unexpected was the total resistance of the carbocyclic analogue C-DHPG-cMP to the plant enzyme, as well as to cell extracts, all the more puzzling in that the conformation of its cyclic phosphate ring, determined by NMR spectroscopy, is similar to that for DHPG-cMP. It also did not inhibit hydrolysis of other cyclic phosphates, including cAMP.

Both (R)- and (S)-DHBG-cMP, with 5-membered cyclic phosphate rings, were hydrolyzed by the enzyme at equal rates somewhat lower than that for 2':3'-cAMP, with a K_m of 1 mM. Following selective inactivation of the enzyme, they were still good substrates, albeit with a slightly reduced rate of hydrolysis. Surprisingly, both of these were readily hydrolyzed by beef heart cPDase, with K_m of 1 mM and a V_{max} somewhat lower than that for cAMP, notwithstanding the presumed specificity of the mammalian enzyme for 3':5' 6-membered cyclic phosphate rings. This is being further investigated.

Preliminary trials indicated that the *seco* nucleoside cyclic phosphates were inactive against HSV-1 in cell culture (Y.-C. Cheng, personal communication). Trials with the other cyclic phosphates of acyclonucleosides are under way.

Furthermore, the *seco* nucleoside cyclic phosphates exhibited no detectable binding to cAMP-dependent protein kinases (B. Jastorff, personal communication).

As noted above, Germershausen et al. [10] also observed opening of the cyclic phosphate ring of DHPG-cMP by cellular extracts. Since the resulting product, DHPG-MP, was a good substrate of cellular GMP kinase, they concluded that the product was (S)-DHPG-MP. It should therefore be noted that non-conventional cPDase isolated from mammalian cells [24], with properties similar to our plant cPDase [23] hydrolyzes 3':5'-cyclic phosphates to give uniquely the 5'-phosphate. This could, therefore, be the enzyme responsible for hydrolysis of DHPG-cMP.

4. Intracellular Phosphorylation of Nucleoside Analogues

Nucleosides and many nucleoside analogues, following uptake by cells, are converted to the monophosphates (usually the 5'-phosphates, but see below) by ATP-dependent nucleoside kinases, or/and NMP-dependent nucleoside phosphotransferases (where N is a nucleoside), recently referred to as "cytosolic 5'-nucleotidases". We defer discussion of the latter to a separate section, below.

The principal known nucleoside kinases are as follows:

(a) *TK*: Mammalian cells contain cytosolic TK1 and mitochondrial TK2. The former phosphorylates thymidine, 2'-deoxyuridine and some of their analogues; TK2 partially overlaps the specificity of TK1, but also phosphorylates dCyd and some of its derivatives, hence is really a pyrimidine deoxyribonucleoside kinase. The substrate specificities of the human enzymes have recently been extensively documented [25], particularly with reference to nucleosides with antiviral activity. Relevant to this was the observation that, while TK2 exhibited a broad specifity with regard to the pyrimidine base, it was much more restrictive as regards modification of the deoxyribose moiety.

(b) *dCK*: This enzyme phosphorylates not only dCyd and its analogues, but also dAdo, dGuo, araA. The specifity of the human enzyme was recently described [25]. Noteworthy is its inability to accept as substrates acyclonucleosides such as Acyclovir, DHPG, DHBG.

(c) *AK*: Adenosine kinase will also phosphorylate dAdo, albeit at a much lower rate; as well as such analogues as the antiviral araA and ribavirin.

(d) *UK*: Uridine kinase is really a uridine/cytidine kinase, since it effectively phosphorylates both uridine and cytidine, as well as a number of analogues such as 3-aza and 5-azauridine, and 5-azacytidine. Phosphorylation of the latter accounts for its antitumour activity [27]. The properties of this kinase have been recently described [28].

(e) *dGK*: This enzyme, of mitochondrial origin, apparently has a rather narrow substrate specificity [29]. It will readily phosphorylate araG, a drug currently under investigation as a therapeutic agent for lymphoid cell disorders [30].

(f) *GK*: No guanosine kinase has as yet been identified in mammalian cells. The first such enzyme has just been isolated and purified from the protozooan parasite *Trichomonas vaginalis* [31]. Its preferred substrate is Guo (V_{max}/K_m 120). The corresponding rate constants for Ino (≈ 3), Urd (≈ 2), Ado (≈ 0.5), Cyd (≈ 0.2) and dGuo (≈ 0.1) are quite low, and virtually nil for other deoxynucleosides.

The foregoing is only a general classification. The properties and substrate specificity of a given kinase may vary from different sources. And the ATP phosphate donor may frequently be substituted for by other nucleoside 5'-triphosphates (see, e.g. Ref. 29).

A recent interesting finding is the report by Kong et al. [32] that 5-azacytidine induces a high level of dCK activity in a dCK-deficient HL-60 cell line resistant to araC, and that the induced enzyme exhibited kinetic properties similar to those of the wild-type enzyme.

4.1. *TK of herpesviruses*

Some herpesviruses code for TK activities which differ significantly from the TK of their host cells. The TKs of pseudorabies virus and herpes virus *saimiri* possess only dT kinase activity, whereas those of HSV-1, HSV-2 and VZV, like mitochondrial TK2, additionally phosphorylate dCyd. They also display a much broader tolerance for substrate analogues, extending to acyclonucleosides with a purine ring, such as ACV, DHPG, DHBG. The TKs of HSV-1 and VZV, and to a lesser extent HSV-2, possess associated dTMP kinase activity [33]. Since they phosphorylate dCyd, it would be of interest to determine whether they also exibit dCMP kinase activity.

These viral encoded enzymes also exhibit stereoselectivity or stereospecificity towards chiral antiviral substrates, most extensively investigated by MacCoss et al. [34] and Karkas et al. [18] for the HSV-1 enzyme. For example, DHPG, an effective agent vs HSV-1, is dependent for its activity on prior phosphorylation by the viral, but not host, kinase. DHPG is a pro-chiral molecule, but the monophosphate exists as two enantiomers, R and S (Scheme 8). In fact the HSV-1 TK stereo*specifically* leads to (S)-DHPG-MP, which is then stereo*selectively* phosphorylated by cellular enzymes to (S)-DHPG-TP, which effectively inhibits the viral DNA polymerase. Synthetic (R)-DHPG-MP is poorly phosphorylated by cellular enzymes to its triphosphate which, in turn, is a poor inhibitor of the viral polymerase. Similar stereochemical properties of HSV TK have been documented for other chiral antiviral acyclonucleoside [34] as well as for chiral nucleoside analogues.

5. Inhibitors of Nucleoside Kinases

Bearing in mind the existence of a variety of nucleoside kinases with multiple and/or overlapping specificities, and the utility of establishing which of these is

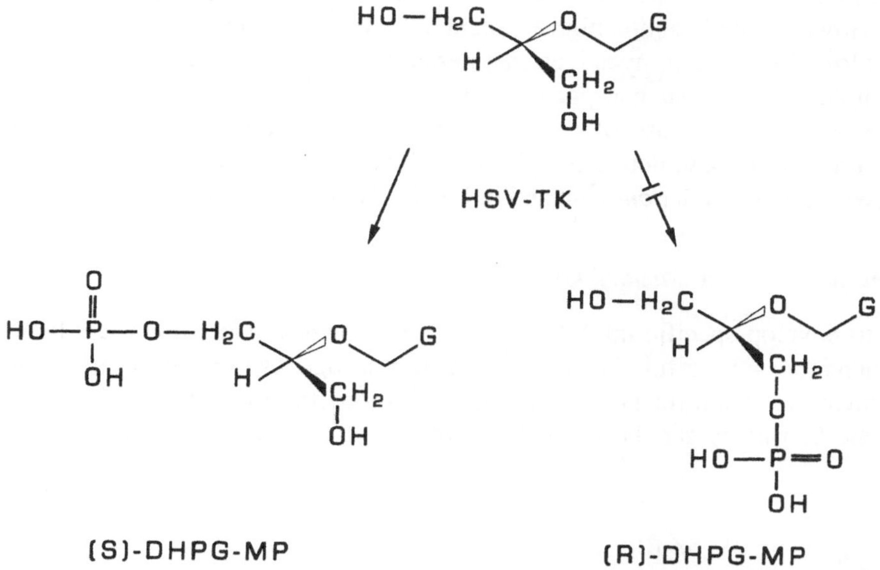

Scheme 8. Showing selective phosphorylation of the prochiral acyclonucleoside DHPG by the TK of HSV. The only product of phosphorylation is (S)-DHPG-MP

involved in the intracellular phosphorylation of a therapeutically active nucleoside analogue, it is surprising that more effort has not been devoted to development of specific non-substrate inhibitors of these enzymes.

Bisubstrate analogue inhibitors, such as Np_nN' (where N and N' are nucleosides linked *via* their 5' positions by n phosphoryl residues) are remarkably effective *in vitro* inhibitors of nucleoside kinases (e.g.[35–37]), but these have been designed primarily as probes for the active site(s) of the enzymes, and for investigating their mechanism of action. Their high negative charge, which restricts their transport across cell membranes, and their susceptibility to PDases, for the moment precludes their use as intracellular inhibitors, although efforts to circumvent these obstacles are currently under way. The rational design of multisubstrate analogue inhibitors has recently been reviewed [38].

Consequently identification of the kinase(s) responsible for phosphorylation of a given nucleoside analogue have been based largely on studies with purified enzymes, supplemented with the use of mutant cell lines deficient in one or more kinases.

The most comprehensive search for a kinase inhibitor is that of Miller et al. [39], who screened 119 nucleoside analogues for substrate/inhibitor properties vs rabbit liver AK. Amongst those nucleosides which were not substrates, but exhibited appreciable binding, the most effective as inhibitor was 5'-amino-5'-

deoxyadenosine, competitive with respect to Ado, and with a K_i at neutral pH of 20 nM. However, despite this highly potent activity, it has apparently not as yet been exploited in *in vivo* investigations; perhaps because of its susceptibility to other cellular enzymes, such as phosphorylases or deaminases.

One example of the use of an AK inhibitor (5-iodotubercidin, with a K_i of 10 μM at neutral pH), to determine phosphorylation routes of several nucleoside analogues in *Leishmania donovani* and in L cells, is that of LaFon et al. [40].

5.1. Inhibitors of viral-encoded kinases

Efforts to develop specific inhibitors of the TK enzymes of herpesviruses have been much more successful, due in part to the less stringent substrate specifities of these enzymes. Several such inhibitors, and their properties [41–43], are displayed in Scheme 9. But by far the most effective embrace a series of 5'-substituted

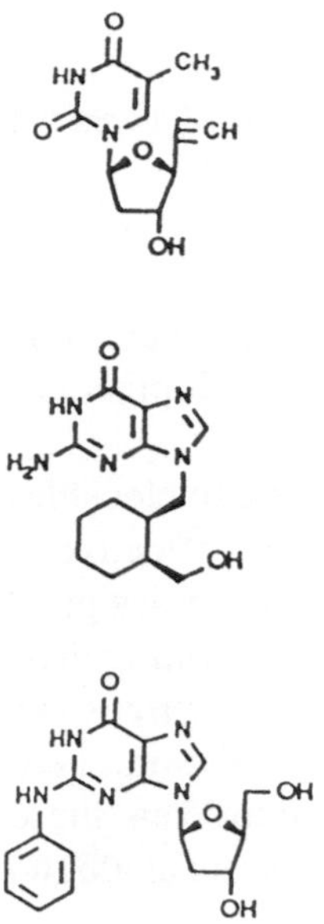

K_i vs HSV-1 TK ~ 0.09 μM
vs HSV-2 TK ~ 0.4 μM.
No inhibition of cytosolic
TK1 at 150 μM [41].

IC_{50} vs HSV-1 TK ~ 0.17 μM.
Not a substrate, but cyclopropyl
and cyclobutyl derivatives are.
No inhibition of cytosolic TK1
at 70 μM [42].

K_i vs HSV-1 TK ~ 0.3 μM.
Not a substrate.
No inhibition of human
cell TK at 1 mM [43].

Scheme 9. Properties of three specific inhibitors of HSV TK

5-ethyl-2'-deoxyuridines, the parent nucleoside of which was long ago synthesized and shown to inhibit HSV replication [44]. Several of the more potent compounds in this series, reported by Martin et al. [45], are shown in Scheme 10.

The very low cytotoxicity of the foregoing testifies to their high selectivity for HSV, relative to host cell, kinases. They have been shown to possess the ability to

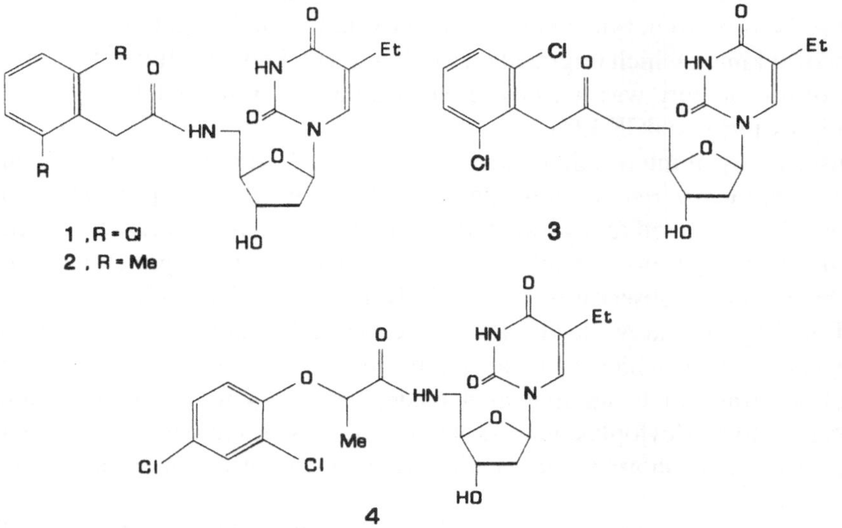

Scheme 10. 5′-substituted 5-ethyl-2′-deoxyuridines, specific inhibitors of HSV TK: IC_{50} of **1** and **4** vs HSV-2 TK ≈ 3nM; IC_{50} of **3** vs HSV-1 TK ≈ 30μM; IC_{50} of all compounds vs HeLa and Vero host cells > 200 μM

reverse the anti-HSV activity of nucleoside analogues which are dependent on viral TK for their activation. It is to be anticipated that such inhibitors will prove useful in establishing whether antiviral activity of new nucleoside analogues is dependent on activation by viral TK, and in further studies on the properties and mechanism of action of the viral kinases. From this standpoint, it would clearly be desirable to examine whether the above compounds also inhibit the dCK activity of the HSV kinases, and the thymidylate kinase activity associated with the HSV-1 TK.

Some clinical utility for these inhibitors may also be envisaged, e. g. it has been inferred that viral TK is not required for the establishment of latency by herpesviruses, but is essential for "reactivation" of latent viruses (responsible for recurrences of infection following therapy). A specific HSV TK inhibitor may therefore be clinically advantageous in special circumstances, e.g. in immumosuppression treatments, during chemotherapy of AIDS, etc., where HSV reactivation may lead to morbidity or mortality [46].

6. Cytosolic 5′-Nucleotidases, Nucleoside Phosphotransferases

The antiviral activities of acyclonucleosides, such as Acyclovir (ACV) and Gancyclovir (DHPG), are dependent on their initial intracellular "activation" (phosphorylation by viral-encoded nucleoside kinases). No acyclonucleoside has been found to be a substrate for a cellular ATP-dependent kinase. Datta & Pagano [47] found that ACV is phosphorylated in activated Burkitt somatic cell hybrids by an ATP-dependent kinase which migrated on a column with known cellular dCK, but the nature of this activity was not identified. Neither ACV nor DHPG are substrates of highly purified dCK [25].

It is consequently of interest that both of these are phosphorylated to a low, but significant, extent in a variety of non-infected cells. Keller et al. [48] first showed that this could be accounted for by a so-called cytoplasmic 5'-nucleotidase purified from rat liver. It hydrolyzes nucleoside-5'-phosphates with a preference for purines. The additional observation that hydrolysis of 5'-IMP is inhibited by Ino [49] was found by Worku & Newby [50] to be due to formation of an enzyme-phosphate intermediate which can transfer the phosphoryl moiety back to the hydrolyzed substrate, or to another nucleoside, instead of to water. Although widely referred to as "cytoplasmic 5'-nucleotidase", such enzymes are similar to the nucleoside phosphotransferases identified much earlier in bacteria and plants.

The purified enzyme phosphorylates ACV, at high concentration, at a rate of 2–5% that for hydrolysis of 5'-IMP used as the phosphate donor; and the level of its activity in Vero cells, widely used as a host cell for viral infection, accounted for the extent of phosphorylation of ACV in such non-infected cells [48]. DHPG is also substrate, and is phosphorylated in non-infected Vero and HeLa cells to the mono-, di- and triphosphates twice as rapidly as ACV [51].

The foregoing recalls an earlier observation that ACV is an excellent acceptor substrate for wheat shoot nucleoside phosphotransferase, and, with p-nitrophenyl-phosphate as phosphate donor, is converted to the nucleotide in 50–60% yield [52].

Nucleoside phosphotransferases, ubiquitous in higher plants, catalyze the following reaction sequence:

$$NMP + E \rightleftharpoons E \cdot NMP \rightleftharpoons E \cdot P_i + N$$
$$E \cdot P_i + N' \rightleftharpoons E \cdot P_i \cdot N' \rightleftharpoons E + N'MP$$
$$E \cdot P_i + H_2O \rightarrow E + P_i$$

where N and N' are nucleosides, E is the enzyme, and $E \cdot P_i$ a covalently-bound enzyme-phosphate intermediate [53].

Direct evidence for obligatory formation of the covalently bonded phosphoryl-enzyme intermediate has been adduced for the enzyme from several plant sources [53, 54]. Note that, following hydrolysis of the donor NMP, the resultant $E \cdot P_i$ intermediate may transfer its phosphoryl moiety to another nucleoside N', or back to the N liberated from the donor NMP (phosphotransferase activity), or to water (nucleotidase activity). Activated phosphomonoesters, such as phenylphosphate or p-nitrophenylphosphate, may frequently replace NMP as donors.

The donor and acceptor specificities of the enzyme from plant sources are not absolute. The highly purified enzyme from carrots partially phosphorylates the 3′-OH of acceptor nucleosides. With the barley enzyme, and adenosine as acceptor, the products include 85% 5′-AMP and 15% 3′-AMP [53]. With the malt sprout enzyme, it has been reported that either a 5′-, or to a lesser extent a 3′-, but not a 2′-, NMP may serve as donor [54].

By contrast, the crude wheat shoot enzyme is strictly specific for the primary 5′-CH$_2$OH of an acceptor nucleoside. An essential requirement is the presence of a primary hydroxyl; e.g. with p-nitrophenylphosphate as the phosphate donor, the enzyme will efficiently phosphorylate 5′-homonucleosides [55] and, albeit to a lower extent, the primary hydroxyls of 1-(β-hydroxyethyl)cytosine and 5-(β-hydroxyethyl)uracil [56, 57], both of which may be considered as examples of acyclonucleosides. It is of interest, in this context, that $E. coli$ K-12 contains an ATP-dependent hydroxymethylpyrimidine kinase which efficiently phosphorylates the primary hydroxyl of 2-methyl-4-amino-5-hydroxymethylpyrimidine, and which exhibits properties similar to the enzyme previously characterized as pyridoxal kinase [58].

The wheat shoot enzyme will quite efficiently phosphorylate primary hydroxyls ef antiviral and cytotoxic nucleosides [56, 57] and acyclonucleosides [21, 22]; and has been widely employed, both in our and other laboratories, for small-scale (≈1 mg) and large-scale (≈1 gm) phosphorylation of such nucleosides [57].

6.1. Nucleoside phosphotransferases in mammalian systems

The existence in chick embryo extracts of activity which catalyzes the transfer of phosphate from a 5′-NMP to the 5′-hydroxyl of a nucleoside was reported almost 30 years ago. Subsequently Kit et al. [59] partially purified the enzyme from cytosolic extracts of chick and duck embryo cells, and reported activities exceeding up to 30-fold the activity of TK. The finding, reported in the previous section, that acyclonucleosides may be phosphorylated by such activities in non-infected cells to a low, but significant, extent, has further stimulated interest in these enzymes, now frequently referred to as cytosolic 5′-nucleotidases. Not all cytoplasmic 5′-nucleotidases exhibit phosphotransferase activity (see Refs. 60 and 61 for discussion). Particularly interesting is the finding of Fridland et al. [42] that a cytoplasmic 5′-nucleotidase exists in both wild-type and AK-deficient mutants of human lymphoblastoid cells, and that this activity is involved in the intracellular phosphorylation of the nucleoside analogue tiazofurin (see also below).

The pH optima of these enzymes are in the alkaline range as compared to the acid pH optima of the plant enzymes. Their ability to use phenylphosphate or p-nitrophenylphosphate as donors is also rather limited or non-existent. It is also not known whether they exhibit stereospecificity, e.g. whether, like some plant enzymes, they can transfer the phosphate to the 2′- or 3′- hydroxyl of a nucleoside.

The finding (see above) that DHPG is phosphorylated in non-infected cells raises the question as to whether it is phosphorylated to the active (S)-enantiomeric phosphate (as it is by HSV TK), or to a racemate of the S and R enantiomers (by the plant enzyme). At least in mouse liver extracts, it is the (S)-enantiomer which is formed [70].

7. Site(s) of Enzymatic Phosphorylation of Nucleosides

The known nucleoside kinases transfer the γ-phosphoryl moiety of a nucleoside triphosphate donor to the primary 5'-hydroxyl of a nucleoside. Since nucleoside phosphotransferases exist in mammalian cells, and some of them of plant origin may transfer the phosphoryl moiety of a monophosphate donor to the secondary 3'-OH of an acceptor nucleoside (see above), it becomes desirable to verify the site(s) of phosphorylation of a nucleoside incubated in intact cells or cell extracts, as illustrated by the following example.

The nucleoside analogue 5,6-dichloro-1-(β-D-ribofuranosyl)benzimidazole (DRB) is a known inhibitor of mRNA transcription in mammalian systems. In order to determine whether *in vivo* inhibition is due to DRB itself, or some phosphorylated derivative, the metabolism of DRB was examined in the salivary gland cells of *Chironomus tentans,* a model system for following *in vivo* transcription. It was found that the nucleoside is converted uniquely to the monophosphate, and it was tacitly assumed that this was the DRB-5'-monophosphate (which was found not to be an inhibitor of transcription) [63].

A subsequent more detailed analysis, with digestion of the labeled phosphorylated DRB with 3'- and 5'-nucleotidases, followed by thin-layer electrophoresis in borate buffer demonstrated the presence of 2'- and 5'- monophosphate isomers in proportions of 30–45% and 55–70%. No 3'-isomer was detected [64].

Since adenosine partially mimics the inhibitory effect of DRB in this system, its intracellular metabolism was then investigated in the same way. This led to the finding that the monophosphorylated adenosine consisted of a mixture of 2'-AMP (74%), 3'-AMP (23%) and 5'-AMP (3%) [65]. The low level of 5'-AMP may be due to the fact that it is further converted to ATP. The enzyme(s) involved in formation of these phosphorylated products remain to be identified. Clearly one potential candidate is a nucleoside phosphotransferase (cytosolic 5'-nucleotidase). The acceptor specificities of such enzymes of mammalian origin have not been extensively investigated as those from higher plants and microorganisms.

Can we exclude the existence of some novel non-conventional nucleoside kinase? An example of such an enzyme is one of the known sulfate-activating enzymes widely distributed in Nature, APS kinase. This enzyme catalyzes the transfer of phosphate from ATP to the 3'-hydroxyl of adenosine-5'-phosphosulfate (APS) to give APS-3'-phosphate (PAPS) and its properties have been well documented [66]. We are presently engaged in attempts to characterize the phosphorylating activities of extracts of *Chironomas* cells.

8. Phosphorylation Routes and Activities of Phosphorylated Species

Since the discovery that ACV is phosphorylated to a low, but significant, extent by cytosolic 5'-nucleotidase (nucleoside phosphotransferase), considerable progress has been made, largely by Fridland and coworkers, in pin-pointing the role of this class of enzymes in phosphorylation of nucleoside analogues. For example, the antitumour agent, tiazofurin, is phosphorylated in human lymphblastoid cells by either AK and/or cytosolic 5'-nucleotidase [62]. The anti-HIV agent ddIno, which is not phosphorylated by AK or dCK in human lymphoid cells, is phosphorylated to a low, but sufficient, extent to the active nucleotide, as is also ddGuo [67]. With the use of human lymphoid cells deficient in nucleoside kinases, it was also demonstrated that the anti-HIV agent carbovir (carbocyclic 2',3'-didehydro-2',3'-dideoxyguanosine) is not initially phosphorylated by AK, dCK or TK2, but is by cytosolic 5'-nucleotidase [68].

By contrast, the pathway for activation of ddAdo, also an antiretroviral agent, is somewhat more complex, and may occur *via* three pathways in human T-lymphocytes, viz. direct phosphorylation by AK or dCK to ddAMP, deamination of ddAdo to ddIno and its phosphorylation to ddIMP, and reamination of the latter to ddAMP by adenylosuccinate synthetase/lyase [69]. A detailed investigation of the kinetics of phosphorylation of antiviral purine and pyrimidine necleoside analogues by dCK indicated that the purine nucleosides are much less efficient as substrates, and that their phosphorylation by dCK is under regulation by nucleotide pools [70]. The activity against herpesviruses of the carbocyclic analogue of dGuo, C-dGuo. has been shown to be dependent on initial phosphorylation by the viral-coded TK. This compound may exist in two enantiomeric forms, and it is the D-enantiomer which resembles the natural configuration of nucleosides. But, while antiviral activity was closely associated with the D-enantiomer, the L-enatiomer also exhibited low, but significant antiviral activity, and low levels of the phosphorylated racemate were detected in non-infected cells [71], in line with earlier observations that L-nucleosides are phosphorylated in mice [72]. It is very likely that such phosphorylation is due to cytoplasmic 5'-nucleotidase. A rather curious, but unexplained, finding was the similar antiviral activity of the D-enantiomer and the racemate.

The purine nucleoside analogue araG, considered to function in a fashion analogous to the antileukemic agent araC, has been considered to undergo phosphorylation by dCK. The recent demonstration that it is an excellent substrate for the mitochondrial enzyme dGK [30] suggests that both the foregoing enzymes may be responsible for such phosphorylation.

Apart from its intrinsic fundamental interest in relation to cellular metabolism, it is of obvious importance to establish the pathways of phosphorylation of nucleoside analogues, and the mode of action of the various phosphorylated species, for the rational development of chemotherapeutic methods. Efforts in this direction have been gaining momentum, and we limit ourselves here to a few illustrative examples.

Particularly striking in this regard is the human protozoan parasite *Trichomonas vaginalis,* which is incapable of *de novo* synthesis of purine or pyrimidine nucleotides. It is apparently unique in that it possesses no ribonucleotide reductase activity, hence is incapable of converting ribonucleotides to deoxyribonucleotides for DNA synthesis. It turns out that this organism possesses, in addition to a nucleoside phosphotransferase, a second such enzyme quite specific for deoxyribonucleosides as acceptors [73]. It is clear that this enzyme is a potentially excellent target for anti-trichomonial chemotherapy.

The mode of action of the anti-AIDS agent AZT (3′-azido-3′-deoxythymidine) is considered to involve its intracellular phosphorylation to AZT-MP by cellular TK, to AZT-DP by dTP kinase, and nucleoside diphosphate kinase to AZT-TP. The latter is a potent inhibitor of the HIV reverse transcriptase, but not of the host cell DNA polymerase; in addition AZT-MP may undergo incorporation, leading to chain termination and inactivation of the template/primer. Another potential target is the RNase H activity of the HIV reverse transcriptase; and it has been shown that AZT-MP is such an inhibitor, although its IC_{50} is in the millimolar range in the presence of Mg^{2+}, it is reduced to 50 µM in the presence of Mn^{2+}. Bearing in mind that the level of AZT-MP in AZT-treated cells may attain millimolar concentrations, it is possible that such inhibition may contribute to sensitvity of the virus to AZT [74]. It is conceivable that the monophosphates of other dideoxynucleoside analogues active against HIV may owe part of their activity to more effective inhibition of the HIV RNase H activity.

The clinical utility of AZT is frequently limited by cytotoxic side effects, and numerous efforts have been devoted to determine the sources of these effects. In one such study [75], the ability of phosphorylated derivatives of AZT to inhibit cellular ribonucleotide reductase was eliminated as a possible source of toxicity. In another study [76], AZT-MP was examined as a potential inhibitor of host thymidylate synthase, and was discounted as such in cells in which accumulation of AZT-MP is not high. However, in cells where the level of AZT-MP may attain a millimolar range, by no means rare, inhibition of thymidylate synthase may be of significance.

The antiviral activities of acyclonucleosides, such as ACV and DHPG are dependent on their monophosphorylation by viral-encoded TK, followed by conversion by cellular kinases to the triphosphates, which are selective inhibitors of the viral DNA polymerases, as well as chain terminators. However, many acyclonucleosides are good inhibitors of purine nucleoside phosphorylase (PNP, see Refs. 77 and 77a, and references cited). The most potent such *in vitro* inhibitors of the human enzyme include the pyrophosphates of ACV and DHPG, with K_i values in the nanomolar range [78]. Their high negative charge precludes their use as exogenous inhibitors in cellular systems; but attention has been drawn to the fact that their formation intracellularly may readily be a significant source of side effects of both these druge [78], as well as other acyclonucleosides, including those which undergo low, but significant, phosphorylation in non-infected cells (see above).

We conclude this section with brief reference to two very recent perspicuous accounts of the mode of action of two nucleoside analogues with potent cytotoxic, and antitumour, activities.

The deoxycytidine analogue 2',2'-difluorodeoxycytidine is a specific inhibitor of DNA synthesis with promising therapeutic activity in several murine tumour models. It is initially phosphorylated by dCK, and then converted to the di- and tri-phosphates, the latter of which competes with dCTP for incorporation into DNA by DNA polymerases α and δ. However, a meticulous analysis of the metabolism of the nucleoside in CCRF-CEM cells, coupled with its effects on intracellular dNTP pools, strikingly similar to the effects produced by the known ribonucleotide reductase inhibitor hydroxyurea, showed that it is the 5'-diphosphate (and to a much lesser extent the 5'-triphosphate) which is primarily responsible for inhibition of DNA synthesis *via* potent inhibition of ribonucleoside reductase [79].

Halogenated dAdo analogues have proven useful in tumour chemotherapy, e.g. 2-fluoro-araA (2-F-araA) and 2-chloro-2'-deoxyadenosine (2-Cl-dAdo) (Scheme 11), with the latter under phase I trials vs T-cell neoplasms and phase II trials vs chronic lymphocytic leukemia. In K562 cells it is phosphorylated initially, but perhaps not exclusively, by dCK. The mechanism of growth inhibition by 2-F-araA is inhibition of DNA polymerase α, ribonucleoside reductase and DNA primase; that of 2-Cl-dAdo is somewhat similar, but is a more potent inhibitor of ribonucleotide reductase than of DNA polymerase α (Ref. 80, and references cited). The properties of these two analogues prompted Parker et al. [80] to

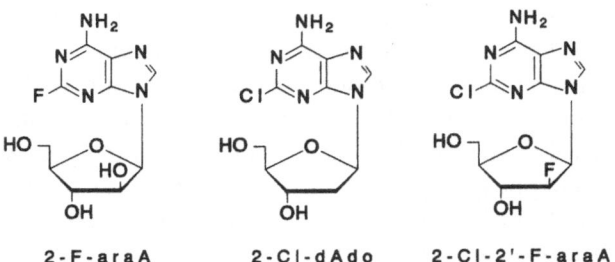

Scheme 11. Three structurally related cytotoxic halogenated deoxyadenosine analogues with promising therapeutic activity against a variety of tumours

synthesize 2-Cl-2'-F-araA (Scheme 11), bearing in mind that substitution of a fluorine at C(2') in the arabino configuration enhances resistance to cleavage of the glycosidic bond by purine nucleoside phosphorylase. In K562 cells, competition studies with natural nucleosides indicated initial phosphorylation by dCK (but this does not necessarily exclude involvement of other kinases). The 5'-triphosphate was a potent inhibitor of DNA polymerase α, but much less so of DNA polymerases β and γ, and of DNA primase. Incorporation into DNA of 2-Cl-2'-F-araAMP inhibited chain elongation, like 2-F-araAMP, more effectively than incorporation of 2-Cl-dAMP. Studies of dNTP pools pin-pointed ribonucleotide reductase as a major target; and this, together with inhibition of DNA chain elongation, indicated that this analogue combines the optimal inhibitory properties of the two parent analogues 2-F-araA and 2-Cl-dAdo.

9. Oxetanocin Nucleosides and Analogues

A relatively recent, and novel, development was the isolation, from a *Bacillus megaterium* strain, of an unusual nucleoside antibiotic wih potent activity against Gram-positive bacteria, and identified as oxetanocin-A (see Scheme 12). This is a most unusual analogue of adenosine, with the pentose ring replaced by a four-membered oxetane ring [81]. The corresponding oxenatocins of Guo, Hx, Xao and 2,6-diaminopurine were then prepared by chemical and/or enzymatic transformations of oxetanocin-A [82].

Notwithstanding its structural dissimilarity from the dideoxynucleosides active vs HIV, oxtenocin-A proved to be an active anti-HIV agent, and its triphosphate a selective inhibitor of HIV reverse transcriptase. Oxetanocin-A, and analogues with other base moieties, also display activities vs other viruses, e.g. oxetanocin-G is as active as DHPG (Scheme 12) vs human cytomegalovirus (HCMV) [85], and several oxetanocin-A analogues, in which the 2'-substituent on the oxetane ring has been modified or deleted, are more active vs HIV than the parent compound (see Ref. 84, and references cited).

Daikoku et al. [85] have chemically synthesized [1]H-labeled oxetanocin-GTP, with the triphosphate moiety at the 4'-position of the oxetane ring, which corresponds to the 5-'triphosphate of a natural nucleoside. This proved to be a preferential inhibitor of HCMV DNA polymerase. It was further shown that oxetanocin-G is converted to its mono- and diphosphates 2-fold, and to its triphosphate 8-fold, more effectively in HCMV-infected cells than in mock-infected cells. In the absence of a known HCMV-coded nucleoside kinase, it is possible that initial enhanced monophosphorylation may be due to the increased induced levels of cellular dT and dG kinases in infected cells. Since purified cellular dT and dG kinases are now available, this may be readily checked. Another potential candidate is a cellular nucleoside phosphotransferase (cytosolic 5'-nucleotidase). Furthermore, Daikoku et al. [85] did not establish unequivocally

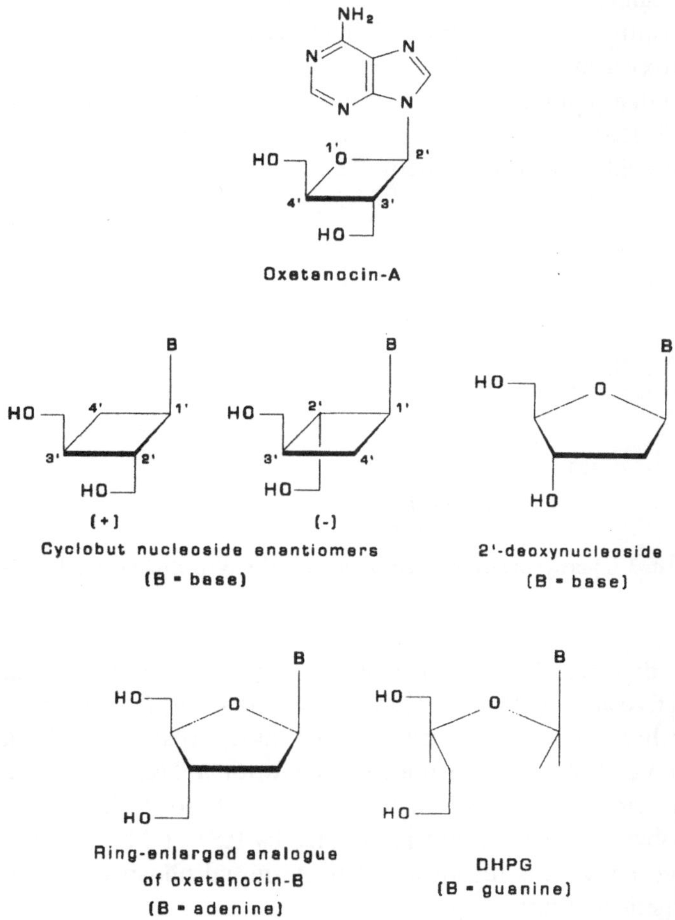

Oxetanocin-A

(+)

Cyclobut nucleoside enantiomers

(B = base)

(-)

2'-deoxynucleoside

(B = base)

Ring-enlarged analogue

of oxetanocin-B

(B = adenine)

DHPG

(B = guanine)

Scheme 12. Oxetanosin and enantiomeric cyclobut nucleosides. Note that the oxetanocin nucleoside more closely mimics the 3'-hydroxymethyl-2'-deoxynucleoside (bottom left, see also Scheme 13) and the acyclonucleoside DHPG (bottom right) than the 2'-deoxynucleoside (middle right)

that intracellular phosphorylation of oxetanocin-G occurs uniquely at the 4'-position. It would therefore have been useful if they had also synthesized oxetanocin-GMP and oxetanocin-GTP with the phosphates on the positional isomer at the 3'-position, since their synthetic procedure yielded a mixture of nucleoside isomers blocked at the 4'- or 3'-positions. The potential significance of this will become clearer when considered relative to the behaviour of 3'-branched analogues of 6-thio-dGuo (see below).

The unabating interest in the oxetanocins has stimulated the synthesis of the corresponding carbocyclic analogues, in which the oxetane ring is replaced by a cycylobutane ring. These analogues, referred to as cyclobut-B (where B is the base residue), exist as two enantiomeric forms. From Scheme 12 it will be seen that the relationship of the hydroxy groups and the base moiety in the (+)-enantiomer *appears* to correspond to that in a natural 2'-deoxynucleoside. More accurately it corresponds to that in a 3'-hydroxymethyl-2', 3'-dideoxynucleoside (Scheme 13) or in an acyclonucleoside such as DHPG (Scheme 12).

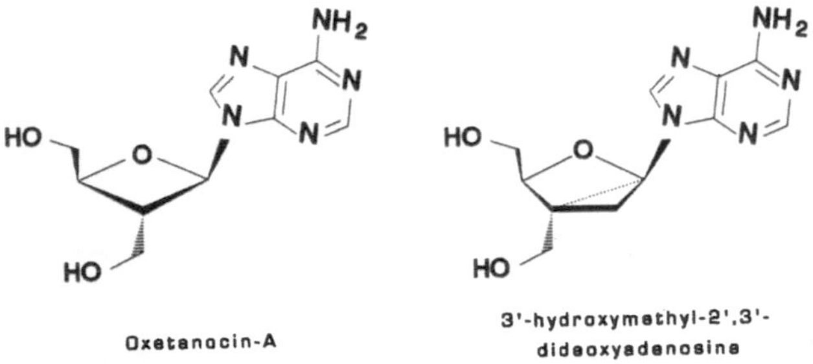

Oxetanocin-A

3'-hydroxymethyl-2',3'-
dideoxyadenosine

Scheme 13. Oxetanocin-A and its 3'-hydroxymethyl ring-enlarged analogue, 3'-hydroxymethyl-2',3'-dideoxyadenosine

It was initially shown that cyclobut-G, consisting of a racemic mixture of the two enantiomers, is as active as Acyclovir vs HSV-1 and HSV-2 *in vitro*, and was more effective than Acyclovir against TK⁻ strains of the same viruses. It is also a useful therapeutic agent vs HIV-1 (which does not code for a TK), and some herpesviruses, including cytomegalovirus, which does not code for a TK. It was then found that (±)-cyclobut-G is readily phosphorylated by HSV-1 TK, suggesting that the viral-encoded enzyme is responsible for the initial phosphorylation step essential for antiherpetic activity.

However, extension of the foregoing, based on synthesis of the individual enantiomers, led to the surprising finding that, whereas only the (+)-enantiomer exhibits anti-HSV-1 activity, it is the (−)-enantiomer which is preferentially phosphorylated by the HSV-1 TK, the difference in rates being 20-fold. It was, furthermore, shown with the aid of ^{1}H NMR that only the 3'-hydroxyl in both enantiomers (corresponding to the 5'-hydroxyl of a nucleoside) is phosphorylated, thus ruling out that differences in sites of phosphorylation may account for the differences in rates of phosphorylation and differences in antiviral activity [86]. This also underlines the distinct stereoselectivity of HSV-1 TK, as in the case of the prochiral DHPG by phosphorylation of a prochiral center to yield a single enantiomer of the chiral monophosphate product (see above).

In separate studies (±)-cyclobut-G was reported to be 5- to 25-fold less potent vs TK⁻ mutants of HSV (see Ref. 6, and references cited). This implies that sufficient levels of (+)-cyclobut can be generated by HSV TK to have an observable effect on antiviral activity, suggesting that (+)-cyclobut-G phosphorylated metabolites are exceptionally potent, and that cellular enzymes are not implicated. It is clearly essential to examine the residual kinase activity in HSV TK⁻ mutants; and to determine what role, if any, may be ascribed to the more efficiently phosphorylated derivatives of (−)-cyclobut-G.

The triphosphates of the two enatiomers of cyclobut-G and cyclobut-A have been synthesized by Bisacchi et al. [87], who found that (+)-cyclobut-GTP is a selective inhibitor of HSV-1 DNA polymerase relative to HeLa DNA polymerase, whereas (−)-cyclobut-G is much less inhibitory against both enzymes. Both (+)-cyclobut-G and (+)-cyclobut-A were also efficacious in the therapy of a lethal mouse CMV infection.

An additional point of interest which must be taken into account is the possible effect of intracellular enzymes on the oxetanocin and cyclobut nucleosides. Shimada et al. [82] found that oxetanocin-A is deaminated by both mammalian and bacterial adenosine deaminases to the corresponding oxetanocin-Hx. Actinomyces organisms converted oxetanocin-A to oxetanocin-Xao *via* oxetanocin-Hx (presumably by deamination, followed by oxidation by xanthine oxidase). And chemically synthesized 2-amino-oxetanocin-A was deaminated by adenosine deaminase to oxetanocin-G.

9.1. *Ring-enlarged oxetanocin nucleosides*

The advent of oxetanocin nucleosides, and their promising antiviral activities, has in turn stimulated interest in 3'-branched-chain analogues of 2'-deoxynucleosides, also referred to as ring-enlarged or ring-expanded oxetanocin nucleosides.

For example, 3'-hydroxymethyl-2',3'-deoxyadenosine was prepared as a ring-expanded analogue of oxetanocin-A (see Scheme 13) by Tseng et al. [84]. When tested against HIV in ATH8 cells, it exhibited an activity profile comparable to that of oxetanocin-A itself, although neither conferred full protection. The corresponding isomeric 2'-hydroxymethyl-2',3'-dideoxyadenosine, also synthesized, proved totally inactive.

Bamford et al. [88] have reported the preparation of a series of 3'-hydroxymethyl nucleosides of uracil and 5-substituted uracils, the sugar rings of which were 2'-deoxyribose, arabinose or lyxose. When tested against a number of viruses, the only active compound was the 3'-deoxy-3'-hydroxymethyl derivative of (E)-5-bromovinyl-(β-D-arabino-pentofuranosyl)uracil, at concentrations well below the cytotoxicity threshold.

It is to be anticipated that carbocyclic analogues of the foregoing will shortly be forthcoming, bearing in mind the known enhanced activities of numerous carbocyclic nucleoside analogues relative to the parent nucleosides.

Highly relevant to the foregoing compounds, and their mechanism of action, is an earlier study by Acton et al. [89] on the antitumour properties of anomeric and 3'-branched-chain homologues of 6-thio-2'-deoxyguanosine. Although attention was previously drawn to the significance of these findings [90, 91], they have hitherto not been appreciated.

9.2. The α- and β-anomers of 6-thio-2'-deoxyguanosine (TGdR)

It is generally assumed that the substrates of nucleoside kinases are the β-anomeric nucleosides. There are, however, examples of naturally occurring α-anomeric nucleosides, e.g. 1-(α-D-ribofuranosyl)-5,6-dimethylbenzimidazole, the 3'-phosphate of which is a key substituent of vitamin B_{12}. Attention was long ago directed to the fact the α-anomers of some synthetic nucleoside analogues exhibit appreciable antiviral and antitumour activities [92]. Since these activities are frequently dependent on prior intracellular phosphorylation, it is pertinent to inquire what enzyme(s) are involved.

β-TGdR is a potent antitumour agent, but its rapid cleavage by intracellular purine nucleoside phosphorylase (PNP) mitigates its potential advantages over that of the base, 6-thioguanine. Nakai & Le Page [93] noted that α-TGdR (which is resistant to PNP) is phosphorylated, albeit less efficiently than β-TGdR, by extracts of normal tissues, but not by purified dCK from these tissues, for which the β-anomer was a substrate. However, in an araC-resistant mouse L1210 cell line, both anomers were phosphorylated and incorporated into DNA, to an extent proportional to the carcinostatic effect. Since the α-anomer is not cleaved by PNP, and is not phosphorylated in bone marrow (accounting for its low toxicity), it appeared to be a promising candidate for antitumour activity. Further studies showed that in tumours α-TGdR is incorporated into DNA at the terminus of short chains, whereas β-TGdR undergoes internal incorporation. The authors postulated the existence of a kinase differing from dCK and capable of phosphorylating both anomers.

Based on this assumption, Acton et al. [89] suggested that the commonly accepted postulate of structural dissimilarity between the two anomers (1 and 2 in Scheme 14) required reexamination. The purine moiety of each may occupy the same site on the enzyme if 2 is redrawn as 2a, to mimic 1. This requires that the exocyclic CH_2OH of 1 be replaced on the enzyme by the CH_2OH of 2a, and vice versa (see Scheme 14). Further comparison of structures 2a and 1 suggested that the α-anomer would better mimic the natural β-anomer 1 if the 3'-OH of 2a is replaced by a 3'-CH_2OH to give 4a, the 3'-branched analogue 4 of α-TGdR. If this reasoning is correct, 4 should be a better substrate than 2.

The 3'-branched analogues of the two anomers (3 and 4) were synthesized by Acton et al. [89]. Both exhibited comparable growth inhibition of WI-12 human lymphoblastoid cells (more effectively than the parent α-TGdR). Both were also

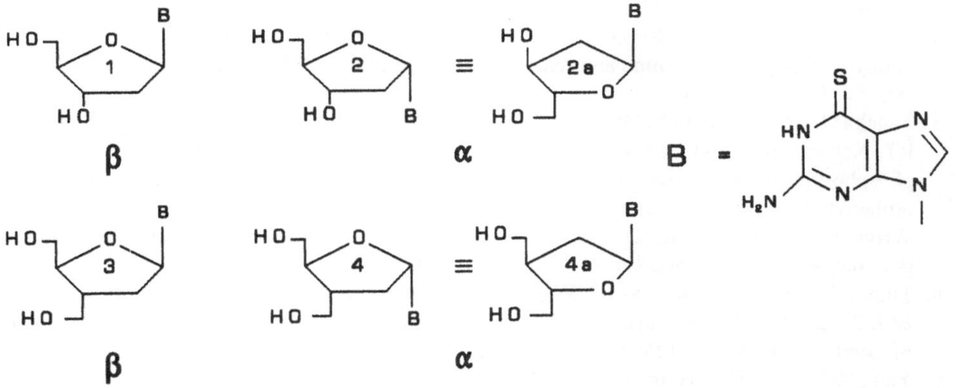

Scheme 14. The β– and α-anomers of 6-thio-2′-deoxyguanosine (TGdR) and their 3′-hydroxymethyl congeners (see text for details)

phosphorylated more effectively than **2**, and incorporated terminally into short chains in the DNA of Mecca lymphosarcoma in mice. However, unlike **2**, **4** was now phosphorylated in bone marrow, as well as in tumour, cells and incorporated into the DNA of both these cells [94], thus dashing the hope initially entertained for better selectivity of **4** vs tumours cells.

The kinase(s), or other phosphorylating enzyme(s) responsible for phosphorylation of **2**, **3** and **4** remain to be identified. Although **2** was not a substrate for purified dCK, the authors apparently did not test **3** and **4** as potential substrates or inhibitors. More recently Park & Ives [29] reported that β-TGdR (**1**), which is a substrate for dCK (see above), is not an inhibitor of highly purified dGK, hence not a substrate.

Acknowledgements. I am indebted to Jarek Cieśla for assistance in preparation of the figures for this text. Experimental results described profited from the support of the Ministry of Higher Education and the Polish Cancer Research Program.

References

1. Kuśmierek JT and Shugar D, Nucleotides, nucleoside phosphate diesters and phosphonates as antiviral and antineoplastic agents — an overview. In: *Antiviral Mechanisms in the Control of Neoplasia* (Chandra P, ed), pp. 481–498, Plenum Press, NY, 1979.
2. Martin JC (ed), *Nucleotide Analogues as Antiviral Agents.* ACS Symp Ser No. 401, American Chemical Society, Washington, 1989.
3. De Clercq E, Broad-spectrum anti-DNA virus and anti-retrovirus activity of phosphonylmethoxyalkylpurines and pyrimidines. *Biochem. Pharmacol.* **42:** 963–972, 1991.
4. Balzarini J and De Clercq E, 5-Phosphoribosyl 1-pyrophosphate synthetase converts the acyclic nucleoside phosphonates 9-(3-hydroxy-2-phosphonylmethoxypropyl)adenine and 9-(2-phos-

phonylmethoxyethyl)adenine directly to their antivirally active diphosphate derivatives. *J. Biol. Chem.* **266**: 8686–8689, 1991.

5. Rose WC, Crosswell AR, Bronson JJ and Martin JC, *In vivo* antitumour activity of 9-[(2-phosphonylmethoxy)ethyl]guanine and related phosphonate nucleotide analogues. *J. Nat. Cancer Inst.* **82**: 510–512, 1990.

6. Tanaka H, Baba M, Saito S, Miyasaka T, Takashima H, Sekiya K, Ubasawa M, Mitta I, Walker RT, Nakashima H and De Clercq E, Specific anti-HIV-1 acyclonucleosides which cannot be phosphorylated—Synthesis of some deoxy analogues of 1-[(2-hydroxyethoxy)methyl]-6(phenylthio)thymine. *J. Med. Chem.* **34**: 1508–1511, 1991.

7. Wright GE and Brown NC, Deoxyribonucleotide analogs as inhibitors and substrates of DNA polymerases. *Pharmacol. Ther.* **47**: 437–497, 1990.

8. Tolman RL, Field AK, Karkas JD, Wagner AF, Germershausen J, Crumpacker C and Scolnick EM, 2'-nor-cGMP: a seco cyclic nucleotide with powerful anti-DNA-viral activity. *Biochem. Biophys. Res. Commun.* **128**: 1329–1335, 1985.

9. Yang ZH, Lucia HL, Tolman RL, Colonno RJ and Hsiung GD, Effect of 2'-nor-cyclic GMP against guinea pig cytomegalovirus infection, *Antimicrob. Agents Chemother.* **33**: 1563–1568, 1989.

10. Germerhausen J, Bostedor R, Liou R, Field AK, Wagner AF, MacCoss M, Tolman RL and Karkas JD, Comparison of the modes of antiviral action of 2'-nor-deoxyguanosine nad its cyclic phosphate, 2'-nor-cyclic GMP. *Antimicrob. Agents Chemother.* **29**: 1025–1031, 1986.

11. Tihon C and Green M, Cyclic AMP-amplified replication of RNA tumour virus-like particles in Chinese hamster ovary cells. *Nature New. Biol.* **244**: 227–231, 1973.

12. Stanwick TL, Anderson RW and Nahmias AJ, Interaction between cyclic nucleotides and herpes simplex viruses: Productive infection. *Infect. Immun.* **18**: 342–347, 1977.

13. Robbins SJ and Rapp F, Inhibition of measles virus replication by cyclic AMP. *Virology* **106**: 317–326, 1980.

14. Miller CA and Carrigan DR, Reversible repression and activation of measles virus infection in neural cells. *Proc. Natl. Acad. Sci. USA* **79**: 1629–1623, 1982.

15. Robbins SJ, Stimulation of measles virus replication by cyclic guanosine monophosphate. *Intervirology* **32**: 204–208, 1991.

16. Cho-Chung YS, Clair T, Tortero G and Yokozaki H, Role of site-selective cAMP analogs in control and reversal of malignancy. *Pharmacol. Ther.* **50**: 1–33, 1991.

17. Prisbe EJ, Martin JC, McGee DPC, Barker MF, Smee DF, Duke AE, Matthews TR and Verheyden JPH, Synthesis and antiherpes virus activity of phosphate and phosphonate derivatives of 9-(1,3-dihydroxy-2-propoxy)methyl guanine. *J. Med. Chem.* **29**: 671–675, 1986.

18. Karkas JD, Germershausen J, Tolman RL, MacCoss M, Wagner AF, Liou R and Bostedor R, Stereochemical considerations in the enzymatic phosphorylation and antiviral activity of acyclonucleosides. I. Phosphorylation of 2'-nor-2'-deoxyguanosine. *Biochim. Biophys. Acta* **911**: 127–135, 1987.

19. Lassota P, Kazimierczuk Z, Zan-Kowalczewska M and Shugar D, 2',3'-*seco* pyrimidine nucleosides and nucleotides, including structural analogues of 3':5'-cyclic CMP and UMP, and their behaviour in several enzyme systems. *Biochem. Biophys. Res. Commun.* **137**: 453–460, 1986.

20. Stolarski R, Kazimierczuk Z, Lassota P and Shugar D, Acyclo nucleosides and nucleotides: Synthesis, conformation and other properties, and behaviour in some enzyme systems, of 2',3'-seco purine nucleosides, nucleotides and 3':5'-cyclic phosphates, analogues of cAMP and cGMP. *Z. Naturforsch.* **41c**: 758–770, 1986.

21. Stolarski R, Lassota P, Kazimierczuk Z and Shugar D, Solution conformations of some acyclonucleoside and nucleotide analogues of antiviral acyclonucleosides, and their substrate/inhibitor properties in several enzyme systems. *Z. Naturforsch.* **43c**: 231–242, 1988.

22. Stolarski R, Cieśla JM and Shugar D, Monophosphates and cyclic phosphates of some antiviral acyclonucleosides: Synthesis, conformation and substrate/inhibitor properties in some enzyme systems. *Z. Naturforsch.* **45c**: 293–299, 1990.

23. Zan-Kowalczewska M, Cieśla JM, Sierakowska H and Shugar D, Potato tuber cyclic-nucleotide phosphodiesterase. Selective inactivation of activity vs nucleoside cyclic 3′,5′-phosphates and properties of the native and selectively inactived enzyme. *Biochemistry* **26**: 1194–1200, 1987.

24. Helfman DM and Kuo JF, A homogenous cyclic CMP phosphodiesterase hydrolyzes both pyrimidine and purine cyclic 2′:3′- and 3′:5′-nucleotides. *J. Biol. Chem.* **257**: 1044–1047, 1982.

24a. Chiatante D, Newton RP and Brown EG, Properties of a multifunctional 3′,5′-cyclic nucleotide phosphodiesterase from *Lectuca* cotyedons: Comparison with mammalian enzymes capable of hydrolysing pyrimidine cyclic nucleotides. *Phytochemistry* **26**: 1301–1306, 1987.

25. Eriksson S, Kierdaszuk B, Munch-Petersen B, Oberg O and Johansson NG, Comparison of the substrate specificities of human thymidine kinases 1 and 2 and deoxycytidine kinase toward antiviral and cytostatic nucleoside analogs. *Biochem. Biophys. Res. Commun.* **176**: 586–592, 1991.

26. Munch-Petersen B, Cloos L, Tyrsted G and Eriksson S, Diverging substrate specificity of pure human thymidine kinases 1 and 2 against antiviral dideoxynucleosides. *J. Biol. Chem.* **266**: 9032–9038, 1991.

27. Plagemann PGW, Behrens M and Abraham D, Metabolism and cytotoxicity of 5-azacytidine in cultured Novikoff rat hepatoma and P388 Leukemia cells and their enhancement by preincubation with pyrazofurin. *Cancer Res.* **38**: 2458–2466, 1978.

28. Payne RC, Cheng N and Traut W, Uridine kinase from Ehrlich ascites carcinoma: Purification and properties of homogenous enzyme. *J. Biol. Chem.* **260** : 10242–10247, 1985.

29. Park I and Ives DH, Properties of a highly purified mitochondrial deoxyguanosine kinase. *Arch. Biochem. Biophys.* **266**: 51–60, 1988.

30. Lewis RA and Link L, Phosphorylation of arabinosyl guanine by a mitochondrial enzyme of bovine liver. *Biochem. Pharmacol.* **38**: 2001–2006, 1989.

31. Miller WH and Miller RL, Guanosine kinase from *Trichomonas vaginalis*. *Mol. Biochem. Parasitol.* **48**: 39–46, 1991.

32. Kong X-B, Tong WP and Chou T-C, Induction of deoxycytidine kinase by 5-azacytidine in an HL-60 cell line resistant to arabinosylcytosine. *Mol. Pharmacol.* **39**: 250–257, 1991.

33. Fyfe JA, Differential phosphorylation of (E)-5-(2-bromovinyl)-2′-deoxyuridine monophosphate by thymidylate kinases from herpes simplex viruses types 1 and 2 and varicella zoster virus. *Mol. Pharmacol.* **21**: 432–437, 1982.

34. MacCoss M, Tolman RL, Ashton WT, Wagner AF, Hannah J, Field AK, Karkas JD and Germershausen JI, Synthetic, biochemical and antiviral aspects of selected acyclonucleosides and their derivatives. *Chemica Scripta* **26**: 113–121, 1986.

35. Ikeda S, Chakravarty R and Ives DH, Multisubstrate analogs for deoxynucleoside kinases. *J. Biol. Chem.* **261**: 15836–15843, 1986.

36. Bone R, Cheng Y-C and Wolfenden R, Inhibition of thymidine kinase by P^1-(Adenosine-5′)-P^5-(thymidine-5′)-pentaphosphate. *J. Biol. Chem.* **261**: 5731–5735, 1986.

37. Davies LC, Stock JA, Barrie SE, Orr RM and Harrap KR, Dinucleotide analogues as inhibitors of thymidine kinase, thymidylate kinase and ribonucleotide reductase. *J. Med. Chem.* **31**: 1305–1308, 1988.

38. Broom AD, Rational design of enzyme inhibitors: Multisubstrate analogue inhibitors. *J. Med. Chem.* **32**: 2–7, 1988.

39. Miller RL, Adamczyk DL, Miller WH, Koszalka GW, Rideout JL, Beacham III LM, Chao EY, Haggerty JJ, Krenitsky TA and Elion GB, Adenosine kinase from rabbit liver. II. Substrate and inhibitor specificity. *J. Biol. Chem.* **254**: 2346–2352, 1979.

40. LaFon SW, Nelson DJ, Berens RL and Marr JJ, Inosine analogues: Their metabolism in L cells and in *Leishmania donovani*. *J. Biol. Chem.* **260**: 9660–9665, 1986.

41. Nutter LM, Grill SP, Dutschman GE, Sharma RA, Bobek M and Cheng Y-C, Demonstration of viral thymidine kinase inhibitor and its effect on deoxyribonucleotide metabolism in cells infected with herpes simplex virus. *Antimicrob. Agents Chemother.* **31**: 368–374, 1987.

42. Ashton WT, Meurer LC, Tolman RL, Karkas JD, Liou R, Perry HC, Czelusniak SM and Klein RJ, A potent selective non-substrate inhibitor of HSV-1 thymidine kinase: (±)–9-[{(Z)-2-(hydroxymethyl)cyclohexyl}methyl]guanine and related compounds. *Nucleosides & Nucleotides* **8**: 1157–1158, 1989.

43. Hildebrand C, Sandoli D, Focher F, Gambino J, Ciarrochi G, Spadari S and Wright G, Structure-activity relationships of N^2-substituted guanines as inhibitors of HSV-1 and HSV-2 thymidine kinases. *J. Med. Chem.* **33**: 203–206, 1990.

44. Swierkowski M and Shugar D, A nonmutagenic thymidine analog with antiviral activity: 5-Ethyldeoxyuridine. *J. Med. Chem.* **12**: 533–534, 1969.

45. Martin JA, Duncan LB, Hall MJ, Wong-Kai-In P, Lambert RW and Thomas GJ, New potent selective inhibitors of herpes simplex virus thymidine kinases. *Nucleosides & Nucleotides* **8** : 753–764, 1989.

46. Spadari S and Wright G, Antivirals based on inhibition of herpesvirus thymidine kinases. *Drug News & Perspect* **2**: 333–336, 1989.

47. Datta AK and Pugano JS, Phosphorylation of Acyclovir in activated Burkitt somatic cell hybrids. *Antimicrob. Agents Chemother.* **24**: 10–14, 1983.

48. Keller PM, McKee SA and Fyfe JA, Cytoplasmic 5′-nucleotidase catalyzes acyclovir phosphorylation. *J. Biol. Chem.* **260**: 8664–8667, 1985.

49. Itoh R, Purification and some properties of cytosolic 5′-nucleotidase from rat liver. *Biochim. Biophys. Acta* **657**: 402–410, 1981.

50. Worku Y and Newby AC, Nucleoside exchange catalyzed by the cytoplasmic 5′-nucleotidase. *Biochem. J.* **205**: 503–510, 1982.

51. Smee DF, Boehme R, Chernow M, Binko BP and Matthews TR, Intracellular metabolism and enzymatic phosphorylation of 9-(1,3-dihydroxy-2-propoxymethyl)guanine and acyclovir in herpes simplex virus-infected and uninfected cells. *Biochem. Pharmacol.* **34**: 1049–1056, 1985.

52. Birnbaum GI, Cygler H, Kusmierek JT and Shugar D, Structure and conformation of the potent antiherpes agent 9-(2-hydroxyethoxymethyl)guanine (Acycloguanosine). *Biochem. Biophys. Res. Commun.* **103**: 968–974, 1981.

53. Prasher DC, Carr MC, Ives DH, Tsai T-C and Frey PA, Nucleoside phosphotransferase from barley. Characterization and evidence for ping-pong kinetics involving phosphoryl enzyme. *J. Biol. Chem.* **257**: 4931–4939, 1982.

54. Billich A, Stockhowe U and Witzel H, Nucleoside phosphotransferase from malt sprouts. I. Isolation, characterization and specificity of the enzyme. *Biol. Chem. Hoppe-Seyler* **367**: 267–278, 1986.

55. Lassota P, Kusmierek JT, Stolarski R and Shugar D, Pyrimidine homoribonucleosides: Synthesis, solution conformation and some biological properties. *Z. Naturforsch.* **42c**: 589–598, 1987.

56. Giziewicz J and Shugar D, Preparative enzymatic synthesis of nucleoside-5′-phosphates. *Acta Biochim. Polon.* **22**: 87–98, 1975.

57. Giziewicz J and Shugar D, Nucleoside 5′-phosphates: Enzymatic phosphorylation of nucleosides to the 5′-phosphates. In: *Nucleic Acid Chemistry*, Part II (Townsend LB & Tipson RS, eds), pp. 955–961, John Wiley & Sons, New York, 1978.

58. Mizote T and Nakayama H, Purification and properties of hydroxymethylpyrimidine kinase from *Escherichia coli. Biochim. Biophys. Acta* **991**: 109–113, 1989.

59. Kit S, Leung W-C, Trkula D and Dubbs DR, Characterization of nucleoside phosphotransferases and thymidine kinase activities of chick embryo cells and chick-mouse somatic cell hybrids. *Arch. Biochem. Biophys*.**169**: 66–76, 1975.

60. Zekri M, Harb J, Bernard S and Mefla K, Purification of bovine liver cytosolic 5'-nucleotidase: kinetic and structural studies as compared to the membrane isoenzyme. *Eur. J. Biochem*. **172**: 93–99, 1988.

61. Hoglund L and Reichard P, Cytoplasmic 5'(3')-nucleotidases from human placenta. *J. Biol. Chem*. **265**: 6589–6595, 1990.

62. Fridland A, Connelly MC and Robbins TJ, Tiazofurin metabolism in human lymphoblastoid cells: Evidence for phosphorylation by adenosine kinase and 5'-nucleotidase. *Cancer Res*. **46**: 532–537, 1986.

63. Egyhazi E and Shugar D, 5,6-Dichlororibofuranosylbenzimidazole (DRB) is phosphorylated in salivary gland cells of *Chironomus tentans*. *FEBS Lett*. **107**: 431–435, 1979.

64. Egyhazi E, Ossoinak A, Holst M, Rosendahl K and Tayip U, Kinetic analysis of uptake and phosphorylation of 5,6-dichlororibofuranosylbenzimidazole (DRB) by salivary gland cells of *Chironomus tentans*. *J. Biol. Chem*. **255**: 7807–7812, 1980.

65. Holst M and Egyhazi E, Transport and metabolism of adenosine in relation to the transcriptional activity of hnRNa genes in *Chironomus* salivary glands. *Eur. J. Biochem*. **147**: 631–636, 1985.

66. Renosto F, Seubert PA and Segel IH, Adenosine-5'-phosphosulfate kinase from *Penicillium chrysogenum*: Purification and kinetic characterization. *J. Biol. Chem*. **259**: 2113–2123, 1984.

67. Johnson MA and Fridland A, Phosphorylation of 2',3'-dideoxyinosine by cytosolic 5'-nucleotidase of human lymphoid cells. *Mol. Pharmacol*. **36**: 291–295, 1989.

68. Bondoc LL, Shannon WM, Secrist JA III, Vince R and Fridland A, Metabolism of the carbocyclic nucleoside analogue carbovir, an inhibitor of human immunodeficiency virus in human lymphoid cells. *Biochemistry* **29**: 9839–9843, 1990.

69. Johnson MA, Ahluwalia G, Connelly MC, Cooney DA, Broder S, Johns DG and Fridland A, Metabolic pathways for the activation of the antiretroval agent 2',3'-dideoxyadenosine in human lympoid cells. *J. Biol. Chem*. **263**: 15354–15357, 1988.

70. Sarup JC, Johnson MA, Verhoef V and Fridland A, Regulation of purine deoxynucleoside phosphorylation by deoxycytidine kinase from human leukemic blast cells. *Biochem. Pharmacol*. **38**: 2601–2607, 1989.

71. Bennett Jr LL, Shealy YF, Allan PW, Rose LM, Shannon WM and Arnett G, Phosphorylation of the carbocyclic analog of 2'-deoxyguanosine in cells infected with herpes viruses. *Biochem. Pharmacol*. **40** : 1515–1522, 1990.

72. Jurovcik M and Holy A, Metabolism of pyrimidine L-nucleosides. *Nucleic Acids Res*. **3**: 2143–2154, 1976.

73. Wang CC and Cheng HW, The doexyribonucleoside phosphotransferase of *Trichomonas vaginalis*: A potential target for anti-trichomonial chemotherapy. *J. Exp. Med*. **160**: 987–1000, 1984.

74. Tan CK, Civil R, Mian AM, So AG and Downey KM, Inhibition of the RNase H activity of HIV reverse transcriptase by azidothymidylate. *Biochemistry* **30**: 4831–4836, 1991.

75. Harrington JA, Miller WH and Spector T, Effector studies of 3'-azidothymidine nucleotides with human ribonucleotide reductase. *Biochem. Pharmacol*. **36**: 3757–3761, 1987.

76. Dzik JM, Bretner M, Kulikowski T, Cieśla J, Cieśla JM, Rode W and Shugar D, Interaction of the 5'-phosphate of the anti-HIV agent 3'-azido-3'-deoxythymidine and 3'-azido-2',3'-dideoxyuridine with thymidylate synthase. *Biochem. Biophys. Res. Commun*. **155**: 1418–1423, 1988.

77. Stein JM, Stoeckler JD, Li S-Y, Tolman RL, MacCoss M, Chen A, Karkas JD, Ashton WT and Parks Jr RE, Inhibition of human nucleosides and nucleotides. *Biochem. Pharmacol*. **36**: 1237–1244, 1987.

77a. Bzowska A, Kulikowska E, Shugar D, Bing-yi C, Lindborg B and Johansson NG, Acyclonucleoside analogue inhibitors of mammalian purine nucleoside phosphorylase. *Biochem. Pharmacol.* **41**: 1791–1803, 1991.

78. Tuttle JV and Krenitsky TA, Effects of acyclovir and its metabolites on purine nucleoside phosphorylase. *J. Biol. Chem.* **259**: 4065–4069, 1984.

79. Heinemann V, Xu Y-Z, Chubb S, Sen A, Hertel LW, Grindey GB and Plunkett W, Inhibition of ribonucleotide reduction in CCRF-CEM cells by $2',2'$-difluorodeoxycytidine. *Mol. Pharmacol.* **38**: 567–572, 1990.

80. Parker WB, Shadix SC, Chang C-H, White EL, Rose LM, Brockman RW, Shortnacy AT, Montgomery JA, Secrist III JA and Bennett Jr LL, Effects of 2-chloro-9-(2-fluoro)-D-arabinofuranosyl)adenine on K562 cellular metabolism and the inhibition of human ribonucleotide reductase and DNA polymerases by its $5'$-triphosphate. *Cancer Res.* **51**: 2386-2394, 1991.

81. Shimada N, Hasegawa S, Harada T, Tomisawa T, Fuji A and Takita T, Oxetanocin, a novel nucleoside from bacteria. *J. Antibiotics* (Tokyo) **39**: 1623–1625, 1986.

82. Shimada N, Hasegawa S, Saito S, Nishikjori T, Fujii A and Takita T, Derivatives of oxetanocin: Oxetanocins H, X and G, and 2-aminooxetanocin A. *J. Antibiotics* (Tokyo) **11**: 1788–1790, 1987.

83. Nishiyama Y, Yamamato N, Takahashi K and Shimada N, Selective inhibition of human cytomegalovirus by a novel nucleoside, oxetanocin G. *Antimicrob. Agents Chemother.* **32**: 1053–1056, 1988.

84. Tseng CK-H, Marquez VE, Milne GWA, Wysocki RJJr, Misuya H, Shirasaki T and Driscoll JS, A ring-enlarged oxetanocin A analogue as an inhibitor of HIV infectivity. *J. Med. Chem.* **34**: 343–349, 1991.

85. Daikoku T, Yamamoto N, Saito S, Kitagawa M, Shimada N and Nishiyama Y, Mechanism of inhibition of human cytomegalovirus replication by oxetanocin G. *Biochem. Biophys. Res. Commun.* **176**: 805–812, 1991.

86. Kohlbrenner WE, Carter CD, Fesik SW, Norbeck DW and Erickson J, Efficiency of phosphorylation of the cyclobut-G (A-69992) enantiomers by HSV-1 thymidine kinase does not correlate with their anti-herpesvirus activity. *Biochem. Pharmacol.* **40:** R5-R10, 1990.

87. Bisacchi GS, Braitman A, Cianci CW, Clark JM, Field AK, Hagen ME, Hockstein DR, Malley MF, Mitt T, Slusarchyk WA, Sundeen JE, Terry BJ, Tuomari AV, Weaver ER, Young MG and Zahler R, Synthesis and antiviral activity of enantiomeric forms of cyclobutyl nucleoside analogues. *J. Med. Chem.* **34**: 1415–1421, 1991.

88. Bamford MJ, Coe PL and Walker RT, Synthesis and antiviral activity of $3'$-deoxy-$3'$-C-hydroxymethyl nucleosides. *J. Med. Chem.* **33**: 2494–2501, 1990.

89. Acton EM, Goerner RN, Uh HS, Ryan KJ and Henry DW, Improved antitumour effects in $3'$-branched homologues of $2'$-deoxythioguanosine: synthesis and evaluation of thioguanine nucleosides of 2,3-dideoxy-3-(hydroxymethyl)-D-*erythro*-pentofuranose. *J. Med. Chem.* **22**: 518–525, 1979.

90. Shugar D, New antiviral agents: some recent developments. In: *Medicinal Chemistry Advances* (De Las Heras FG and Vega S, eds), pp. 225–238, Pergamon Press, Oxford, 1981.

91. Shugar D, Antiviral agents —some current developments. *Pure & Appl. Chem.* **57**: 423–440, 1985.

92. Shugar D, Progress with antiviral agents. *FEBS Lett.* **40** (Suppl.): S48–S62, 1974.

93. Nakai Y and LePage GA, Characterization of the kinases involved in the phosphorylation of α- and β-$2'$-deoxythioguanosine. *Cancer Res.* **32**: 2245–2451, 1972.

94. LePage GA, Banks PA, Noujaim MJ and Buzzell GR, *In vivo* effects in mice produced by a $3'$-branched homologue of α-$2'$-deoxythioguanosine. *Cancer Chemother. Pharmacol.* **5**: 127–131, 1980.

Subject Index